Chemistry, A Human Concern

Chemistry, A Human Concern

Jay A. Young

Professor of Chemistry, Emeritus,
King's College

Macmillan Publishing Co., Inc.
New York

Collier Macmillan Publishers
London

Macmillan Publishing Co., Inc.
866 Third Avenue, New York, New York 10022

Collier Macmillan Canada, Ltd.

Library of Congress Cataloging in Publication Data

Young, Jay A
  Chemistry, a human concern

  Includes bibliographies and index.
  1. Chemistry. I. Title
QD31.2.Y68   540   76–30624
ISBN 0–02–431160–X

Printing: 1 2 3 4 5 6 7 8   Year: 8 9 0 1 2 3 4

He looks with favor upon their hearts
    And shows them his glorious works
That they may describe
    The wonders of his deeds
And praise his holy name.

Sirach: 17, 7–8

Preface

For students with major interests in areas distantly removed from chemistry, we have three choices for the content and presentation of a single, introductory course in chemistry. One, "What is good for the chemistry major is therefore good for everyone," is both arrogant in its assertion and all but impossible to execute. The second, "Give a general survey of chemistry, and try to cover all the bases," is in my opinion little better. Good students from other disciplines are insulted by a superficial approach, and students of lesser ability are also not stimulated to find much that is useful when such a presentation is used.

There are medians between these extremes. Of these, the one chosen for this book attempts to identify several topics which, in themselves, are of interest to most students (whatever their major curricular goals might be) and to relate these to chemistry. I have tried to write to the student, seeking to begin where the student is, and to lead from there into a bit of chemistry. Some of the chemistry is, indeed, not very sophisticated (there is no real molecular orbital theory in this book); some is reasonably challenging (the relation between structure and properties of polymers, for example). Many of the topics, though not the manner of exposition, are the same as those in "standard" introductory courses; several other topics are not.

Broadly, the book is divided into two parts, expositional chapters that discuss topics from introductory chemistry courses, and commentary chapters intended to extend the content of chemistry into areas that are already perceived to be meaningful (for nonchemical reasons) by the students. Although the chapters are sequentially designed, most of them can be taken in any other order, and some omitted, without significant loss of continuity. This is discussed further in the "Mezzologue," which follows Chapter 7.

It is suggested that the course be taught in a manner so as to rely on short and long term papers, oral discussions, and prepared presentations as the means of evaluating students performance. For this purpose, each of the expositional chapters present discussion questions rather than the more familiar chemistry-oriented questions, at the end of each; the commentary chapters end with annotated bibliographies intended to encourage different students with differing interests to search in a variety of directions for further information that relates a portion of the chemistry studied to their own concerns. In my experience, the device is successful, and enjoyable for both professors and students.

Some may wish to consider this book for use as the text in a course which prepares students for more rigorous, or standard, introductory chemistry instruction. Of the two most direct ways to so prepare students, I have chosen to motivate rather than to engender awe. Even without mathematics, beginning chemistry is complicated enough; but without motivation it is impossible. The rigor, I feel, can come later; initially let us try to help students become sympathetic toward the reasons why rigor is ultimately necessary.

It is a distinct pleasure to acknowledge the chapter contributions by Dr. C. A. McAuliffe and by Dr. and Mrs. John W. Moore, and the helpful comments and suggestions of Professors Mary Berry, Harry Day, and Paul Melius. Two graduate student associates and friends, Francis McCullough and Patrick Morgan, were equally helpful. Mr. Wallace Shows stimulated me to write this book from my notes; his successor at Macmillan, Mr. James L. Smith, "inherited" this book when he assumed his present position as editor and has been most patient with a procrastinating author. Alphabetically last, but not otherwise so characterized, I acknowledge the assistance and personally expressed interest of Dr. Robert E. Varnerin, Manager of Education, Manufacturing Chemists Association. None of these friends is responsible for the errors which surely exist in this book.

Other friends are cordially encouraged to identify these to me for incorporation into a second edition, should this work prove to be that useful in its present form to my colleagues and to our students, to all of whom this book is herewith respectfully dedicated and now presented.

Washington, D.C.   Jay A. Young

To the Instructor

This text is based upon the principle that every student taking the course is not expected to carry away the same knowledge and attitude as every other student in the course. In effect this amounts to an overt, explicit recognition of what has always been the case anyway, despite our attempts to induce some kind of conformity by administering the same exams and quizzes to each member of the class and our utilization of objective grading standards to the extent that this is possible.

However, the open suggestion that each student is different upon entry into the course and is expected to maintain his/her unique characteristics *because* of this course does imply the use of different teaching strategies. Some colleagues for good reason will reject the assertions made here. To these I recommend their adoption of some of the analogies and examples presented here and there in this text. For those who are still with me it may be useful to address three important matters:

> What might I do in the classroom, if I adopt this text?
> How does one arrive at a grade for students in the course?
> What about the laboratory, if time and other factors permit?

These possibilities suggest themselves for incorporation into the classroom: Generate, lead, and participate in student discussions based either on questions at the ends of some chapters or on cited references at the ends of other chapters. Add your own favorite chemical stories and instruction at appropriate places; to encourage this, some important and interesting concepts have been deliberately omitted: asymmetric carbon and optical isomerism, entropy and the tendency toward randomness as a driving force, reaction rates and mechanisms, surface (colloid) chemistry, the chemistry of many common substances including soaps and detergents, waxes, fertilizers, cooking, cosmetics. Make liberal use of lecture demonstrations and lecture experiments. Much of the chemistry in this text is assertive and descriptive; add your own comments on the process of science, the logic and imagination between observed fact and accepted theory. Insert historical anecdotes and developments. Include your own illustrations of the liberal, humanizing aspects of chemistry. Select one or more of the references at the end of a chapter, building that discussion into a lecture or series of lectures.

Grading criteria are personal; the following comments are intended to suggest my own approach as a place for other instructors to start. I like to convey to the students that they are expected to show what they have obtained from the course. Early on, each student is asked to prepare a list of expectations and is aided to phrase these meaningfully. Both of us keep a copy. At the end of the course, students are asked to present or to refer to evidence that demonstrates their goals have or have not been achieved to some degree. Usually, new goals become evident during the course and students are asked to present or refer to evidence indicating accomplishment of these. The grade is based upon the degree of excellence in phrasing the goals initially, in identifying the degree of accomplishment in the stated and in the discovered goals, and on their contributions to classroom discussions. Alternatively, grades can be based upon the quality of short or long, frequent or infrequent, term papers coupled with performance on quizzes and examinations based upon the content of specified chapters and lectures. If this alternative is chosen, I recommend the use of performance objectives as the most efficient way to inform students for what they will be responsible in the quizzes and examinations.

If the latter alternative is chosen, the grading of term papers can become arduous. One solution is to require a short abstract and to form committees of students who then evaluate the term papers for the professor using as their primary criteria the degree of consonance and completeness between the term paper and the abstract. If this procedure is followed, there must be provision for appeal from the student-committee-assigned grade. Quizzes and examinations have another purpose, of course: to keep the class on its toes during the weeks of the semester or quarter. Sometimes, depending upon the class, and varying from year to year, such a technique is either necessary or insulting. When it is necessary, I prefer to use short answer (one to five or six words) questions and administer the quizzes almost daily. The questions should be heavily dependent upon the content of the assigned chapter, and the quizzes should be announced in advance.

The conduct of the associated laboratory is usually a matter of local option. For large classes, the logistic problems will often preclude any laboratory instruction. In these cases extensive use of lecture experiments, perhaps by adding one more hour of scheduled lecture time to make room, seems advisable. In any event, suitable laboratory or lecture experiment work can be centered around three schemata:

> The chemistry around us
> Esthetics of chemistry
> Practical and applied chemistry

As is obvious from many of the following examples, I consider laboratory work to be an opportunity to extend, rather than to review, what happened in the classroom, although there is certainly nothing wrong in emphasizing the latter instead of the former.

For the chemistry around us, an examination of the chemical and physical properties of polymers of various sorts, the analysis of vinegar, colorimetric analysis, electrolysis of water, the burning of steel wool in air and in oxygen, photographic chemistry, and qualitative analysis either in the traditional sense, or more briefly, presented instead as a search for clues in a hypothetical detective story, or as a study of the utility of solubility rules.

Esthetics of chemistry implies to some degree the fun of watching chemical phenomena. Examples include luminescent reactions, the iodine clock, reactions of some of the alkali metals with water or halogens, or other compounds and elements, various gas–gas reactions such as SO_2 and H_2S or SO_2 and HI, the thermal decomposition of $(NH_4)_2Cr_2O_7$, ligand exchanges which involve color change such as exchanging chloride ion for water on the copper ion, and a host of the other possibilities.

Practical and applied chemistry includes some applications of theory for its own sake, such as the Dumas determination of the molecular weight of a volatile liquid, and directly practical analogies, such as the reduction of copper oxide with illuminating gas or the reduction of lead oxide with charcoal, as analogous to the manufacture of pig iron. Other possibilities include the "nylon rope trick," generating nylon at the interface between immiscible liquid solutions of the two monomers, chromatographic separations, introductory studies of radioactivity, and so on.

Jay A. Young

Foreword to the Student

The very nature of our being human drives us, each and severally, to modify the conditions of our existence. Thus, one writes a poem, another invents a machine to pit cherries, someone else discovers oxygen, still another works to change the rules of football; none are satisfied with things as they are. Whether we are successful or not in these attempts, all of us are driven to pursue forever those changes that we hope will be effective.

This might have been a book about poetry or the art of making football rules. Both are important—perhaps equally, perhaps unequally, at least to some of us. As it happens, this is a book about chemistry, an attempt to relate how and why chemistry too can be used to modify the conditions of our existence.

It is necessary for someone to write good poetry. The same applies to pitting cherries, and to playing football and discovering oxygen. It is desirable that we nonpoets, cherry pie eaters, oxygen breathers, and football watchers recognize the drives that make others into poets, cherry pitter inventors, oxygen discoverers, and football players.

This book is about oxygen discoverers (in a metaphoric sense at least). It has been written to all poets, cherry pie eaters and cherry pitter inventors, to football players, rule makers and watchers, for information, and, I hope, for your enjoyment and response. As human beings, chemistry is a subject for our mutual concern, whether we are more, or less, interested in chemistry than we are in other important matters.

Jay A. Young

Contents

chapter 1

This Earth and Its Surroundings

Let's begin with where we are, on the Earth, one of nine planets of our solar system, traveling in a distorted circular, elliptical, orbit around the Sun. Actually, the Sun is a star, more or less average, not the biggest or smallest, not the brightest or the least bright, not as hot or as cold (if a star can be really cold) as some. This star, our Sun, is one of 100,000,000,000 stars (more or less) in our galaxy, which we call the Milky Way galaxy.

In the observable universe, there are approximately 100,000,000 galaxies, each with about 100,000,000,000 stars. Figure 1–1 is taken from a

Figure 1–1.   The Andromeda galaxy; the X indicates an approximate location in this galaxy that corresponds to our location in the Milky Way galaxy. [Courtesy of the Lick Observatory.]

photograph of the Andromeda galaxy, a "near" neighbor of our galaxy. The Andromeda galaxy is called a spiral galaxy; you can see a spiral like structure in Figure 1–1. As nearly as can be determined, the Milky Way galaxy is also a spiral galaxy of about the same size and general shape as the Andromeda galaxy. If our Sun were a member of the Andromeda galaxy and at the same location in that galaxy as it is in the Milky Way, it would be more or less at the spot marked X in Figure 1–1, rather far out from the center. This is probably a good thing. We cannot see the center of our galaxy directly because between us and the center there are massive clouds of "cosmic dust," but there is some evidence that in the center of our galaxy tremendous explosions frequently occur. Life there would not be likely to be tranquil and perhaps could not exist at all.

**NUMBERS,
SIZE,
DISTANCE,
AND WEIGHT**

Let's get back to the Sun, and also undertake a little mental effort. Your author, that's me, thinks he is a scientist, a chemist, and that you would like to know a little about science, particularly chemistry— if someone can show you that it is worth the trouble. The problem for an author is that before he can make a subject interesting, the reader must do a little work. The best I can do is to promise that the result will be worth the effort, and then hope that you will agree later on that it was.

Now, to the first bit of work. Our Sun has a mass of 2×10^{33} grams. The 33 up above the ten is called an exponent. Most people are familiar with the exponent 2. For example we say that 10^2 is "ten squared," or 10×10, or 100. In the same way, 10^3 is "ten cubed," or $10 \times 10 \times 10$, or 1000. Notice the trick: 10^2, with an exponent of 2, is the same as 100, with two zeros; and 10^3 with an exponent of 3 is the same as 1000, with three zeros. So, you can guess that 10^{33} is the same as

$$1,000,000,000,000,000,000,000,000,000,000,000$$

with 33 zeros. Two times 10^{33} is easy; it is

$$2,000,000,000,000,000,000,000,000,000,000,000$$

of course. Now, if we knew how big a gram was, we would have some idea of the mass of the Sun. A gram, which will be abbreviated "g" from now on,* is not much. There are 28.3495 g in 1 ounce (oz). Round this off to 30, since we really don't care about details at this point, and multiply by 16 to get 480 g to 1 pound (lb). It is easier for most people to remember 500 than it is to remember 480, so we can say that roughly there are 500 g in 1 lb.

Never ask a person who is trying to diet what his or her weight is in grams. A 135-lb girl would have to reply, "Sixty-seven thousand five hundred grams," and that is a discouragingly large number. (That number of grams is calculated from our "rounded off" number, no pun intended, of 500 grams per pound (g/lb). Using the more precise information of 28.3495 grams per ounce (g/oz), a 135-lb person has a mass of 61,235 g, if we forget about decimals.)

Now let's get back to the Sun. If it has a mass of 2×10^{33} g, the equivalent number of pounds is, roughly, found by dividing by 500. This comes out to be 4×10^{30} lb.

What is the size of our Sun? As everyone knows, it is shaped like a sphere, pretty much, and it has a radius of 7×10^{10} centimeters (cm). (Here we go again!) The reason for the emphasis on grams and centimeters, and other

---

* In chemistry units of measure, such as grams, milliliters, centimeters, and so on, are abbreviated when they appear with numerals. In each case the abbreviation we will use will be shown in parentheses. Many of these will be familiar to you; you will soon get used to the others.

things not yet mentioned, is partly selfish and partly altruistic. First, it is easier for an author who has been thinking about science for 30 years, or longer, to use these terms. But in this country during the next several years we will be changing to this system, called the "International System," and it will be helpful for you to get a bit of a jump on other people.

A centimeter (cm) is fairly small, too; it takes exactly 2.54 cm to make 1 inch (in.). If you figure this one out, it comes to about 10 cm for 4 in. The span of a typical human hand, opened up as much as possible, from the tip of the thumb to the tip of the middle (longest) finger, is a little shorter than 8 in. So, for some idea of centimeters, your hand span is approximately 18 or 20 cm. If you are fairly tall, more than 6 feet (ft), that would be 183 cm, then your hand span is typically more than 20 cm, maybe 22 cm or a bit more.

Let's come back to the Sun again. A radius of 7×10^{10} cm works out to be 2.3×10^9 ft, or 4.3×10^5 miles, rounding things off a bit as we divide by 2.54, and then by 12, and then by 5280. I want to be sure everyone is still with me, so let's translate the 2.3×10^9-ft radius, to be sure. It is the same as

$$2,300,000,000 \text{ ft}$$

The exponent 9 on the 10 is the count for moving decimal places. Earlier, it was described as the number of zeros, and this is the same as counting how much to move the decimal place. For example, if you only have one digit, like 2×10^9 that *would* be 2,000,000,000, since a decimal point just on the right of the 2 (2×10^9) is understood to be there. In the same way, 4.3×10^5 miles is

$$430,000 \text{ miles}$$

as you have probably already decided. The radius of the Earth is about 4000 miles, so you can see that the Sun is incredibly larger than the earth.

Almost everyone has heard that the Sun is 93 million miles distant from the Earth; its distance in centimeters is 1.5×10^{13} cm, roughly. It depends upon what season of the year you think of in this case. In our summer in the Northern Hemisphere the Sun is a little farther away than it is in winter; the opposite in the Southern Hemisphere. However, the Earth's orbit around the Sun is close enough to a circle for us to be able to say, approximately, the distance is 1.5×10^{13} cm.

Until a person gets used to working with numbers like 7×10^{10} and 1.5×10^{13}, they can be a bit confusing. Here, for example, the radius of the Sun, 7×10^{10} cm and 1.5×10^{13} cm for its distance from us, might seem to be about the same. At a first glance, without experience, these two numbers might not seem to be very different. However, it turns out (if you bother to divide 7×10^{10} into 1.5×10^{13}) that 1.5×10^{13} is more than two hundred times bigger. Between us and the Sun, that is, there is room to put about 214 other suns, all the same size, in a line. So, although the Sun is large, our distance from it is a great deal larger.

TEMPERATURE The Sun is hot, as everyone knows. Its surface temperature is approximately 5000° Celsius. (Another new term, naturally!) Degrees Celsius (°C) is not so bad if the temperature is high. Merely double it to get the equivalent in degrees Fahrenheit (°F); the temperature of the Sun is about 10,000°F.

This doubling to change from degrees Celsius to degrees Fahrenheit will give you a fair approximate result when the Celsius temperature is higher than about 80°C. For example, 100°C (which you would have approximated at 200°F) is exactly the same as 212°F. At more ordinary temperatures, this does not work as well; room temperature, say 20°C, is the same as 68°F. But at −40°, they are identical; −40°C is exactly −40°F.

However, any way you say it, the surface of the Sun is not cool. How do we know that this is the temperature of the Sun's surface? No one has been there; no satellite from Earth has probed this part of our solar system directly. The answer is that we are not absolutely sure. The same doubts apply to whether or not our Milky Way galaxy actually would look like the Andromeda galaxy, or whether our Sun is really in the approximate position in our galaxy that I said it was in locating that X in Figure 1–1. So, now is it that we know these things, or perhaps think that we know them?

The full answer to this question even for the few details that have been discussed up to this point in the first chapter would require a much larger book than this one. But we can get some idea of why we think we know the temperature of the surface of the Sun. To see this, it would be helpful for you to perform a quick experiment. You will require some matches, a paper clip, a very dark enclosure (such as a closet), a little patience, close observation, and asbestos fingers, or a willingness to suffer a small, minor burn which will heal in less than a day. Straighten out the paper clip, so that you can heat the tip of one end in the flame of a match. Do so, in the darkened space, and closely observe the color of the heated tip.

Usually, it takes two or three matches all burning together to heat the tip enough so that it will glow visibly in the dark. And this is what is needed. After the tip is hot enough so that it is glowing, extinguish the match flame and watch the tip closely. It will cool pretty fast. But, and this is the whole point, notice the change in color as it cools. It becomes less and less orange, more and more reddish, finally fading away to a dull red, and then to no visible glow, at all.

If we knew the relation between the color of the glow and the temperature, and could measure the color, we could tell the temperature from the color. This can be done. The orangish glow that you were probably able to achieve is equivalent to a temperature of about 900°C; the dull red glow just before no glow can be seen indicates a temperature of about 500°C. Above 900°C, the color changes from organgish to a yellowish then to white, and still hotter yet, to a bluish white. From the color of the surface of the Sun, it can be estimated that its temperature is in the neighborhood of 5000°C. However this is true if, and only if, we assume that the same color-temperature relationship that can be observed here on Earth is equally applicable to conditions at the surface of the Sun.

There are several reasons why we can make this assumption about similarity in conditions for color and temperature here and there, but it would take several pages to describe them and would involve a rather prodigous effort by the student, as well. Unfortunately we must conserve our space and effort for other subjects, so, if you are interested, the place to begin is in the index of other books under the entry, "black body radiation."

While we are still thinking about the surface of the Sun, another semiphilosophical point can be considered. What is a surface? A surface is usually a limiting area. On one side we have one kind of thing; on the other side, we have something else. This is one problem with the Sun. There is no sharp limit for the edge, or surface, of the Sun. Over a distance of several miles, as we would go from outside toward the Sun's center, the substance of the Sun, mostly hydrogen atoms and ions (we will get to these later) become more and more numerous. But there is no place anywhere near the Sun where there are no hydrogen atoms and ions and, a little bit nearer, where there are a lot. Hydrogen ions, from the Sun, often shoot past the Earth, and on farther out, for example.

So when we speak of the surface of the Sun, we are actually talking about a fairly thick shell. On the outside of this shell, there are not very many hydrogen atoms and ions around; on the inside, several miles away, toward the center of the Sun, there are many many more. Still further inside the Sun, the great majority of particles that we would find (according to deductions from what man can actually see) would be hydrogen ions and electrons, at a temperature of $5 \times 10^6 \,°C$.

THE HYDROGEN ION

What is a hydrogen ion? Since no one has ever seen one, any description is bound to be false to some extent. Nonetheless to think about what a hydrogen atom might look like, try this idea.

Think of an incredibly small object, a very, very tiny ball, with a diameter of 1×10^{-13} cm. Notice the negative exponent, *minus* 13. This signifies a 1 followed by 13 zeros in the bottom part (denominator) of a fraction, with a 1, on the top, like this:

$$\frac{1}{10,000,000,000,000} \text{ cm}$$

There is no way really to describe such a small object as though it had a shape;

so, to think of it as a ball is probably not correct, but it is as good as any other way. This tiny object has a positive electrical charge. Almost everyone has had the experience of accidentally receiving an electrical "shock." When that happened, several millions of electrical charges (probably both positive and negative) passed through your body, or some part of your body.

In the interior of the Sun, you can think of these hydrogen ions, or **protons** which is another name for the same thing, as zipping about in all directions, first this way, then that, after two of them collide; and after another collision with another proton, darting off in some still different direction, until the next collision. On the average, though, the distance from one proton to the next in the interior of the Sun is about 1×10^{-8} cm. This sounds as though they were close together,

$$\frac{1}{100,000,000} \text{ cm apart}$$

but compared to the size of the protons themselves, they are much farther apart from each other than the Earth is from the Sun. So, in their terms we could say that, being so far apart, they do not collide with one another until they have traveled a very long distance, on the average.

ENERGY FROM FUSION AND ITS RESULTS

Even so, by a complicated process that is not well understood, every once in a while (several millions of times a second considering the whole of the interior of the Sun) four protons interact in what is called a fusion reaction to form one helium ion and two positrons. As a result of this, energy is released because (we could say) the mass of the one helium ion and the two positrons is slightly less than the mass of the four protons. (This would be something like putting four oranges which weighed $\frac{1}{2}$lb each in a sack so that the total weight of oranges was 2lb. Then, an instant later, the sack would eject two cumquats, of about $\frac{1}{2}$oz each, and now would contain one very, very hot grapefruit which weighed $1\frac{1}{2}$lb).

To put it differently, as the Sun emits energy, some of which we receive here on Earth, its mass becomes less. There is no cause for concern, however, because in 10^9 years (that's 1 billion years) only 1% of the hydrogen ions (protons) in the Sun will have been converted into helium ions.

The energy released in this fusion process in the Sun is important to us for another reason. The Sun is big and heavy, as we know. It is so heavy that if it were not for the continuous release of energy, which keeps the protons moving at very high velocities and continuously bumping into each other, they would instead be crushed into a much smaller space by the gravitational forces between and among all the protons and all the other particles, in the Sun. The Sun would collapse to a much smaller size, if it was not continuously releasing energy.

Eventually of course, almost all the protons in the central region of the Sun will be consumed, changed by fusion into helium ions. At that time, far in the future, when less energy is produced, the Sun will start to collapse. When anything this big shrinks, it gets hotter. As the Sun gets hotter and somewhat smaller, the temperature of the interior will reach perhaps 10^8 °C, or higher. In the interior at that time, remember, there will be mostly helium ions. At that very high temperature, helium ions will undergo fusion, forming carbon ions, mostly, which have less mass than the original helium ions. So, the Sun will emit energy, get still hotter, and expand, which will cool it a little.

This expansion is expected in about 5×10^9 years from now. The Sun's light will be more reddish than it is now (it will be cooler), and it will be very large, more than 200 times larger in diameter than it is now, or more than 8 million times larger in volume. If the Earth still exists, it will be inside the Sun! Eventually, many eons after that, when the helium has mostly been consumed by fusion into other ions, the Sun will shrink again, to a much smaller size than it ever had been, to a sphere about the same size as the Earth is now. In this shrinking it will of course become much hotter, and the color of the light it emits will be less reddish, whiter. The Sun will have become a "white dwarf" star. It will still have almost the same mass, only a few percent less, than it has now. Hence its density, the amount of mass in a given volume, will be much greater than it is now. At present, the average density of the Sun is only a little more than the density of water. A teaspoonful of Sun today is only half again as heavy as a teaspoonful of water. A teaspoonful of clay has about twice the mass of a teaspoonful of today's Sun. But then, as a white dwarf, a teaspoonful of Sun will be more massive than a Rolls Royce car.

The prediction of the eventual fate of our Sun is based upon astronomical observations of the stars. In the sky now, there are many white dwarfs which once, probably, were like our Sun is now. Stars that are more or less about the size of our Sun, or smaller, will probably evolve into white dwarfs eventually. Larger stars often have a different end. For example, stars that are more than twice the size of our Sun, and near the end of their life, will expand to several million times their previous size, just as our Sun will do, at about the time when all the hydrogen ions in their interior have fused into helium ions. That is, these large stars, like our Sun, will first shrink and get much hotter. Helium ions will fuse to form carbon ions, energy will be released, and the star will expand enormously. Then, when all the helium ions, or many of them, have been exhausted, the very large reddish colored star will shrink, like our Sun, on its way to becoming a white dwarf.

But, because the star is more massive than our Sun, when it becomes more or less the size of a white dwarf, its gravitational pull will still be very great, great enough to cause that star to become even smaller. The compressive forces in such an object are terrific. The shrunken star will become very, very hot, so hot that it will explode, sending debris at high velocities all over its part of the universe. It will emit a torrent of light during the explosion that will last for several days, and it will become one of the brightest objects around, much brighter than any star you or I have seen. (The last supernova, as such stars are called, was seen in our galaxy in 1604. On the average, there is a supernova in each galaxy about once every hundred years, so the few supernovas in our galaxy since 1604, if any, have occurred in a distant part and were not noticed.

After the explosion, only a small fragment of the original star will remain. The remainder is called a neutron star. For a star, a neutron star is quite small, perhaps 5 to 10 miles in diameter, although it is quite dense. A teaspoonful of neutron star would weigh more than all the Rolls Royces ever made.

But let's get back to the explosion; without that explosion, you and I could not exist. The details are not clear, but in that explosion all the elements except hydrogen, helium, carbon, and perhaps a couple of others are formed. As you perhaps know, man is mostly water, which is made of two elements, hydrogen and oxygen. In addition, a typical living body contains nitrogen, carbon, and more hydrogen, as well as lesser amounts of other elements, iron, copper, calcium, phosphorus, cobalt, manganese, sodium, chlorine, potassium, and others. With the exception of hydrogen, all of the elements of which you are composed were either generated in a nice well behaved star, like our Sun which makes helium (there is not much helium in living tissue); or in a "red giant," which our Sun will one day become, where carbon is generated; or from a supernova explosion of a larger star as it shrinks to a size less than a white dwarf, where, in a very short time, all the other elements are formed.

The question is, how did those elements scattered all over the place in explosions of supernovas ever manage to get collected into one location, on this earth, and some of them anyway, get into you and me? Is it conceivable that we are indeed the stuff of which stars are made? We will try to get at the answer to that question in the rest of the pages in this chapter.

THE HOT BIG BANG AND THE UNIVERSE

Probably it will be best to go back to the very beginning. Well perhaps not to the actual beginning, but back about 10^{10} years ago. At that point in time, all of the matter that now composes the universe was in the same place, very dense, very compact. What it was like before that is anyone's guess. For our purposes we can say that time began about 10^{10} years ago, at time zero.

The entire universe, then, at time zero, was in what can be called a singular state. What this means in detail is not understood, even if it is correct. A pertinent disclaimer must be made here. Science deals with the material universe, since this is all that can be observed. As you know, other versions of the beginning of the universe are offered by man's religious beliefs. When all of these versions, the several religious and the scientific, are examined closely, many differing details can be noted. Yet, the broad scope of each version is, surprisingly perhaps, not inconsistent with the other versions. It is reasonable to suggest that the versions attributable to religious beliefs were developed from traditions that depended upon the cultures of the peoples who related them, parent to children, for thousands of generations. In the transmission, details that were probably never clearly understood became altered.

There is nothing factual in the scientific version presented here. It is theoretical; it too reflects details that are not now clearly understood. It differs from other versions only because it is based upon observed phenomena, interpreted by the imagination and intellect of mankind and subjected to thoughtful criticism as it has been developed during the past few decades. Here then, is the current scientific version, subject to change as more observations are made and as more careful criticisms are made:

Probably, at time zero, the primordal cosmos, or egg, or fireball, the whole universe collected in one place, consisted of either neutrons, or protons and electrons. A **neutron** is a tiny particle, about the same size and mass as a proton, but it has no charge. An **electron** is about 2000 times lighter than a proton or neutron, with a single negative electrical charge. Anyway, neutrons can sometimes spontaneously change into protons, electrons, and antineutrinos. An **antineutrino** is something else again. It has no charge; it has no mass (as far as you can tell); it travels at a high velocity through any kind of matter without leaving (well, hardly) a trace of its passage behind.

So, if the cosmos at time zero was not composed of protons and electrons, it soon became composed of protons, electrons, and antineutrinos and about 100 seconds later, some of those protons had interacted with other protons to form helium ions. Most of the helium in the universe today was formed in those first few seconds. As nearly as we can tell, there are approximately 10^{80} (and *that* is a very,very large number) atoms of all kinds in the entire universe. About 8 out of every 100 of these are helium, and about 92 out of every 100 are hydrogen. The rest, of the hundred or so different kinds of elemental atoms, are present in the universe in only trace amounts, about one atom in every thousand is an atom of the other elements.

Now, when you have about 10^{80} atoms (or ions) in one place, and about 10% of them fuse to form helium ions instead of the hydrogen ions they were before, about 100 seconds ago, an awesome amount of energy is released. The temperature goes up terrifically; there is a big, big explosion. In words that are inadequate, there occurs a "hot big bang." This theory of the beginning of the universe is called the hot big bang theory.

The energy that was released was in the form of photons. **Photons** are particles of light, for example. To see an object, you need photons. An ordinary source of light, any light, emits photons, and if some of those photons strike an object and bounce off from it, entering your eye, you can see that object. Back in the time when the universe was perhaps only a few seconds old, there were a tremendous number of photons. Traces of these photons can be observed today. Sensitive equipment is necessary, which was first available after the end of World War II (though some earlier observations, we now know, were made as long ago as 1928). In fact, these recent observations of the remnants of those photons, which were produced 10^{10} years ago, are one of the "proofs" that the hot big bang theory might be correct.

Meanwhile, as a result of the hot big bang explosion, the protons and electrons were blasted away from the cosmos-egg into what was empty space just before that. When a proton and an electron get together, they form a hydrogen atom. This is a pretty good description of a **hydrogen atom**: a proton with an associated electron.

By about 10^5 years after the hot big bang, things had cooled off somewhat, cool enough so that the electrons could become associated with the protons, one for one, to form hydrogen atoms. For all practical purposes, 10^5 years (100,000 years) after the hot big bang is still 10^{10} years ago. For example, to compare 10^5 years with 10^{10} years is like comparing $10 to $1,000,000. For all practical purposes, if you subtract $10 from $1,000,000, you would still have $1,000,000 left.

So, almost 10^{10} years ago this was the situation: hydrogen atoms and helium atoms were hurtling through space away from the center of that explosion. (**Helium atoms** are helium ions with two electrons associated; one ion for each two electrons.) Naturally, you would expect some turbulence as these atoms traveled onward. Here and there small eddies, whirls and whorls, would form and disappear. Once in a while, though, a larger disturbance would form, and remain. Those hydrogen and helium atoms would then travel together at high speed away from the center of the explosion.

Now, any assembly of matter has some mass. Anything that has mass will attract, by gravity, any other object that has some mass. If the assembly of matter is big enough, its attraction for another object that has mass will be fairly large, large enough to trap the other object. You and I are trapped (thank goodness) on Earth because it is an assembly of matter big enough to hold us by gravitational attraction. So, as this larger whirl, or whorl, or whatever kind of a disturbance it was, continued, it attracted other hydrogen and helium atoms to it. It kept growing in size.

This kind of thing, probably, was happening in many different spots all over the expanding cloud of hydrogen and helium atoms. Stars were being born!

As a whirl or whorl grew by adding more atoms of hydrogen and helium (mostly hydrogen because about 90% was left over as hydrogen from the hot big bang), it began to shrink under the force of its own gravity. As it became smaller, it became hotter. When it was big enough, about 10^{33} g (remember, this stands for grams) or more, and hot enough, about 10^7 °C, the protons in the interior lost their electrons, some of which sort of hung around inside the newly born star and some of which escaped. But the protons at this temperature can fuse to form helium ions, releasing energy, just as our Sun does, now. Eventually some of those stars decayed to white dwarfs, some to supernovas, which exploded leaving a neutron star as a residue, and, some theorists think, other stars decayed and became black holes. It is supernovas which interest us.

In the final moments, just before a supernova erupts, there is thought to be a surplus of neutrons in the outer parts of the star. In the confusion of the catastrophic implosion of the interior of that star, as it becomes a supernova, the outer regions form other elements, iron, californium, molybdenum, barium, calcium, phosphorus, and so on, in an instant. And then, an instant later, those atoms are blasted all over the place, traveling at high velocities.

If this kind of thing happens often enough, after a while, after zillions of years, there will be a large number of atoms of other elements, in addition to the hydrogen and helium, all moving away from that center of the hot big bang, all forming whirls and whorls, and sometimes new stars, which finally decay as the others did before them. Many of these whirls or whorls developed not into a single star but into solar systems. And one of these solar systems is our own. Our system was formed about 5.4×10^9 years after the hot big bang, or about 4.6×10^9 years ago. Our solar system had its beginning a little more than $4\frac{1}{2}$ billion years ago, when the universe was about half as old as it is now.

THE SOLAR SYSTEM

Our solar system was unique from the first moments. For example, often as these collections of atoms develop into a whirling cloud, the speed of rotation is relatively high. Then, as gravitational effects cause the collection to shrink, its rotational speed increases and the collection splits approximately in two, ultimately forming two stars. In our case, the initial rotation of the collection of atoms was rather slow so that, as it shrank and increased its rotational speed, the main central part of the cloud did not split into parts; we have one star for our Sun and a number of other condensed parts of the cloud that are not big enough or hot enough to be stars.

Of course, as this collection of atoms in the main part became smaller in size and the atoms were forced closer together by gravity, the whole mass increased in temperature. Eventually the hydrogen to helium fusion began (and continues to this day) and energy was continuously emitted. Some of this energy swept out from the new star in the form of a "solar wind," protons moving at high velocity. The solar wind reached distances further from the Sun than the Earth now is, even out beyond the orbit of Mars. The lighter atoms, hydrogen and helium, were swept out beyond these limits, and eventually condensed to form the planets Jupiter, Saturn, and those further away. The heavier debris that now existed in the collection of atoms was not swept away as extensively. This remained in the vicinity of the Sun and condensed to form the inner planets, Mercury, Venus, our Earth, and, further out, Mars and the Asteroids (just beyond Mars). This debris consisted mostly of iron, magnesium, silicon, oxygen, sulfur, nitrogen, and carbon, along with the hydrogen and helium that happened not to be swept away. As a result, the inner planets, Mercury, Venus, the Earth and its moon, Mars, and the Asteroids, all have about the same composition. They were formed from the accretion of the heavier debris. Our Earth is composed mostly of iron, in the center, and of silicon and oxygen in the outer portions, for example.

The accretion of the debris that formed our Earth was built up over a long period. It is still going on, at the rate of several tons each day. For example, every one has seen a "falling star." Probably not all falling stars have the same origin, but some of them are pieces of debris falling into our Earth after their original formation many, many years ago. Probably in the early history of our Earth, this accretion of debris was more frequent, and no doubt some of the pieces were quite large. The crater in Arizona, Figure 1–2, is evidence of a relatively recent large piece and its collision with our Earth.

Figure 1–2.   The Meteor Crater, Arizona. [Photograph by Joseph Muench.]

THE
ORIGINAL
ATMOSPHERE
AND THE
BEGINNING
OF LIFE

Not all of the matter that collected to form our Earth was solid. Water, a compound of hydrogen and oxygen; methane, a compound of hydrogen and carbon; and ammonia, a compound of hydrogen and nitrogen; where collected. Methane and ammonia are gases under conditions familiar to us today. Methane is used today as a fuel; it is a component of some natural and artificial fuel gases. Ammonia, the compound of hydrogen and nitrogen, should not be confused with the "ammonia" used in household cleaning. This household cleaner is made from the compound, ammonia, dissolved in water, usually with a little soap or detergent and perhaps some perfume added. The compound, ammonia, is responsible for the penetrating odor of household ammonia; as a gas, it tends to leave the water in which it is dissolved.

These substances, steam (from the water), methane, and ammonia, formed the early atmosphere of Mercury and Venus, also. These planets are close enough to the Sun to be much hotter than the Earth (The surface of Venus, for example, is at the temperature of molten lead.) Mars is sufficiently far from the Sun to be much cooler than the Earth; the little water that is there is mostly in the form of ice. Only the Earth is at the delightfully proper distance for water to exist predominately in the liquid state.

The atmosphere of the inner planets is not the same now as it was initially. Venus, for example has lost almost all of its methane, water, and ammonia; today its atmosphere is largely nitrogen and carbon dioxide. How did this happen?

Venus is close to the Sun and subject to more intense radiation than the Earth. This intense energy from the Sun ripped apart the steam molecules, breaking them up into hydrogen and oxygen atoms. It did the same to the methane, producing carbon and hydrogen atoms; and the same to the ammonia, producing nitrogen and hydrogen atoms. Now Venus is about the same size as the Earth. As on the Earth, any object that is going fast enough, about 7 miles per second (mi/sec) or faster, will leave the planet if it is going straight upward. This figure, 7 mi/sec straight upward, is the approximate "escape velocity" for the Earth and for Venus. Hydrogen atoms are light. At the high temperatures that exist on Venus, they travel at an average speed much greater than the escape velocity. Therefore, as the steam, methane, and ammonia molecules were broken apart by the radiation from the Sun, the hydrogen atoms tended to leave, especially when they happened to be going straight up. This was bound to happen sooner or later, since gaseous atoms bat about in all directions. This left carbon, oxygen, and nitrogen atoms in the gaseous atmosphere of Venus. The carbon and oxygen atoms joined together over the years, to form carbon dioxide; and the nitrogen atoms joined with each other, forming pairs of nitrogen atoms (which we would identify as "nitrogen molecules" today).

Much the same happened on our Earth. It too developed an atmosphere consisting of carbon dioxide and nitrogen as the major constituents. The major difference, however, between the Earth and Venus was that, on our planet, most of the water was in the liquid state, or solid state, as ice. Because Earth was cooler, we did not lose all of our water.

It's a good thing our Earth was as massive as it was, at least for life as we know it. Mars is cooler than the Earth, since its is further than we are from the Sun, and Mars has lost most of its water. Mars is smaller than the Earth; the escape velocity is about half that of the Earth. When the original compounds in the atmosphere of Mars were ripped into atoms by radiation from the Sun, the hydrogen atoms escaped. Even though Mars is cooler, it is still generally warm enough for light-weight atoms to achieve the lesser required escape velocity. If the Earth were nearer the Sun, or further away, if it were heavier, or lighter, our atmosphere would have been different than it was then. Probably our atmosphere would not have evolved into what it is today. Life would be different here, or nonexistent as a consequence.

So, as of about 3 billion years ago, the Earth had an atmosphere which was different from the rest of the inner planets. It contained carbon dioxide and nitrogen, with some water vapor, and with liquid water on the surface. It was the carbon dioxide and liquid water that caused a big change. Carbon dioxide dissolves in water; this solution is responsible for the pleasantly sharp taste of "carbonated" beverages, including the popular cola and noncola soft drinks, beer, and sparkling wines, For you and for me, carbonated water is harmless (and pleasant); for stones and rocks, it is disaster. Given enough time, say several millions of years, carbonated water is very corrosive. You can get a rough idea of this corrosive power by noting that carbonated water is called carbonic acid by chemists. Acids are corrosive. As the carbon dioxide in our Earth's atmosphere dissolved in the water, the nitrogen molecules were left. Eventually, our atmosphere was practically all nitrogen.

However, not quite *all* was nitrogen. Although radiation from the Sun did indeed rip apart many of the methane, ammonia, and water vapor molecules in our early atmosphere, some were not completely torn apart. Instead of only atoms, fragments of molecules were formed. Some idea of the character of these fragments can be obtained by looking at the chemists' name for them: free radicals.

A free radical is a reactive species. Saying it differently, these free radicals will join with each other to make larger molecules.

According to one theory, not widely accepted, the crude oil deposits were formed in this way. The substances in deposists of crude oil are large molecules, for the most part.

Some of the other molecules formed included the amino acids, today best known as the constituents of protein and, probably, the basis of life. Also, some sugars, some carbon–nitrogen ring compounds (so-called because of the ring shape of the molecules) were formed, along with many other varieties. In general, these molecules were washed out of the atmosphere where they were formed into the primordal oceans by rain.

Then, a little over 3 billion years ago, somehow, by processes that have not yet been reproduced in the laboratory, these molecules joined in a unique way so that the assembly of matter was able to reproduce itself. Simple cellular organisms developed next, again by processes that are not understood as yet. About all that we can be sure of is that oxygen, if it were in the atmosphere then, would have been poisonous to these early forms of life.

THE OXYGEN ATMOSPHERE AND THE NEXT STEPS IN EVOLUTION

About one billion years later, however, a major step forward (for us), occurred. Chelated magnesium became involved in living cells. In words that everyone can understand, the first forerunners of chlorophyll appeared in living cells. Chlorophyll contains magnesium in a special way; the magnesium is said to be chelated because, we think it is held in the chlorophyll molecule something like a crab's claw might hold an object. (The word, chelate, is derived from the Greek word for claw or hoof.) The hemoglobin in our blood contains chelated iron, but this appeared later in the evolutionary development of living forms.

With the chelated magnesium incorporated into living tissue, a new source of energy became available. These living organisms were capable of using the light from the Sun which filtered into the depths of the ocean water where they lived. This energy from the Sun was used to change carbon dioxide, dissolved in the ocean water from the atmosphere above, into sugars and other molecules. But, in doing so, the living organism discards oxygen, which finds its way into the atmosphere. Now the first ecological crisis occurred; some living forms were ejecting poisonous oxygen into the atmosphere!

Fortunately, there was enough iron dissolved in the oceans to avert this crisis temporarily and, incidentally, do us some good today. That dissolved iron reacted with the oxygen and formed the solid, iron oxide. Iron oxide is insoluble in water; it precipitated, that is, drifted to the bottom of the oceans. And it piled up. (Much later, geological processes lifted those iron oxide deposits above the water level, forming mountains of iron oxide. We obtain our iron today from those deposits, using it to make varieties of steel for man's use.) Over several millions of years, as the dissolved iron in the oceans was consumed, the quantity of oxygen in the atmosphere increased. This happened so slowly, however, that some living organisms had evolved to forms that could endure oxygen. As time went on, some organisms were able to use the oxygen, as you and I do, to support life. According to the theory of evolution, then, you and I would be quite different, or not here at all, had it not been for that iron dissolved in the primordal ocean water.

Another interesting development also took place as the quantity of oxygen increased in the atmosphere. The radiation from the Sun which rips molecules apart, or makes the free radical fragments, could also make mincemeat of us and of other modern living forms. Fortunately, we are protected by a special form of that same atmospheric oxygen. High above the Earth, at what could be thought of as the outer part of our atmosphere, the dangerous radiation from the Sun interacts with the large amounts of oxygen present. It rips oxygen molecules apart, forming oxygen atoms which recombine into ozone molecules. These ozone molecules form a relatively thick layer in our upper atmosphere and are able to absorb a large part of radiation from the Sun which would be inimical to life. Water will also absorb this kind of radiation.

Until the atmosphere contained enough oxygen to form ozone which would protect life, no life could exist on dry land, or in the top foot or so of the oceans. All living forms that ventured, or were pushed, into the top foot of the ocean waters, or onto land, perished from the Sun's energetic radiation. Until about 600 million years ago, all life was in the oceans 1 or 2 ft and more below the surface. It was at this time that colonies of cells began to form, also. In these colonies different cells took on different functions, each subgroup contributing to the maintainence of the colony.

As nearly as we can estimate today, up until about 200 million years ago, all the dry land mass of the Earth was accumulated pretty much in one place; the rest of the Earth was covered with water. That is, for the first more or less 3 billion years of the Earth's existence, radioactive heat had been building up, and up, and up. There is more on this in Chapters 6 and 12, but for now we will only note that radioactive processes emit energy. In this case, that energy was emitted deep in the interior of the Earth and it had no way to get out. The interior of the Earth, originally cool, became warmer and warmer and hotter and hotter. Finally, it erupted in what you and I would call a slow process. In terms of the billions of years of the Earth's history, it could be called a rapid process. By that process, slow or rapid according to the point of view, the single land mass began to break up into the continents and smaller islands that we recognize today. This process is still going on. The contintents are moving slowly, accompanied by earthquakes and volcanic eruptions as dramatic evidence of the radioactive heat generated, even now, deep in the interior.

At this time there was a great assortment of life on the Earth; conifer like trees and other plants, all sorts of reptiles some of which we now call dinosaurs, as well as some insects, and many different simpler forms.

About 100 million years ago, azeleas appeared, the hardwood trees developed, grasses began, rhododendrons were first seen. Then, about 70 million years ago, the large dinosaurs became extinct. We could guess that they encountered some kind of ecological crisis (for them), but the details remain unknown at this time. At any rate, at about this time, the mammals appeared. Evolution continued and had we been present about 2 million years ago, we would have seen a green world, teeming with life. Magnificent forests, acre upon acre of grassy plains, herds of animals, birds, insects, fish in abundance.

The Earth was balanced, probably; it was a magic garden. It was about at this time that man appeared.

MAN AND HIS EARTH

The Earth is not a magic garden today, and it is arguable whether it should be returned to that state because this almost certainly would mean the cessation of man's life as we know it now. Few, however, would argue that we should not at least try to get it into a better balance. There does not seem to be any other place nearby to flee to, for example, our Sun is a single star. Many stars exist as pairs, probably with planets in orbits around the pair. Such planets are subject to wide ranges of temperature fluctuation. When the planet is in position to receive radiation from both of its suns, the temperature on the planet is high. When it has moved to a different position so that one sun hides the other, the temperature is much lower. Life as we know it, as least, could not exist on such a planet.

If our Sun were much more massive than it is, say half again larger, it would have a lifetime of about $2\frac{1}{2}$ billion years; then it would expand to a red giant state. Our Earth would long ago have been consumed by the Sun. (Fortunately, given its size, our Sun has another 4 or 5 billion years before it becomes a red giant.) If our Sun were about one quarter more massive than it is, it would be ready to form a red giant at an age of about $4\frac{1}{2}$ billion years; this is the present age of our Sun. If our Sun were a little smaller, say about three fourths of its actual mass, its temperature would be lower, our Earth would be too cold to support life, even the simplest forms that we know of.

If our Earth were a little closer to the Sun, its temperature would be hot enough to boil water on the surface. Water would be a gas, not a liquid. Life as we know it could not exist. Also, at that higher temperature condition, high winds, more forceful than a hurricane, would buffet the surface. On the other hand, if the Earth were a little farther from the Sun, the surface would be mostly ice. Life would be uncomfortable, if not impossible, at least as we know it.

If the Earth was a little heavier than it is, with larger oceans and more iron dissolved in those oceans, oxygen produced by early forms of life would still be trapped by the dissolved iron. Little oxygen, if any, would yet have entered the atmosphere. There would be no ozone layer to protect life as we know it from the intensity of the harmful ultraviolet radiation from the Sun. We could not live on such a planet. If the Earth were a little lighter, more hydrogen would have escaped from the atmosphere in the early history of its development. Advanced forms of life could not develop.

Of course, considering the 10^{11} stars in our Milky Way galaxy alone, it is reasonable to imagine that some of them have planets and that more than a few might have a planet about the size of ours, about the same distance (maybe almost exactly the same distance) from a star almost exactly the size of our Sun, with that sun about as old as ours is. If this is indeed reasonable, then we could conclude that somewhere in this galaxy life somewhat like ours exists.

Our galaxy is a big place in space. It is approximately 1000 parsecs thick (in its thickest spot). A parsec is the distance traveled by light in about 3 years; and light travels at the rate of 186,000 miles a second (mi/sec), or 3×10^{10} centimeters per second (cm/sec). Our galaxy is a disclike object, with a diameter of about 30,000 parsecs. In all that space, the chance of finding a specially sized single star, with a suitably sized planet the right distance away, and more or less near to us as well, is very small. We do not know enough about how stars and planets are formed to make a reliable estimate of how many earthlike planets with friendly single stars for suns there are among the 10^{11} stars in our galaxy. Given the very special conditions that we enjoy and given the immensity of our galaxy, it is very unlikely that intelligent life at least somewhat similar to ours exists within 100 parsecs, perhaps not within 1000 parsecs (that is, 300 years of travel at the speed of light). For all practical purposes we are alone. No external material help is available, we cannot even communicate with others, since the most rapid communication techniques we know travel at the speed of light. It would take years to get an answer, even if we knew what language to use.

These considerations have caused some people to despair and others to ask questions about the origin of our unique condition. Some people see the hand of a divine Creator, of a paternal and omnipotent Being, who designed our home for us and conclude that we are not alone, not unaided nor faltering without hope. Indeed there are persuasive arguments both favoring and against any view of humans on Earth and the existence of a Creator. In the ultimate, individual determinations are not made by argument but by faith (or a lack of faith) as the predominant factor. This is not a book on theological cosmology, in any event, but it is useful to be aware of the modern basis for questions regarding purpose and creation and why people hold to the answers they have found.

Our excursion into the universe, our galaxy, our solar system, and our Earth has been brief. Many details have been mentioned only briefly or not at all. For your next steps, if you are interested, the references at the end of this chapter are recommended although they too are only a beginning.

It has been necessary to present this story as though it was a kind of myth, somehow known to be true. Only once, with the hot paper clip, did I attempt to give you a suggestion that what has been said is based upon observations, experience, experimental evidence, and deductions. Some of the references at the end of this chapter discuss the detailed nature of these deductions at length, but the going gets mathematically involved rather quickly. It is important, however, to see just what man can deduce from his observations. It is also important to realize that man has been doing this for years and years. Nonetheless, it would be a bit of a relief to get back to familiar ground and to think about smaller systems than galaxies and stars. So, the next chapter comes from a series of lectures on "The Chemical History of a Candle," by Michael Faraday, first presented more than 100 years ago.

Before we get to the perceptible reality of a burning candle, consider a profound question. Take a walk some starry night; look at the stars. What are they for? Why are they there? What good are they? The Psalms give us some answers, other poets have provided their interpretations. But from this chapter, if it takes more than 10^9 years to make elements other than hydrogen, helium, and carbon; if it took 10^8 galaxies, each with 10^{11} stars to set up the conditions for life on one planet (and perhaps on a few others); if this was required for that life to exist, then we could say of the stars, they are there because we are here.

FOR FURTHER INFORMATION

A. J. Zadde and T. A. Smits: *Making Friends with the Stars.* Barnes and Noble, New York, 1963. This paperback book introduces the stars in the constellations, naming them and identifying them. It is not technical. The theme of the author is, Go out: look at the stars.

J. Cornell and E. N. Hayes (eds.): *Man and Cosmos.* W. W. Norton, New York, 1975. A perspective view of the solar system, not the cosmos, in a set of nine lectures addressed to a general audience.

E. Keller: Man and the universe. *Chemistry,* **45**(7):4 (1972). A four part discussion ranging from early ideas about the universe to radio astronomy and quasars. Readable and reliable. It is followed, on pages 23–26, by remarks on the early history of the moon by J. A. Wood.

D. W. Sciama: *Modern Cosmology.* Cambridge University Press, Cambridge, 1971. This is perhaps the most recent and authoritative thorough survey of our understanding of cosmology available in English. The discerning reader will find that I have used Sciama heavily in preparing the manuscript for this chapter in this book.

A. L. Hammond: Stellar old age (II): neutron stars and pulsars. *Science,* **171**:1133 (1971). The weekly publication of the AAAS, *Science,* carries frequently appearing surveys of current topics under the title "Research Topics." As written by Hammond and his colleagues, these are uniformly excellent and informative, well worth the small effort necessary for those not in science who wish to be informed about current developments in science. Thus, for example, see Hammond's remarks on the uniqueness of the Earth's climate, *Science,* **187**:245 (1975), or W. D. Metz: *Science,* **186**:814 (1974) on models of the origin of the solar system.

P. J. Wyllie: *The Way the Earth Works.* Wiley, New York, 1976. Clear and concise, on the dynamics of the Earth, by a practiced author, with a bibliography of suggested readings.

J. F. Dewey: Plate tectonics. *Sci. Amer.,* **226**(5):56 (1972). Once the earth was formed, how did the continents get their present positions? Plate tectonics has one answer. Like all other articles in this magazine, the illustrations alone inform as well as support the printed words. For further, and more technical, details see W. A. Elders et al.: *Science*, **178**:15 (1972) on the spreading of that part of the Earth's crust known as Southern California.

R. Howard: Recent solar research. *Science*, **177**:1157 (1972). This article is technical. Once mastered, the beautiful complexity of our Sun can be better appreciated.

There are several articles on the origin of the universe that may be of interest; for starters, see B. J. Bok: The birth of stars. *Sci. Amer.*, **227**:48 (August, 1972); M. J. Reese and J. Silk: *Sci. Amer.*, **222**:26 (June, 1970) on the origin of galaxies, or a series of articles in *The Science Teacher* from the origins of the Earth to lunar exploration puzzles, **40**:20 (March, 1973), or F. H. Shu: *Amer. Sci.*, **61**:524 (1973) on the process of star formation in galaxies.

V. Trimble: Cosmology: Man's place in the universe. *Am. Sci.*, **65**:76 (1976). There are perhaps four possibilities: An anthropic view (presented in this chapter of this book), the universe is more or less what it is because we are here to observe it; there may be many universes and ours is one of the few with living beings intelligent enough to ask questions about it; perhaps in the distant future we will have learned enough about the nature of matter to comprehend the origin, the then present, and the future of our single universe; or, the complexity of this universe necessarily generated that which we see and are aware of today.

W. A. Fowler: *Chem. Eng. News*, **90** (March 16, 1964) and *Sci. Amer.*, **195**:82 (September, 1956). Two older articles on the origin of the elements.

B. E. Turner: Interstellar molecules. *Sci. Amer.*, **228**:51 (March, 1973). Dr. Turner is a radio astronomer who has studied the composition of the otherwise empty space in the nearby regions of our universe. He and his colleagues have identified more than two dozen different kinds of molecules in space. This article describes the means of detection of those species, with conjectures about their origin. For a related bit of information, see D. M. Rank et al.: *Science*, **174**:1083 (1971).

Isaiah 45: 18, on the point that the universe is not a chaos.

C. Sagan and T. Page: UFO's—a scientific debate. Cornell University Press, Ithaca, N.Y., 1973. Maybe yes, maybe no (it says here). Also take a look at the reviews of this book, one is by P. A. Sturrock in *Science*, **180**:593 (1973), but there are others some pro, some con.

R. N. Bracewell: *The Galactic Club*; *Intelligent Life in Outer Space*. W. H. Freeman, San Francisco, 1975. An effective criticism of *Chariots of the Gods*, remarks on Von Daniken's ideas, a lack of imagination in applying economic constraints when considering the plausibility of space probes from out there, to here, plus a lively assortment of challenging suggestions constitute the content of this well written book.

G. Gatewood: Finding unseen worlds. *Astronomy*, **4**(4):6 (1976). The basis for determining whether there is life on other planets in other star-planet systems is yet to be determined. About all we can do now is to talk about probabilities. This article discusses the kind of solid evidence that we might hope to observe. For the argument that we are indeed alone in the universe, see G. Verschuur: We are alone. *Astronomy*, **3**(12):46 (1975).

For a more down to earth view, so to speak, see the entire issue of the *Scientific American* for September, 1975. The theme of this issue is the solar system; articles include a general discussion, origin and evolution, the sun, the inner planets, our moon, other smaller bodies such as the meteoroids, the outer planets, and interplanetary space.

J. R. Gott et al.: Will the universe expand forever?. *Sci. Amer.*, **234**:62 (March, 1976). Presently, the answer is affirmative. What this means, and why the affirmative answer does not seem to be too tentative (maybe!).

P. Moore and I. Nicolson: *Black Holes in Space*, W. W. Norton, New York, 1976. Maybe they don't exist, maybe they do. If they do, this book tells you what black holes are and why some scientists think that they are. You can read it in two or three hours.

This chapter has presented the story of the Earth from its beginning up to the present (as much of it as can be encompassed in a few pages) but what about the future? For one man's idea see the *Fourth of July*, a commemorative magazine published on that date in 1976 by the *Washington Post*, in an article by Thomas O'Toole, the science editor for this distinguished newspaper. The title is, Inventing the future. You may also wish to read *Science, values and human judgment*, by K. R. Hammond and L. Adelman in *Science*, **194**:389 1976.

chapter 2

A Candle Is More Than Wax, Wick, and Flame

The universe, after all, is a big subject, and most of it is pretty far away. Almost any idea about what the universe really is ought to be considered seriously. It's so large a subject that we can never know more than a fraction of all there is to be known. There is a lot of room for ideas.

A candle would seem to be quite the opposite case. A candle is small. It can be right in front of your nose. There is not much to it, or so it might seem. A candle is nothing but a candle, period. However, to say that is as narrow a view as to say the Sun goes around the Earth because it looks that way. In its own way, and in more than the fact that it too, like a star, emits light, a candle is as complex as the universe itself. Of course, we do know more about candles, because we can get closer to them, and we can control them to some extent.

Any student who undertakes the study of a subject like chemistry when their personal interests are elsewhere has the right to ask pointed questions about why this study has been imposed upon them. There can be no really valid universal answer. Generally speaking the study of chemistry helps us to understand the physical world in which we live. How it helps and why it is important to understand our physical surroundings are different for each student. This is another way of saying that each of you will have to find your own personally unique answer.

To help you find your own answer to the question "Why is chemistry important to me?" you could read this whole book, but that is a bit much to ask right now. Instead, try this chapter and react to it and to what you have read in the first chapter. The two combined are intended to provide you with the essentials of your personal answer to the question. The rest of the book, from Chapter 3 on, is intended to fill in some of the details.

THE LARGE ASPECTS OF A SMALL OBJECT

OK, let's get started. This chapter is Lecture II on the Chemical History of a Candle, first delivered during the Christmas holidays in the winter of 1860–61 to a group of young adults by Professor Michael Faraday of the Royal Institution, in London. As it happened, Professor Faraday was 69 years old when he delivered this series of lectures and he thought his memory was failing. So, he very carefully made copious notes, in advance. As nearly as can be determined, the printed version here is an almost verbatim reproduction of his remarks.

Figure 2–1.   A candle flame.

In Lecture I Faraday described the fabrication of a candle; he pointed out that the solid wax melts and is transported up the wick, where it burns. He called attention to the rounded bottom of the flame and to the gap between the bottom of the flame and the top of the candle wax. He explained why the flame is rounded and why the flame does not reach down all the way to the wax (Figure 2–1). Further, he discussed the observation that the wax of the candle melts in the center but not around the edges so as to form a cup in which the liquid melted wax is retained. He identified the different parts of the flame of a candle by examining the shadow cast by the flame when it is illuminated by a strong light and when a scorch pattern is formed on a piece of paper held vertically in the flame but withdrawn before it caught fire.

Now, we are ready for Lecture II. Enjoy!

Where Does a Candle Go to?
A Candle: Brightness of the Flame.—
Air Necessary for Combustion.—
Production of Water.

We were occupied the last time we met in considering the general character and arrangement as regards the fluid portion of a candle, and the way in which that fluid got into the place of combustion. You see, when we have a candle burning fairly in a regular, steady atmosphere, it will have a shape something like the one shown in the diagram (Figure 2–2), and will look pretty uniform, although very curious in its character. And now I have to ask your attention to the means by which we are enabled to ascertain what happens in any particular part of the flame; why it happens; what it does in happening; and where, after all, the whole candle goes to; because, as you know very well, a candle being brought before us and burned, disappears, if burned properly, without the least trace of dirt in the candle stick; and this is a very curious circumstance.

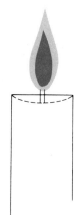

Figure 2–2.   A candle flame, schematic.

Vapor from the Interior of a Flame

In order, then, to examine this candle carefully, I have arranged certain apparatus, the use of which you will see as I go on. Here is a candle; I am about to put the end of this glass tube into the middle of the flame—into that part which old Hooker has represented in the diagram as being rather dark, and which you can see at any time if you will look at a candle carefully, without blowing it about. We will examine this dark part first.

Now I take this bent glass tube, and introduce one end into that part of the flame, and you see at once that something is coming from the flame, out at the other end of the tube; and if I put a flask there, and leave it for a little while, you will see that something from the middle part of the flame is gradually drawn out, and goes through the tube, and into that flask, and there behaves very differently from what it does in the open air (see Figure 2–3). It not only escapes from the end of the tube, but falls down to the bottom of the flask like a heavy substance, as indeed it is. We find that this is the wax of the candle made into a vaporous fluid—not a gas. (You must learn the difference between a gas and a vapor: a gas remains permanent; a vapor is something that will condense.) If you blow out a candle, you perceive a very nasty smell, resulting from the condensation of this vapor. That is very different from what you have outside the flame; and, in order to make that more clear to you, I am about to produce and set fire to a larger portion of this vapor; for what we have in the small way in a candle, to understand thoroughly, we must, as philosophers, produce in a larger way, if needful, that we may examine the different parts.

Combustible Vapor from Wax

And now Mr. Anderson will give me a source of heat, and I am about to show you what that vapor is. Here is some wax in a glass flask, and I am going to

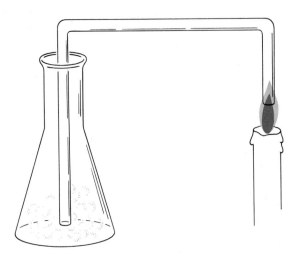

Figure 2–3.   Collecting vapors from the center of a candle flame.

make it hot, as the inside of that candle-flame is hot, and the matter about the wick is hot. (The lecturer placed some wax in a glass flask, and heated it over a lamp. See Figure 2–4.) Now I dare say that is hot enough for me. You see that the wax I put in it has become fluid, and there is a little smoke coming from it. We shall very soon have the vapor rising up. I will make it still hotter, and now we get more of it, so that I can actually pour the vapor out of the flask into that basin, and set it on fire there (Figure 2–5). This, then, is exactly the same kind of vapor as we have in the middle of the candle; and that you may be sure this is the case, let us try whether we have not got here, in this flask, a real combustible vapor out of the middle of the candle. (Taking the flask into which the tube from the candle proceeded, and introducing a lighted taper.) See how it burns (Figure 2–6). Now this is the vapor from the middle of the candle, produced by its own heat; and that is one of the first things you have to consider with respect to the progress of the wax in the course of its combustion, and as regards the changes it undergoes.

I will arrange another tube carefully in the flame, and I should not wonder if we were able, by a little care, to get that vapor to pass through the tube to the other extremity, where we will light it, and obtain absolutely the flame of the

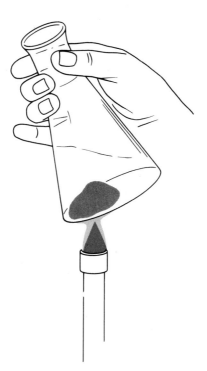

Figure 2–4.   Melting candle wax.

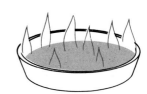

Figure 2–5. Burning vapor from heated candle wax.

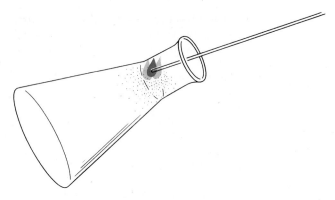

Figure 2–6.   Burning the vapor collected from the center of the flame (see Figure 2–3).

candle at a place distant from it (Figure 2–7). Now, look at that. Is not that a very pretty experiment? Talk about laying on gas—why, we can actually lay on a candle! And you see from this that there are clearly two different kinds of action—one the *production* of the vapor, and the other the *combustion* of it—both of which take place in particular parts of the candle.

Distribution of Heat in a Flame

I shall get no vapor from that part which is already burnt. If I raise the tube (Figure 2–8) to the upper part of the flame, so soon as the vapor has been swept

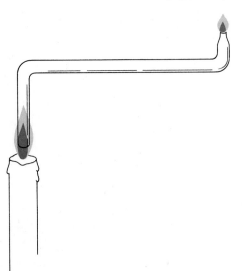

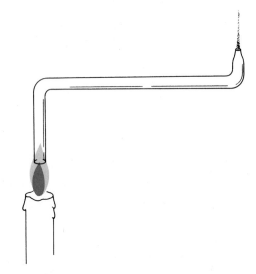

Figure 2–7.   Directly burning vapor from the center of the flame.

Figure 2–8.   Burned vapor from the outer portion of the flame will not burn further.

out what comes away will be no longer combustible; it is already burned. How burned? Why, burned thus: In the middle of the flame, where the wick is, there is this combustible vapor; on the outside of the flame is the air which we shall find necessary for the burning of the candle; between the two, intense chemical action takes place, whereby the air and the fuel act upon each other, and at the very same time that we obtain light the vapor inside is destroyed. If you examine where the heat of a candle is, you will find it very curiously arranged. Suppose I take this candle, and hold a piece of paper close upon the flame, where is the heat of that flame? Do you not see that it is *not* in the inside? It is in a ring, exactly in the place where I told you the chemical action was (Figure 2–9); and even in my irregular mode of making the experiment, if there is not too much disturbance, there will always be a ring. This is a good experiment for you to make at home. Take a strip of paper, have the air in the room quiet, and put the piece of paper right across the middle of the flame—(I must not talk while I make the experiment)—and you will find that it is burnt in two places, and that it is not burnt, or very little so, in the middle; and when you have tried the experiment once or twice, so as to make it nicely, you will be very interested to see where the heat is, and to find that it is where the air and the fuel come together.

Air Necessary for Combustion

This is most important for us as we proceed with our subject. Air is absolutely necessary for combustion; and, what is more, I must have you understand that *fresh* air is necessary, or else we should be imperfect in our

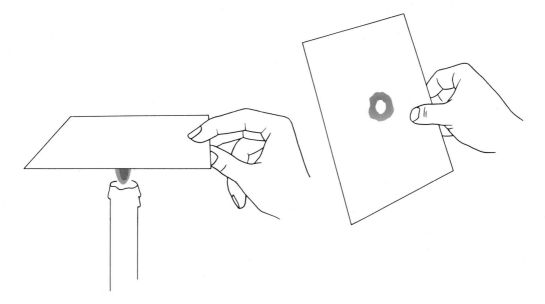

Figure 2–9.   Evidence for the release of heat only from the outer portion of the flame.

reasoning and our experiments. Here is a jar of air; I place it over a candle, and it burns very nicely in it at first, showing that what I have said about it is true; but there will soon be a change. See how the flame is drawing upward (Figure 2–10), presently fading, and at last going out. And going out, why? Not because it wants air merely, for the jar is as full now as it was before; but it wants pure, fresh air. The jar is full of air, partly changed, partly not changed; but it does not contain sufficient of the fresh air which is necessary for the combustion of a candle.

Figure 2–10.   Extinguishing a candle flame.

These are all points which we, as young chemists, have to gather up; and if we look a little more closely into this kind of action, we shall find certain steps of reasoning extremely interesting.

The Argand Lamp

For instance, here is the oil-lamp I showed you—an excellent lamp for our experiments—the old Argand lamp. I now make it like a candle (obstructing the passage of air into the centre of the flame); there is the cotton; there is the oil rising up in it; and there is the conical flame. It burns poorly because there is a partial restraint of air. I have allowed no air to get to it save round the outside of the flame, and it does not burn well. I can not admit more air from the outside, because the wick is large; but if, as Argand did so cleverly, I open a passage to the middle of the flame, and so let air come in there, you will see how much more beautifully it burns. If I shut the air off, look how it smokes; and why? We have now some very interesting points to study: we have the case of the combustion of a candle; we have the case of a candle being put out by the want of air; and we have now the case of imperfect combustion, and this is to us so interesting that I want you to understand it as thoroughly as you do the case of a candle burning in its best possible manner.

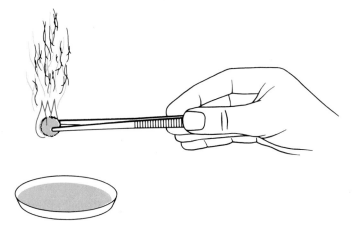

Figure 2–11.   Burning turpentine.

Soot from Imperfect Combustion

I will now make a great flame, because we need the largest possible illustrations. Here is a larger wick (burning turpentine on a ball of cotton. See Figure 2–11). All these things are the same as candles, after all. If we have larger wicks, we must have a larger supply of air, or we shall have less perfect combustion. Look, now, at this black substance going up into the atmosphere; there is a regular stream of it. I have provided means to carry off the imperfectly-burned part, lest it should annoy you. Look at the soots that fly off from the flame; see what an imperfect combustion it is, because it can not get enough air. What, then, is happening? Why, certain things which are necessary to the combustion of a candle are absent, and very bad results are accordingly produced; but we see what happens to a candle when it is burnt in a pure and proper state of air. At the time when I showed you this charring by the ring of flame on the one side of the paper, I might have also shown you, by turning to the other side, that the burning of a candle produces the same kind of soot—charcoal, or carbon.

Combustion with and without Flame

But, before I show that, let me explain to you, as it is quite necessary for our purpose, that, though I take a candle, and give you, as the general result, its combustion in the form of a flame, we must see whether combustion is always in this condition, or whether there are other conditions of flame; and we shall soon discover that there are, and that they are most important to us. I think, perhaps, the best illustration of such a point to us, as juveniles, is to show the result of strong contrast. Here is a little gunpowder. You know that gunpowder burns with flame; we may fairly call it flame. It contains carbon and other materials, which altogether cause it to burn with a flame. And here is some pulverized iron, or iron filings. Now I purpose burning these two things together. I have a little mortar in which I will mix them. (Before I go into these experiments, let me hope that none of you, by trying to repeat them, for fun's sake, will do any harm. These things may all be very properly used if you take care, but without that much mischief will be done.)

Well, then, here is a little gunpowder, which I put at the bottom of that little wooden vessel, and mix the iron filings up with it, my object being to make the gunpowder set fire to the filings and burn them in the air, and thereby show the difference between substances burning with flame and not with flame. Here is the mixture; and when I set fire to it you must watch the combustion, and you will see that it is of two kinds. You will see the gunpowder burning with a flame and the filings thrown up. You will see them burning too, but without the production of flame. They will each burn separately. (The lecturer then ignited the mixture. Figure 2–12.) There is the gunpowder, which burns with a flame, and there are the filings: they burn with a different kind of combustion. You see, then, these two great distinctions; and upon these differences depend all the utility and all the beauty of flame which we use for the purpose of giving out light. When we use oil, or gas, or candle for the purpose of illumination, their fitness all depends upon these different kinds of combustion.

Figure 2–12. Burning gunpowder and burning particles of iron.

Combustion of Lycopodium

There are such curious conditions of flame that it requires some cleverness and nicety of discrimination to distinguish the kinds of combustion one from another. For instance, here is a powder which is very combustible, consisting, as you see, of separate little particles. It is called *lycopodium*, and each of these particles can produce a vapor, and produce its own flame; but, to see them burning, you would imagine it was all one flame. I will now set fire to a quantity (Figure 2–13a), and you will see the effect (Figure 2–13b). We saw a cloud of

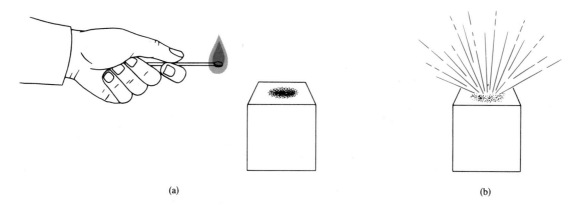

(a)　　　　　　　　　　　　　　　　　　　　(b)

Figure 2–13. **(a) Setting fire to a lycopodium powder. (b) Lycopodium powder burning.**

flame, apparently in one body; but that rushing noise (referring to the sound produced by the burning) was a proof that the combustion was not a continuous or regular one. This is the lightning of the pantomimes, and a very good imitation. (The experiment was twice repeated by blowing lycopodium from a glass tube through a spirit flame.) This is not an example of combustion like that of the filings I have been speaking of to which we must now return.

Soot from a Luminous Flame

Suppose I take a candle and examine that part of it which appears brightest to our eyes. Why, there I get these black particles, which already you have seen many times evolved from the flame, and which I am now about to evolve in a different way. I will take this candle and clear away the gutterage, which occurs by reason of the currents of air; and if I now arrange a glass tube so as just to dip into this luminous part, as in our first experiment, only higher, you see·the result. In place of having the same white vapor that you had before, you will now have a black vapor (Figure 2–14). There it goes, as black as ink. It is certainly very different from the white vapor; and when we put a light to it we shall find that it does not burn, but that it puts the light out. Well, these particles, as I said before, are just the smoke of the candle; and this brings to mind that old employment which Dean Swift recommended to servants for their amusement, namely, writing on the ceiling of a room with a candle. But what is that black substance? Why, it is the same carbon which exists in the candle. How comes it out of the candle? It evidently existed in the candle, or else we should not have had it here.

Cause of the Luminosity of Flame

And now I want you to follow me in this explanation. You would hardly think that all those substances which fly about London, in the form of soots and blacks, are the very beauty and life of the flame, and which are burned in it as

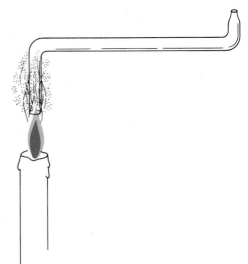

Figure 2–14. The formation of soot (carbon) by a candle flame.

those iron filings were burned here. Here is a piece of wire gauze, which will not let the flame go through it, and I think you will see, almost immediately, that when I bring it low enough to touch that part of the flame which is otherwise so bright, it quells and quenches it at once, and allows a volume of smoke to rise up (Figure 2–15).

I want you now to follow me in this point—that whenever a substance burns, as the iron filings burnt in the flame of gunpowder, without assuming the vaporous state (whether it becomes liquid or remains solid), it becomes

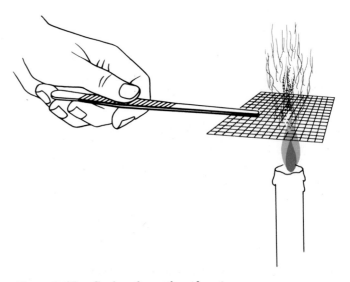

Figure 2–15. Copious formation of soot.

exceedingly luminous. I have here taken three or four examples apart from the candle on purpose to illustrate this point to you, because what I have to say is applicable to all substances, whether they burn or whether they do not burn—that they are exceedingly bright if they retain their solid state, and that it is to this presence of solid particles in the candle-flame that it owes its brilliancy.

Here is a platinum wire, a body which does not change by heat. If I heat it in this flame, see how exceedingly luminous it becomes. I will make the flame dim for the purpose of giving a little light only, and yet you will see that the heat which it can give to that platinum wire, though far less than the heat it has itself, is able to raise the platinum wire to a far higher state of effulgence. This flame has carbon in it; but I will take one that has no carbon in it. There is a material, a kind of fuel—a vapor or gas, whichever you like to call it—in that vessel, and it has no solid particles in it; so I take that because it is an example of flame itself burning without any solid matter whatever; and if I now put this solid substance in it, you see what an intense heat it has, and how brightly it causes the solid body to glow (Figure 2–16). This is the pipe through which we convey this particular gas, which we call hydrogen, and which you shall know all about next time we meet. And here is a substance called oxygen, by means of which this hydrogen can burn; and although we produce, by their mixture, far greater heat than you can obtain from the candle, yet there is very little light.

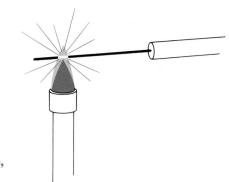

Figure   2–16.   Solid   wire, heated until it glows.

The Lime Light

If, however, I take a solid substance, and put that into it, we produce an intense light. If I take a piece of lime, a substance which will not burn, and which will not vaporize by the heat (and because it does not vaporize remains solid, and remains heated), you will soon observe what happens as to its glowing. I have here a most intense heat produced by the burning of hydrogen in contact with the oxygen; but there is as yet very little light—not for want of heat, but for want of particles which can retain their solid state; but when I hold this piece of lime in the flame of the hydrogen as it burns in the oxygen, see how it glows! (Figure 2–17.) This is the glorious lime light, which rivals the voltaic light, and which is almost equal to sunlight.

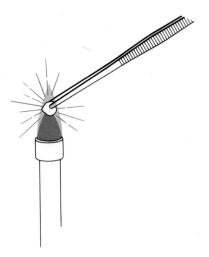

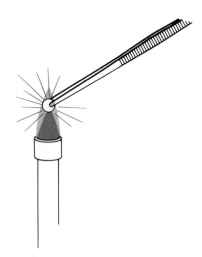

Figure 2–17.   Solid lime, heated to glowing.   **Figure 2–18.   Solid carbon, heated to glowing.**

I have here a piece of carbon or charcoal, which will burn and give us light exactly in the same manner as if it were burn as part of a candle (Figure 2–18). The heat that is in the flame of a candle decomposes the vapor of the wax, and sets free the carbon particles; they rise up heated and glowing as this now glows, and then enter into the air. But the particles, when burnt, never pass off from a candle in the form of carbon. They go off into the air as a perfectly invisible substance, about which we shall know hereafter.

Is it not beautiful to think that such a process is going on, and that such a dirty thing as charcoal can become so incandescent? You see it comes to this—that all bright flames contain these solid particles; all things that burn and produce solid particles, either during the time they are burning, as in the candle, or immediately after being burnt, as in the case of the gunpowder and iron filings—all these things give us this glorious and beautiful light.

Combustion of Phosphorus

I will give you a few illustrations. Here is a piece of phosphorus, which burns with a bright flame. Very well; we may now conclude that phosphorus will produce, either at the moment that it is burning or afterward, these solid particles (Figure 2–19). Here is the phosphorus lighted, and I cover it over with this glass for the purpose of keeping in what is produced. What is all that smoke? That smoke consists of those very particles which are produced by the combustion of the phosphorus.

Here again are two substances. This is chlorate of potassa, and this other sulphuret of antimony. I shall mix these together a little, and then they may be burnt in many ways. I shall touch them with a drop of sulphuric acid, for the purpose of giving you an illustration of chemical action, and they will instantly burn (Figure 2–20). (The lecturer then ignited the mixture by means of sulphuric acid.) Now, from the appearance of things, you can judge for

Figure 2–19. Burning phosphorus, with copious production of solid smoke particles.

yourselves whether they produce solid matter in burning. I have given you the train of reasoning which will enable you to say whether they do or do not; for what is this bright flame but the solid particles passing off?

Combustion of Zinc

Mr. Anderson has in the furnace a very hot crucible. I am about to throw into it some zinc filings, and they will burn with a flame like gunpowder. I make this

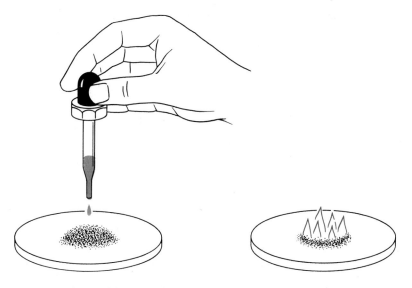

Figure 2–20.   Igniting the mixture.

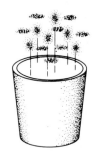

Figure 2–21.   Burning zinc (in the crucible) produces fluffy collections of very light weight smoke particles.

experiment because you can make it well at home. Now I want you to see what will be the result of the combustion of this zinc. Here it is burning—burning beautifully like a candle, I may say (Figure 2–21). But what is all that smoke, and what are those little clouds of wool which will come to you if you can not come to them, and make themselves sensible to you in the form of the old philosophic wool, as it was called? We shall have left in that crucible, also, a quantity of this woolly matter.

But I will take a piece of this same zinc, and make an experiment a little more closely at home, as it were. You will have here the same thing happening. Here is the piece of zinc; there (pointing to a jet of hydrogen) is the furnace, and we will set to work and try and burn the metal. It glows, you see; there is the combustion; and there is the white substance into which it burns. And so, if I take that flame of hydrogen as the representative of a candle, and show you a substance like zinc burning in the flame, you will see that it was merely during the action of combustion that this substance glowed—while it was kept hot; and if I take a flame of hydrogen and put this white substance from the zinc into it, look how beautifully it glows, and just because it is a solid substance.

Luminosity of Coal-gas Flame

I will now take such a flame as I had a moment since, and set free from it the particles of carbon. Here is some camphene, which will burn with a smoke; but if I send these particles of smoke through this pipe into the hydrogen flame you will see they will burn and become luminous, because we heat them a second time. There they are (Figure 2–22). Those are the particles of carbon reignited a second time. They are those particles which you can easily see by holding a piece of paper behind them, and which, while they are in the flame, are ignited by the heat produced, and, when so ignited, produce this brightness. When the particles are not separated you get no brightness. The flame of coal-gas owes its brightness to the separation, during combustion, of these particles of carbon, which are equally in that as in a candle.

I can very quickly alter that arrangement. Here, for instance, is a bright flame of gas. Supposing I add so much air to the flame as to cause it all to burn before those particles are set free, I shall not have this brightness; and I can do that in this way: If I place over the jet this wire-gauze cap, as you see, and then light the gas over it, it burns with a nonluminous flame, owing to its having plenty of air mixed with it before it burns; and if I raise the gauze, you see it

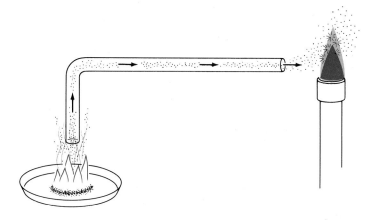

Figure 2–22.   Particles of soot, from burning camphene (camphor) glowing in the hot flame of a gas burner.

does not burn below (Figure 2–23). There is plenty of carbon in the gas; but, because the atmosphere can get to it, and mix with it before it burns, you see how pale and blue the flame is. And if I blow upon a bright gas-flame, so as to consume all this carbon before it gets heated to the glowing point, it will also burn blue. (The lecturer illustrated his remarks by blowing on the gas-light.) The only reason why I have not the same bright light when I thus blow upon the flame is that the carbon meets with sufficient air to burn it before it gets separated in the flame in a free state. The difference is solely due to the solid particles not being separated before the gas is burnt.

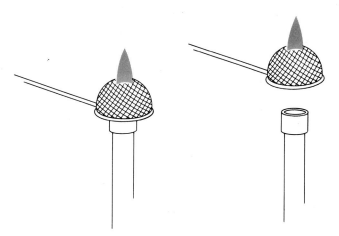

Figure 2–23.   Flowing gas burning above, not below, the wire gauze.

Different Products of Combustion

You observe that there are certain products as the result of the combustion of a candle, and that of these products one portion may be considered as charcoal, or soot; that charcoal, when afterward burnt, produces some other product; and it concerns us very much now to ascertain what that other product is. We showed that something was going away; and I want you now to understand how much is going up into the air; and for that purpose we will have combustion on a little larger scale. From that candle ascends heated air, and two or three experiments will show you the ascending current; but, in order to give you a notion of the quantity of matter which ascends in this way, I will make an experiment by which I shall try to imprison some of the products of this combustion. For this purpose I have here what boys call a fire-balloon; I use this fire-balloon merely as a sort of measure of the result of the combustion we are considering; and I am about to make a flame in such an easy and simple manner as shall best serve my present purpose. This plate shall be the "cup," we will so say, of the candle; this spirit shall be our fuel; and I am about to place this chimney over it, because it is better for me to do so than to let things proceed at random. Mr. Anderson will now light the fuel, and here at the top we shall get the results of the combustion. What we get at the top of that tube is

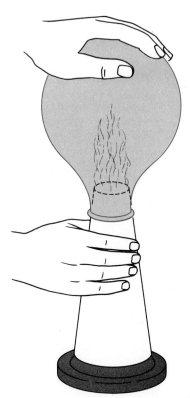

Figure 2–24. Collecting two gas and vapor products of combustion.

exactly the same, generally speaking, as you get from the combustion of a candle; but we do not get a luminous flame here, because we use a substance which is feeble in carbon. I am about to put this balloon—not into action, because that is not my object—but to show you the effect which results from the action of those products which arise from the candle, as they arise here from the furnace. (The balloon was held over the chimney, when it immediately commenced to fill; see Figure 2–24.) You see how it is disposed to ascend; but we must not let it up, because it might come in contact with those upper gas-lights, and that would be very inconvenient. (The upper gas-lights were turned out at the request of the lecturer, and the balloon was allowed to ascend.) Does not that show you what a large bulk of matter is being evolved? Now there is going through this tube (placing a large glass tube over a candle) all the products of that candle, and you will presently see that the tube will become quite opaque. Suppose I take another candle, and place it under a jar, and then put a light on the other side, just to show you what is going on. You see that the sides of the jar become cloudy, and the light begins to burn feebly. It is the products, you see, which make the light so dim, and this is the same thing which makes the sides of the jar so opaque. If you go home, and take a spoon that has been in the cold air, and hold it over a candle—not so as to soot it—you will find that it becomes dim just as that jar is dim. If you can get a silver dish, or something of that kind, you will make the experiment still better; and now, just to carry your thoughts forward to the time we shall next meet, let me tell you that it is *water* which causes the dimness, and when we next meet I will show you that we can make it, without difficulty, assume the form of a liquid.

THE REMAINING FARADAY LECTURES

In the remaining four lectures, Professor Faraday discussed the properties and composition of the water formed when a candle burns, the nature of the combustion process, oxygen and nitrogen in the air, the carbon dioxide that is also produced when a candle burns, how to weigh a gas, the compressibility of air, the combustion of charcoal, and the analogy between respiration and combustion, all in the same delightfully informal style. For the complete set of lectures in a currently available publication, see *The Chemical History of a Candle* by Michael Faraday published by Macmillan, New York, 1962.

With the background presented by these first two chapters, we are ready to consider an extended definition of chemistry, the topic of Chapter 3.

chapter 3

Chemistry Defined

A definition is a sentence, or paragraph, that describes something in such a way that the reader and the writer of the definition both have the same, or nearly the same idea about the thing defined. For example, what is a pencil? But when the thing to be defined is more complex, the definition becomes longer and we become less sure that the writer and the reader both end up with the same understanding. For example, what is love? Define: Poetry. How would you describe good music? What is a candle? Your definition of a candle would be quite different before and after reading Chapter 2. The study of literature is acknowledged to be important for an educated person; define good literature in such a way as to show that it is an important part of your total education. How can we define chemistry?

THE DEFINITION OF CHEMISTRY

In any task involving the definition of a complex concept, one simply begins. So, first, why is there such a subject as chemistry? Answer: because there are people, called chemists, who find that this kind of study of nature fulfills for them three major needs of man: satisfaction of curiosity (Why does a candle burn?), of intellectual excitement (What specific details can I imagine which might be said to occur within the flame of a burning candle?), of esthetic delight (Look closely and enjoy the examination of the different colors and shapes of the candle flame.) From this we can get an initial definition of chemistry: The study of sensual, delightful, facts that demand intellectually imaginative explanations. In this chemists are not different from others. Almost everyone likes to solve a puzzle, for example; and chemists even take some pleasure in the occasional foul odors generated in their kind of work (in this, they are rather different from some others).

Chemistry is the study of change

Our first definition of chemistry is correct in a sense, but too broad to be useful; it would also apply to mathematics or physics, or biology, or almost any science, and even to many aspects of other quite different human activities, such as art, poetry, law, gardening, in fact to almost any subject. We need a more specific definition.

Chemistry is the study of changes. Not all changes, of course. We can change our opinion, we can change our position, or the location of a piece of furniture (move that chair over there, please, Charles); we can change our financial condition (your money or your life, Bud!) or have it changed for us. None of these kinds of changes are chemical changes.

Wood burns and ash forms, along with the less obvious formation of water and carbon dioxide. This is one example of a chemical change. The nature of the substance changes; this is a chemical change. Wood floats; it is fibrous. Ashes sink; they are powdery. Water is different from wood. Carbon dioxide, a gas, is also different from wood. The nature of wood is different from the nature of what is formed when wood burns. Burning therefore is a chemical change.

Ordinarily, wood will not burn in the absence of air. In the chemical change that we call burning, or combustion, two substances change their nature: the wood and the oxygen in the air; though the change in the nature of the wood is more obvious, the nature of the oxygen also changes. Strictly speaking, the wood does not change to ashes; the wood simply ceases to exist as wood. In the same strict sense we could say that the oxygen is consumed also and ceases to exist as oxygen. In the place of the wood and the oxygen, new substances begin their existence, ashes, carbon dioxide, and water. Ordinarily however chemists would say somewhat loosely, that the wood and oxygen did change into ashes, carbon dioxide, and water. Two substances changed into, in this case, three other substances. A change of substance is called a **chemical change**.

Chemical change

Now we can develop a fuller statement about chemical change. A chemical change usually involves two or more original substances. During the change, these substances cease to exist. That is, we can no longer identify their presence. At the same time new substances, which were not present before, can be identified. We can say that the new substances were formed from the original substances by a chemical change.

Man is surrounded by continual chemical changes. Respiration, digestion of food, muscle contraction and extension; baking a cake, frying a steak or hamburger; the process that causes the flow of electricity from a battery for a flashlight, a portable radio, or from a storage battery to start the motor in a car; the hardening of cement, growing plants, decaying organic matter; lighting a match (as distinguished from burning, already mentioned), pyrotechnic displays, detonation of explosives, the ageing of motor oil while it is used as a lubricant; the manufacture of glass, winning metals from their ores, the manufacture of dyes and their use as well in coloring fabrics, the manufacture of fabrics other than cotton, wool, silk, and linen, the growing of these natural fibers; the manufacture of gasoline from crude oil, of paper from wood or rags; the action of a laxative in the body; the manufacture of drugs and their action in the body. Though the whole list does have an end, the portion cited here is a mere beginning.

Chemistry affects life

Chemistry can now be defined more rigorously as the study of chemical changes, as an activity undertaken by man in order to find out why substances change into other substances, how these changes take place, and to apply this new information both to the further study of chemical changes and to the preparation and use of new substances in ways determined by society or individuals for the good or ill of mankind.

The subject of chemistry is then an intellectual and practical concern about chemical changes. No one in his right mind could suggest that the subject of chemistry should be interesting to every person, but it is at the minimum closely related to the life of every person. All of us are concerned about the deterioration of our environment. In each instance, some new substance (or more than one) has been formed as the result of a chemical change. It then acts further causing other, often undesirable, chemical changes. Nitrogen oxides come into the atmosphere from the burning of fuel oil and gasoline, sulfur dioxide from the burning of coal and oil; there are residues of insecticides, phosphates in natural waters, and fertilizer run off into streams and lakes; solid wastes accumulate, from household refuse to junked automobiles and the tailings and washings from mining processes or other industrial processes. In each case a desirable goal has been sought by men and the undesirable side effects have followed as a consequence. Chemical changes, when used by man for his own benefit, sometimes have included detrimental effects as well. The solution to the problems that have been caused will obviously include different decisions about what we desire and how much we are willing to pay, but they will also include more and better controlled chemical changes to alter the present problems.

THEORIES AND MODELS

The ability to control chemical changes requires a great deal of information. Some of this information will be factual; that is, what actually happens, what we can see with our own eyes when our observation is exceptionally detailed. Some of this information will be imaginary; it will describe possible explanations that account for what we observe. The combination of the two, facts and imagination, will then help us to predict what might happen under some new circumstances. This combination is called a theory, or "model".

Let's take as an example the problem of Noah Return and Molly Cule. Every time Noah Return sees Molly Cule he asks her for a date, and every time she refuses. From these facts, we can imagine that Noah rarely takes a bath. Our model says that Noah stinks. We can then predict that if Noah washes himself, Molly will say yes the next time she is asked. Another model, equally valid for the few facts we have, would identify some other characteristic of poor Noah. Perhaps he is very short, whereas Molly is quite tall. We might predict that by stretching Noah, or shortening Molly, all would be well between the two. In either case for our two models, we can help settle the matter by making more observations, by smelling Noah or measuring both Noah and Molly. If Noah was found factually to be well washed and about the same height as Molly, then we would need a third model to account for the total facts now available.

In a manner of speaking, this is a book about models that attempt to explain the reasons for chemical changes. We have already noted some of the features of the so-called oxygen model of combustion. To illustrate this idea of models further, let us consider an earlier model of combustion. Remember, our major purpose is to develop models for the use of man. For example, notice the chemical changes around you: iron rusts, plants grow, cement hardens. Almost everything changes in some way or other sooner or later. In some cases the change can be reversed. Rust can be converted into iron again. Even easier, ice can be changed back into water. Rotted tree leaves and twigs eventually are taken up by other plants, and new leaves and branches are formed. All of these changes are explained by models. The model for living processes is not yet well developed, more facts and more imagination are needed. Eventually this model for living processes will be good enough for man to generate life from nonliving precursors, although there are some who question the wisdom of such an act.

A false model for combustion

To get at that earlier model of combustion, consider once more the combustion of wood. As wood burns you can hear a hissing sound, from the cracks in the wood flames come out. As the wood burns, we could imagine, some substance is forced out. According to this model, that substance is phlogiston.

When wood burns, phlogiston is released and ashes remain behind; that is, wood is composed of ashes and phlogiston, if this model is correct. Paper burns, releasing phlogiston and leaving a different sort of ash behind. The difference between wood and paper according to this model is due to the difference in the ashes of the two.

A good model not only accounts for the detailed observations we make but it also predicts new facts that we should be able to observe. When a candle burns, since it releases phlogiston, it could be expected to become lighter. This fact is of course observable. A candle leaves no ash when it burns. So, according to the phlogiston model a candle is composed of pure phlogiston (except for the wick). The phlogiston model predicts that we could make wood by somehow mixing or combining wood ashes and candle wax. No one has been able to do this, yet, so the model is of questionable validity.

When iron burns the ash which is left would be the part of the iron which is not phlogiston. (Iron will burn, though not everyone knows this. You can see iron burn by lighting some very fine steel wool with a match. Do it outdoors, or over the kitchen sink—otherwise you might start a conflagration.) When charcoal burns very little ash is left; charcoal is mostly phlogiston, we could say. If this is true, the model predicts that by mixing the ash left from the burning of iron with charcoal we could get our iron back again. This can be done. The phlogiston model has made a valid prediction. (Iron ore is equivalent to the ash left when iron burns. Iron ore is mixed with charcoal* in blast furnaces every day of the year throughout the world, and tons of iron are produced.)

As iron burns, and phlogiston is released, we would expect that the ashes would weigh less than the iron we started with. In fact we find that the ashes weigh more. So we modify the phlogiston model to include the idea that phlogiston has negative weight. After all, this is consistent with the observed upward movement of flames that issue from a burning substance. The escaping phlogiston has negative weight; it is repelled by gravity and goes upward. But now, our modified phlogiston model would predict that wood ashes weigh more than the original wood. This prediction is false.

---

* Actually coke from coal, not charcoal from wood, is used in blast furnaces.

Apparently the phlogiston model, though correctly predicting some new facts, does not do so well with others. When a model shows too many of these weaknesses, it is rejected if another imaginative guess, a new model, can be conceived that does better with predicting new observations correctly. The phlogiston model was abandoned about 200 years ago in favor of the oxygen model of combustion that was used earlier in this chapter. At present, this new model predicts new facts correctly and is not likely to be abandoned in the near future.

Constructing a model   Models are the product of the imagination working on observed facts. Some students might enjoy these exercises:

1. Put two glasses of water side by side, one quite hot, the other very cold. Carefully, without stirring, put a drop of ink into each. Imagine an explanation, think of a model, that accounts for your observations.
2. Put about a teaspoonful of baking soda on a saucer and then add some vinegar slowly, drop by drop. Imagine a model that accounts for the fizzing that you observe.
3. Pour a little household ammonia in a saucer in one corner of a room. What model can you think of that accounts for the observation you make later that the odor of ammonia is present in the opposite corner?
4. Wave your hand through the air and observe that it is easy to do. Then, with care, extend your hand out the window of an automobile that is traveling rapidly and note that it is much more difficult to move your hand in that air. What model can account for these observations?
5. Get an old tin can lid and put a bit of sugar on one place and a bit of salt at another place. Heat the lid on the burner of a stove. You will observe that the sugar appears to liquify a little and then turns black but the salt does not, although it will melt if the burner is hot enough. What model will account for these facts?
6. Glass shatters when it is sharply struck; wood merely gets dented. What model accounts for this?

In considering these questions we come to one of the big differences among people. Generally speaking, students who are interested in science for its own sake rather like these kinds of challenges. Others find them dull, or worse; they forsee little enjoyment in thinking about why this or how that. Instead, they would much rather read a good book, or enjoy walking in the rain, or writing a poem, or listening to some good music, or talking to friends. People are different, and should severally prefer different activities. The point is that science, and in particular chemistry, affects our world and it is useful for those who really don't care too much for science to be aware of the kinds of interests science specialists enjoy. It is equally necessary for those oriented toward science to be aware of the valid enjoyment others can have, of course.

THE ATOMIC THEORY

This, however, is a book about chemistry, and we do have to get on with scientific models. To build a good model requires many facts, first. The model we are building here, and which satisfies the observations you might have made about ink in hot and cold water, ammonia odor, scorched sugar, broken glass, and all the rest, is the atomic model or atomic theory of matter. What are some of the other facts on which this model is based?

Salt has the same properties, no matter where it comes from. Purified salt from a Siberian salt mine melts at 801°C; so does the salt that is mined from under the ground in Michigan, near Detroit. The composition of pure salt is always the same, 39.34% sodium and 60.66% chlorine, by weight, no matter where it comes from. The same is true for other pure substances. Water, for example, contains various dissolved components but these can be removed and the pure water that we obtain always has, by weight, 8 parts oxygen and 1 part hydrogen. (These numbers are rounded off slightly; strictly it is 7.999 parts by weight oxygen and 1.008 parts hydrogen.) Carbon dioxide, whether it comes from burning wood, or respired air from your lungs, or from dry ice, is (rounded off) 3 parts by weight carbon and 8 parts by weight oxygen. Carbon monoxide, a different substance, is always 3 parts carbon and 4 parts oxygen, by weight. Ammonia, the gas which has the sharp odor, not the cleaning liquid called ammonia water, contains nitrogen and hydrogen always in the proportion of 14 parts by weight nitrogen and 3 parts by weight hydrogen (rounded off a bit).

More facts: When a chemical change occurs, the mass of the products formed is always exactly the same as the mass of the reacting substances that have disappeared, within the limits of man's ability to measure mass. The ability is pretty good, incidentally; it is routine, though a bit tedious, to measure the mass of the ink in a dot one thousandth the size of the period at the end of this sentence.

The composition of matter

These facts, and others, support the imaginative model used today, the atomic model. This model states that all matter is composed of **atoms**, very small particles; that in many substances the atoms are somehow bound together into groups, called **molecules**, or are otherwise bound together in other kinds of groups.

Iron is composed of iron atoms only. Salt is composed of sodium and chlorine, only, but the atoms are in a different condition in the salt than the iron atoms are in iron. Diamond is composed entirely of carbon atoms bonded together in a still different way. Ammonia is composed of hydrogen and nitrogen atoms, in groups called molecules. Water also exists in molecules, except that water molecules are composed of hydrogen and oxygen atoms, one or more than one of each (we'll get to this in a moment).

Now, to be any good at all, a model must have predictive character. The atomic model has been found to be particularly useful as a predictor. For example,

1. Even though atoms are, according to the model, quite small, we ought to be able to count them, somehow.
2. Although atoms are small, never the less thay have some size, and we ought to be able to also say how big they are.
3. As nearly as we can tell today, there are about 100 different kinds of atoms. The model says that different atoms can form groups of the same atoms, of two different atoms, of three or more different atoms, and so on. So, predictively, there ought to be "quillions" of different substances, each consisting of different combinations of a few, or more, of the hundred or so different atoms.

This third prediction has not worked out, only about 2 million different combinations are known today, and some combinations cannot be obtained. Hydrogen, for example, will not form any kind of atomic combination with helium, or with gold. Iron atoms will not combine with aluminum atoms, but they will with gold. Since it was first suggested about 150 years ago the atomic model has been modified to take care of this kind of discrepancy.

The weight of atoms

What about the first two predictions? We can count atoms, we think, and we can say how big they are—if our counting is correct, or vice versa. To see how this is done we can use an analogy to begin.

Imagine that you have a can of marbles, all identical, and you wish to know the weight of one marble. Suppose further that it is impossible to count the marbles. How could some reasonable idea of the weight of one marble be determined? (Remember, it is against the rules of the analogy to take one marble and weigh it because that would be counting "one marble." Similarly, it is not allowed to take say ten marbles and weigh them, and then divide that total by ten.) The way to do this without counting is to reach in the can, grab a handful of marbles, and weigh them. Then, put those marbles back in the can, grab another handful, trying to make the second handful a bit more or a bit less than the first, and weigh those. And so on. Eventually, there is a good chance that two different handfuls will actually be different by only one marble, if enough handfuls are weighed. Here is an imaginary listing of the data obtained for the first several handfuls of marbles:

First handful	16.9 g
Second handful	22.1 g
Third handful	19.5 g
Fourth handful	7.8 g
Fifth handful	19.5 g

Notice, between the second and third handfuls the weight difference is 2.6 g. Perhaps the second and third handfuls were different by one marble; if so, then one marble weighs 2.6 g. Further, the difference between the first and the third handfuls (or fifth) is also 2.6 g; this lends some support to the suggestion that one marble weighs 2.6 g. Actually, a more detailed inspection of the data will show that one marble cannot weigh 2.6 grams, but to make it a bit easier, we will imagine that a sixth handful was withdrawn from the can and weighed. The result:

Sixth handful	26.0 g

Now look at the differences between the sixth handful and the others. For example, comparing the sixth at 26.0 g and the fourth at 7.8 g, the difference is 18.2 g. *If* one marble weighs 2.6 g, then there are seven more marbles in the sixth handful than in the fourth (since $7 \times 2.6 = 18.2$). But compare the sixth handful at 26.0 g with the second at 22.1 g. The difference is 3.9 g. So far, it has seemed that one marble might weigh 2.6 g and now we see that this cannot be correct. If the sixth and second handfuls differed by one marble, the difference in weight would be 2.6 grams; if they differed by two marbles, the difference in weight would be twice that, 5.2 g. Instead, the difference in weight between the sixth and the second handfuls is 3.9 g. This is consistent with a single marble weight of 1.3 g. And *that* weight, 1.3 g, is consistent with all of the data, for all of the handfuls. It is certainly correct to say that maybe, if we weighed still more handfuls, we would get a result, finally, that would be inconsistent with a weight of 1.3 g for one marble. Actually, the corresponding experiments, where it was impossible to count atoms (since they are so small) have been carried out hundreds of thousands of times in various indirect ways so that we think today that anyone who tries it once more would obtain data that are consistent with what we now believe. In any case, this analogy is useful for us because it does illustrate the point that science arrives indirectly at information about the essence of matter. No scientist doubts that we do know the weight of an iron atom to be 9.27×10^{-23} g, for example, and yet at the same time realizes the assumptive indirect basis of this "knowledge."

Numbers of atoms

If we are willing to say that we know the weight of a single iron atom, then we can count the total atoms in a piece of iron by weighing the piece. This is the way that pennies are counted by some bank tellers, as you may know. (In other banks, the pennies are counted by stacking them into piles of certain heights.) Let us return to our marble analogy and say that one marble weighs 1.3 g Now we can count the marbles in, say, the fourth handful, which weighed 7.8 g. At 1.3 g for each marble, there were six marbles in the fourth handful ($6 \times 1.3 = 7.8$; or $7.8/1.3 = 6$).

So, by weighing marbles without counting, we can get a reasonable number for the weight of one marble. Then, by weighing a bunch of marbles and assuming the weight of one to be acceptable, we can count marbles with a bit of arithmetic. We can do the same for atoms. For example, a medium sized iron nail weighs 7.5 g. Since one iron atom weighs 9.27×10^{-23} g, we can calculate (or count) the number of iron atoms in that nail by dividing. Simply divide 9.27×10^{-23} into 7.5; the answer is a lot of iron atoms—8.1×10^{22}, as a rounded off answer. For exercise, how many copper atoms are there in a penny? A penny is made of practically pure copper and weighs about 3 g; one copper atom weighs 1.1×10^{-22} g.

The size of atoms

To continue, if we can count atoms, we can get some idea about their size. From the calculated number of iron atoms in a nail, or copper atoms in a penny, it is easy to see that atoms are very small. In scientific work, however, "very small" is not much use. More precision is desirable. To see how this precision is obtained, we can return to our marble analogy, this time with a cube shaped can full of marbles. The marbles in this cubical can have been packed in neatly, layer on top of layer, with each layer as full as it can get, as illustrated in Figure 3–1. The can-full is weighed and the weight of the empty can subtracted. From this, the weight of the marbles is found to be 665.6 g. We already know that it is reasonable to assume that each marble weighs 1.3 g, so, dividing 1.3 into 665.6, we get 512 marbles in the cube shaped can.

What comes next is simple for some students and difficult for others, equally intelligent, who have a different way of thinking about things. So that

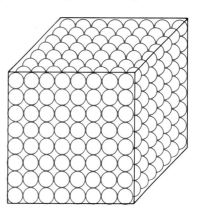

Figure 3–1. Marbles packed in regular array in cubical container.

everyone will follow, we will make a full explanation. The number of marbles in any box, cube shaped or not, can be found by multiplying the number of marbles along a long edge, a thickness edge and a depth edge. For a cube shaped container each edge is the same length so there is the same number of marbles along each of the three edges. The next problem is to identify the number that, multiplied by itself three times (the cube of that number), equals 512. That number is the number of marbles along any edge of the cube shaped can. It turns out to be 8; $8 \times 8 \times 8$ equals 512. Now if we knew the size of the cube shaped can, we could calculate the size of a marble. For example, if we measured the can and found it to be 4 in. along any edge, we would know that each marble was $\frac{1}{2}$ in. in diameter.

Now that we have seen how to measure something indirectly, let us go back to the iron atom example. A cube of pure iron that is 1 cm on each edge always weighs 7.86 g. Since we know one atom of iron weighs 9.27×10^{-23} g, by dividing we find that there are 8.5×10^{22} iron atoms in a 1-cm cube of iron. If we assume that the atoms are spherical and nicely arranged in neat layers in that cube, we can calculate that there are 4.4×10^7 iron atoms along each edge of the cube. [Rounding off, $4.4 \times 10^7 \times 4.4 \times 10^7 \times 4.4 \times 10^7$ equals 8.5×10^{22}; or $(4.4 \times 10^7)^3$ equals 8.5×10^{22}.]

So, if iron atoms are spherical, and if 4.4×10^7 iron atoms, all lined up one after the other have a total length of 1 cm, the diameter of one iron atom can be obtained by dividing (again) 4.4×10^7 into 1; the answer is 2.3×10^{-8} cm.

How big is a copper atom if we assume that it is a sphere also? You know everything you need to know, except for one fact: a cube of copper atoms that is 1 cm on each edge weighs 8.9 g (about the same amount of copper as is in three pennies).

Summary of the atomic theory

To keep things more or less straight, let's review briefly. We have at this point an atomic model, based on these kinds of facts:

Salt is salt, water is water. No matter what the source, pure salt has the same properties, always. The same for water. No matter what the source, the same substance always has the same composition, the same percentage of each component. When a chemical change occurs, such as vinegar reacting with baking soda, or the burning of a candle or piece of paper or wood or iron, or the digestion of food, the weight of the original substances (which disappear in the process) is equal to the weight of the new substances that form.

The model that attempts to account for these kinds of facts is the atomic model. According to that model:

1. All matter is composed of very small particles, called atoms. The atoms themselves maintain their integrity, they are not destroyed in a chemical change.

2. The atoms of any single element are all alike; the atoms of any other element are themselves alike (of course) but different from the atoms of every other element.
3. To a first approximation at least, atoms are probably spherical.

This model permits a few predictions:

1. Atoms can be counted, in principle. They are so small that they cannot be counted directly, but they can be "counted" by using weight relationships.
2. The size of atoms can be measured, in principle. They are too small to measure directly (with a ruler and magnifying glass), but their size can be determined if we know how many (by counting) occupy a volume that we can measure with a ruler (or other volume measuring device).
3. Atoms can be expected to fasten to each other in combinations with the same types of atoms or of different types of atoms.

The first two predictions have been fulfilled, but the third, as we know, is only partially correct. Let us examine these facts in some detail since the model must be changed to fit all facts, if this is possible.

THE COMBINATION OF ATOMS INTO MOLECULES

For our first example, consider water. A detailed extension of the atomic model suggests very strongly that a unit of water, called a molecule, consists of two atoms of hydrogen and one atom of oxygen fastened together somehow. Never more nor less than two atoms of hydrogen to each molecule of water, and only one atom of oxygen.

We express this as H_2O_1, using the subscripts to signify the number of each atom, and "H" to symbolize a hydrogen atom and "O" to symbolize an oxygen atom. By common agreement among all concerned, the "1" subscript is omitted; it is understood to be present. So you will generally see the expression for the number and kind of atoms in a water molecule symbolized as H_2O.

The question is, why this particular combination of two hydrogen atoms and one oxygen atom? Are any other combinations possible? Answer: yes, one other, H_2O_2. This substance is not water, it is explosive, poisonous, and generally disruptive. It is called hydrogen peroxide. You can buy it in a drug store, diluted with water to make it relatively safe. As some blondes know, it is an effective bleach for dark colored hair. (Even in a dilute solution with water, it is disruptive enough to destroy dark hair colors, but not enough to destroy the hair itself; at least not hurt the hair very much.) So, now the question is, why are there only two possible combinations of hydrogen and oxygen atoms, H_2O and H_2O_2?

Let us leave the answer to that question until later while we finish discussing combinations in general. Triple combinations are known, Copper sulfate, $CuSO_4$, with one copper, one sulfur, and four oxygen atoms is an example. The combination is unique; any other number of copper, sulfur, and oxygen atoms, if such were possible (and it is, but not now, please), would not have the properties that copper sulfate has. (In this set of symbols, Cu represents copper, from the Latin for copper, *cuprum*, since C has already been reserved for carbon and Co for cobalt.)

On the other hand, there is no combination of any kind for copper, fluorine, and oxygen atoms. As we know, no one has yet been able to get hydrogen and helium atoms to combine, nor iron and aluminum. Of all the conceivable combinations of atoms of the 100 or so different elements, only a few are known to be possible, perhaps not more than about 3–4 million. It is reasonable to conclude that not all conceivable combinations are possible; the general prediction of the model is incorrect. The model needs changing, and we need more facts and some imagination to do this. Whatever we come up with for a modification must somehow explain why combinations of atoms are restricted. Instead of saying that atoms are spheres, we will endow the modified model with smaller parts that account for a restriction in possible combinations of atoms.

The telltale colors

At this point it is helpful to digress, to reflect upon the remarkable quality of the human imagination, because the new model required a lot of imagination to be generated from the known facts. Human imagination has produced the world around us as it is, with its pleasant and unpleasant aspects. Indeed, lack of imagination, as well as lack of other desirable qualities, is in part also responsible for the problems we face today. At least, imagination is one of the major requisites in order to effect an improvement.

The use of imagination was successful in suggesting a modification of the original atomic model; it was applied to facts dealing with colors, although that would seem to be rather far removed from the question of why only certain combinations of atoms are possible. An analogy suggests itself here. Imagine anything preposterous, such as a teacup, saucer, and spoon covered with fur. In the world of fine arts, Marcel Oppenheim thought of *that* some years ago, and opened up for the delectation of all a delightful resurgence of the fine arts, whether or not we might individually find pleasure in contemplating a fur covered teacup.

Now to the facts we need. A neon sign glows with a reddish color. Other similar signs glow with different colors. Each of these kinds of signs are made of glass tubing bent to form letters or designs and filled on the inside with neon or other gases. When an electric current is passed through the gas in the tubing, it glows with a particular color—a different color for each different gas. (We will ignore other effects here; it is also possible to get different colors by using colored glass tubing.)

Everyone has seen a pyrotechnic display, such as on the Fourth of July. The colors are varied and wonderful. Loosely described, a colored pyrotechnic flame is made by burning rapidly a mixture of black gunpowder, finely powdered magnesium or aluminum metal, and some chemical salt, such as barium nitrate, which produces a pale green colored flame. For a deeper, different shade of green, thallium oxide is used instead of barium nitrate. For a deep red color, calcium chloride is used. The same substances, barium nitrate, thallium oxide, calcium chloride, and others, can be used on burning fireplace logs to produce attractively colored flames.

Each different element, in our example, barium, thallium, and calcium (not the other parts of the names of the substances), imparts a different, unique color to a flame. You may have noticed this in lighting a match. When a match is first lit, if you observe very closely, you can see a faint purplish hue in the flame. This is due to the presence of a potassium compound in the match head composition. After the initial moment of ignition, the color of a match flame becomes yellowish with a faint trace of orange also present. This color is imparted by a sodium compound in the match head composition. After the match head has burned away, the color of the flame changes again to a paler yellow, the color of burning paper or wood.

The question is, why do different elements impart different colors to a flame? And the model that answers this question also answers our prior question, why are only certain combinations of atoms possible?

Now hang on, because this gets a bit hairy, like that fur covered teacup.

The composition of light

Light, the kind we see by, is thought to be composed of particles called photons. Photons are emitted by the billions or more each second from an ordinary light bulb, for example, but the photons are not all the same. Some of the photons have a lot of energy, for a photon, that is. Other photons have less energy than that, and still other photons have still less energy, and so on. Most of the photons from an ordinary light bulb have energy within the range that can be detected by the human eye or skin. The lowest energy photons are called infrared photons; they cannot be seen by the human eye, but you can feel them with sensitive receptors in the skin, particularly the palm of your hand or in your cheek, just below your eye. (You can test this by opening your hand with your palm exposed, reasonably near a light bulb that is turned on.)

Photons with a little more energy can be detected by the human eye; they give rise to the sensation of red. Photons with a bit more energy than those produce an orange sensation; a bit more energy, yellow; still more, green; more yet, green-blue; still more, blue; even more, deep blue or violet. The photons with more energy than those that produce a violet sensation are not detectable by human receptors; they are called ultraviolet photons. (These very high energy photons can cause damage to our skin, called sunburn and, at high dosages, death in a few hours.) When a person receives approximately equal amounts of photons with the entire range of energies that the eye can detect, that is, about equal amounts of photons giving rise to red, orange, yellow, green, green-blue, blue, and violet sensations, the eye interprets the total as a white color. (We could go into this further and learn about color blindness, but that would take us too far from the main thrust of the story.)

How photons are produced

The next question is, where do these photons that are emitted from the light bulb, or the neon sign, or the pyrotechnic flame come from? Simple answer: They are made, generated, inside these sources of light.

The new imaginary model says that an atom is composed of parts. One part is a nucleus, in the middle of an atom. We will learn more about the nucleus in Chapter 6. For now, let us accept without evidence that the nucleus is itself composed of parts called neutrons and protons, crowded together. Buzzing about, outside the nucleus, are the other parts of the atom, the electrons. The nucleus has a positive charge and each electron has a negative charge. The model does not say what "positive charge" and "negative charge" really means. However, we have all experienced what it means when we received a mild "shock" on a cold, dry day from someone who had walked across a thick rug and, after that, touched us. The tingling sensation was due to an electrical charge, positive or negative, depending upon the material in the rug and in the shoe soles of our charged up friend.

Now, the electrons are buzzing about the nucleus in a peculiar way, sometimes near, sometimes far, so dynamically involved that, according to the model it is now (and perhaps will always be) absolutely impossible ever to say where a particular electron is at a particular time. According to the model, it is only possible to say how much energy each electron has. Further, according to this model, an electron can only have certain restricted values of energy; it cannot have any old amount of energy.

Suppose we think of a neon atom in a neon sign. That atom has just received a jolt of energy from the electric current that is passing through the glass tubing. One of the electrons in that neon atom is, therefore, raised to a higher energy condition. Almost immediately afterwards, the electron "falls" to a lower energy level, a lower energy condition. When it does so, a photon is emitted by that atom. The energy of that photon is equal to the difference in energy levels from where the excited electron was compared to the level to which it fell. For a neon atom, the difference in energy levels is fairly small, so a low energy photon is emitted. The energy of the photon is low enough to be detected by a human eye as a red sensation. (Actually, it takes about half a dozen or so of photons for our eye to detect anything, so we would need half a dozen or so different neon atoms with electrons at a high energy level falling within an instant of each other to a lower level, each emitting a photon that happened to be aimed at our eyeball. Then we would say, "Ah ha! a red light." But, in a neon sign there are "zillions" of neon atoms that are excited by an electrical energy jolt each second, so there are "zillions" of photons, all of the same energy, emitted from the neon sign in all directions an instant later.)

In a pyrotechnic flame colored red by the presence of calcium atoms, essentially the same thing happens. Instead of using an electrical stimulus to lift an electron in a calcium atom to a higher energy level, it is the energy from the burning pyrotechnic composition that does the trick. An instant after the electron has been raised to a higher energy level in a calcium atom, it falls to a level a little lower, emitting a photon of low energy, which we detect as a red color.

If you were able to compare carefully the red color of a neon sign with the red color of a pyrotechnic flame, you would notice that the shade of red is slightly different. The red from a neon sign is a little deeper (lower energy) red than the red from a pyrotechnic flame. The two energy levels are a little bit closer together in a neon atom than they are in a calcium atom.

As you can now imagine, the energy levels in a barium atom are considerably farther apart because the presence of barium produces a green colored pyrotechnic flame, and green sensation photons have more energy than red sensation photons.

Energy levels and the combination of atoms

So now we know about the model which explains colored neon signs. What does that have to do with atoms combining or not combining? Just this: The new imaginary model says that the electrons in atoms can exist only at certain energy levels, and, further, each different kind of atom has a different set of possible energy levels for its electrons. No two different atoms have exactly the same set of possible energy levels. However, all atoms of the same element, of course, have the same set of possible energy levels for each of their electrons.

The model further says that atoms can combine, somehow, because of electron interactions between one atom and another. Electrons interact between one atom and another by means of their possible energy levels. So, if the energy levels happen to fit, whatever that means, the atoms can combine, can be attached to each other. This must be the case for hydrogen and oxygen, to form water molecules and hydrogen peroxide molecules. But, according to the new model, the energy levels that are possible for electrons in hydrogen and helium atoms don't fit, so a hydrogen atom and a helium atom cannot join to make a molecule.

Summary of the atomic model for combination

We will look at more details about this model and the combination of atoms in Chapter 5. Here, we can summarize a description of the atomic model:

1. All matter is composed of atoms.
2. These atoms are indestructible.

3. Each atom of each element is identical.
4. For each different element, the atoms are different.
5. The difference in atoms of different elements is in
    (a) their weight,
    (b) their number of electrons, and
    (c) the possible energies of their electrons.
6. The electrons of an atom can possess various amounts of energy within restricted limits; they cannot have any indiscriminate amount of energy.

COUNTING ATOMS

We shall conclude this chapter with a few remarks about counting atoms. Because atoms are so small, they are difficult to count directly; in fact, so far it has been impossible. We must count atoms indirectly, by weighing. Actually, this is not as peculiar as it might seem to be. For example, no one would buy 10,000 grass seeds, or 500 nails, or 800 beans. We buy these kinds of things by weight. Weighing is a kind of counting.

There are all kinds of counting numbers besides 1, 2, 3, 4. We have the number "dozen" for eggs, doughnuts, and oranges. We have the number "ream" for paper (a ream is 500 sheets of paper). The number "score" is sometimes used for years: "Four score and seven years ago...," which amounts to 4 times 20 plus 7 years. For atoms we use the number "**mole**"; the word mole is derived from the word molecule, probably. We use moles as a counting number for atoms, molecules, and other very small particles, such as electrons, for example. Think of a mole as a special kind of dozen, a sort of 'chemist's dozen.' A **mole of atoms is 6.02×10^{23} atoms**, and it is a peculiar number. You would think that a number easier to remember would have been selected.

Actually, the original intention was to have a simple, easy unit of weight for counting atoms. John Dalton, the man we credit with the original atomic model, suspected that hydrogen atoms were the lightest of all atoms, so he suggested that we think of these atoms as having a weight of 1 unit because this would make things about as simple as possible, at least for the lightest atom. However, others were bothered by the point that an atom is awfully small, so that 1 unit of weight is itself too small to be of much use. They suggested that we think instead of 1 g of atoms as a basis; that is, however many atoms of hydrogen would weigh 1 g. Unfortunately, as more and more scientific facts become known, that simple idea of 1 g as a kind of unit for counting the lightest atoms became inaccurate. Scientists discovered they would have to use 1.0080 g instead. The story is a long one and 16.00 g as a counting weight unit for oxygen got involved, along with an international difficulty in communication between chemists and physicists. Finally, a compromise was reached; 12.0000000 grams of a particularly pure form of carbon has been agreed upon as the basic counting unit for a mole of carbon atoms. Today, the idea of a mole as a counting unit, as a "chemist's dozen" is based on this compromise.

One mole of those special carbon atoms weighs exactly 12.0000000 g. One mole of hydrogen atoms, 6.02×10^{23} hydrogen atoms, that is, weighs a bit more than 1.0 g, 1.0080 g to be precise, or 1.01, rounded off to two decimal places. Try to think of a mole as a counting number, like dozen or score, and keep its value in mind from time to time, for exercise. Try this exercise, for instance: 1 mole of gold atoms weighs 196.97 g. If there are approximately 30 g in 1 oz, 1 mole of gold atoms weighs 6.5 oz, rounded off. Find out the current price of gold and then calculate the cost of one gold atom.

SUGGESTIONS FOR DISCUSSION

1. Earlier in this chapter there were several definitions given for chemistry, both explicitly and implicitly. Identify any two of these definitions, indicate which one you prefer, and justify your choice of the preferable definition and your comparative rejection of the other.

2. State the theme, or essence, of this chapter in a short paragraph, or, better yet, in one reasonably short sentence. If possible, compare and criticize your paragraph or sentence with respect to those composed by others in the class.

3. Define or describe some of the following words and phrases: chemical change, property of matter, element, substance, atom, model, combustion, phlogiston, fact, photon, color, glass, wood, stone, iron, copper, gold, sodium, human imagination, energy level, generation of a photon, mole.

4. Respond with a frank yes or no: So far is the study of chemistry interesting to you? If you answered affirmatively, suggest a few reasons why the study of chemistry might not be interesting to others. If you answered negatively, suggest a few reasons why the study of chemistry might be interesting to others. (Notice that the key word is interesting or not interesting; do not address the question of utility or nonutility.)

5. In this chapter, a few experimental investigations were suggested, such as igniting steel wool, observing a drop of ink in water, and so on. Which of these did you do (or watch while someone else did them)? Which did you omit? Justify both the act of participating and of omitting, or one or the other as applicable.

6. Did you calculate the number of copper atoms in a penny? the diameter of a copper atom? and the cost of one atom of gold? If so, state why it was either worth the effort or a waste of time. If you did not do the calculations, explain why it was not interesting, or seemed to be not worth the trouble, or some other reason for not doing the arithmetic.

7. It is correct to say that salt is salt, no matter where it comes from. Is it correct to say a rock is a rock similarly? What about a rose is a rose is a rose is a rose?

8. Perhaps you have recently read a poem or short story or even a novel or perhaps you have composed a poem or written a short story or composed a

piece of music, which seems to you to agree with the theme of this chapter. Perhaps that work seems to you to disagree with the theme of this chapter. Either way, describe that work briefly and then describe as completely as possible how it agrees or disagrees.

chapter 4

Counting Molecules, Drugs, and Molecular Shapes

It is time for a digression from chemistry itself to some of its more interesting and practical results: counting molecules, pharmaceutical products, and the shapes of molecules. Molecular shapes are pretty straight chemistry, of course, and so is any discussion on molecule counting. Hence, there will be some chemistry in this chapter, but mostly it's about drugs.

One reason a study of science is difficult lies in the vocabulary that is necessary; new words appear on almost every other page. The difficulty is compounded because in many cases the words apply to an abstract concept, and the first few times anyone deals with a nonphysical idea it presents difficulties simply because we cannot be sure what it really signifies in a practical sense. The word, rock, for example, is easy to handle; the word, time, is not as simple to grasp and use, although our familiarity with this concept makes it seem less obscure. But to illustrate, think of a definition or description of "rock," and then do the same for the idea of "time."

MODELS FOR THE WAY CHEMICALS ACT
The molecular model for matter

Further, not only do we have new words which apply to abstract concepts, in science we treat those concepts as though they were real, as real as a rock. Take the word, molecule; a molecule is composed of atoms somehow fastened together. But atoms are imaginary, maybe they do, maybe they do not exist. This leaves molecules in a precarious condition, of course. Further, in the preceding chapter we learned that molecules (and atoms) can be counted. This is obviously nonsense if such things do not exist. Nevertheless, we will have to get used to such logical strategies.

Our evidence, the facts about which there can be no doubt, for atoms and molecules is indirect; so their existence is not certain. But from here on we shall act as though a molecule is as real as a rock. We have good reason to take the existence of molecules as real. By assuming that molecules are the basic composition of many kinds of matter, we can make a lot of progress for the benefit of mankind. A good example of this is the development of drugs and the counting of molecules.

A model for the action of alcohol

We can begin with a common drug, ethyl alcohol. Although the details will vary depending upon the size of the person and their recent past history (loss of sleep last night, for example) and the present circumstances (among friends, or alone, perhaps) an average adult will become mildly inebriated by drinking about 20 or 25 g of ethyl alcohol. Usually, this alcohol is diluted with water containing other substances which have special flavors or odors.

One mole of alcohol molecules weighs 46 g, so, rounding off, we see that about $\frac{1}{2}$ mole of alcohol molecules will produce mild inebriation. Since 1 mole of molecules is 6×10^{23} molecules (rounded off a bit), half a mole is 3×10^{23} molecules of alcohol.

Our average adult is composed of cells, biological units that, one way and another, generally cooperate in the process we call living. In an average adult there are about 3×10^{13} cells. Very loosely described, many of these cells can be thought of as a microscopic sized sack, filled with water, with various subcellular components either immersed in the water or dissolved in it. Now, neglecting the water, each cell is composed of about 1×10^{10} molecules, as an average sort of number. So multiplying 3×10^{13} cells by 1×10^{10} molecules per cell we get 3×10^{23} molecules per average person. Conclusion? For mild inebriation, it is necessary to imbibe one alcohol molecule for each body molecule (not counting the water in the body). The question is, how is it that alcohol causes inebriation? The answer is not known, but from the molecule counting we could make a guess that it is necessary to flood the body system with alcohol to produce inebriation. One model for this would say that each body molecule (excluding the water) must pair up with an alcohol molecule, and when this happens the person is inebriated. This model is easily disproved by more facts.

Five moles of alcohol molecules will cause death (in an average instance). When the corpse is examined closely, there is no evidence, even with such an excess of alcohol, that every body molecule has paired up with one or more alcohol molecules. It is more likely that when a person drinks $\frac{1}{2}$ mole of alcohol molecules, some of the molecules in many of the body cells become "saturated" with several alcohol molecules. This "saturation" then produces the symptoms of inebriation by a process for which there is as yet no generally accepted model.

A model for the action of morphine

On the other hand, consider the drug morphine. It, too, is addictive, produces euphoria, deadens sensitivity to pain, and is generally useful under the supervision of a physician. A typical dose of morphine weighs 0.006 g. A morphine molecule is about six times heavier than an alcohol molecule, so 0.006 g of morphine works out to be about 0.00002 mole (that is, 2×10^{-5} mole) of morphine molecules. And, 2×10^{-5} mole times 6×10^{23} molecules per mole is 1×10^{19} morphine molecules, approximately. That number of morphine molecules is 30,000 times smaller than the number of (not counting water) molecules in a typical human body. Clearly, morphine does not produce its effect by flooding, or saturating, as postulated for the alcohol molecules.

The challenge is to suggest a model that would at least hint at the way the drug morphine interacts with the body cells. Several models have been proposed. One says that inside the body the morphine molecules hit or collide with molecules in the body's cells and bounce off, then hit another cell molecule, and so on. It is the number of collisions per second that does the trick, somehow. This model is probably wrong, at least partly so, because it would require 30,000 collisions per second for each cell molecule to be affected once per second. Possibly a less frequent number of collisions per second on a smaller number of "key" molecules in some of the body's cells might be the effective action.

Another model calls attention to the electrons that are part of the atoms in the morphine molecule and suggests that the morphine molecules, somehow, interact electronically with certain key cell molecules to produce the effect. Then the question arises, how is it that the morphine molecule interacts with these special key molecules in a cell? The best accepted answer to this suggests that the morphine molecules all have the same shape, and each fits into a corresponding shaped surface of a key molecule in certain cells, something like a key fitting a lock. In other words, this model suggests that molecules have a shape and that the shape of a molecule is specific and unique. In Chapter 5 we will look at other arguments that lead to this same conclusion. Here, let us simply explore the idea in a general way.

THE SHAPES OF MOLECULES
The shape of a water molecule

As we know, a molecule of water is composed of two atoms of hydrogen and one atom of oxygen. How are these three atoms arranged in space? There are four possibilities, two are linear, that is, on a line:

$$H—H—O \quad and \quad H—O—H$$

and two are angular:

$$
\begin{array}{c}
\quad H \\
\diagup \quad \diagdown \\
H \qquad O
\end{array}
\quad and \quad
\begin{array}{c}
\quad O \\
\diagup \quad \diagdown \\
H \qquad H
\end{array}
$$

Other possibilities, such as O—H—H, or H—H⟍O , and so on, are merely different views of the four structures displayed. For the angular possibilities, there is also a question about the angle, which could be very broad, that is, obtuse, thus making the molecule almost linear; or it could be quite acute, that is, sharp, for example, sharper than shown in the diagramed structures; or the angle could be somewhere in between.

As we shall see in Chapter 5, the best model is angular, with the oxygen atom in the middle, at the apex; the angle is believed to be about 105°, a bit greater than a right angle, 90°, like this:

$$
\begin{array}{c}
O \\
H \quad H \\
105°
\end{array}
$$

The lines in the diagram between the H and O symbols can be thought of as representing pairs of electrons, from the hydrogen and oxygen atoms, which spend most of their time more or less between the two atoms. Each line represents one pair of electrons. As we know, electrons cannot be said to be in a particular place, so the lines in the diagram are not indicative of a concrete "that is where the pair of electrons is at, by golly" statement; instead it is more or less where they are probably at most of the time. We will discuss this in more detail in Chapter 5.

Three-dimensional representations of molecular structures

Many students as well as mature chemists would like to have a more realistic representation of the shape of a water molecule. For this, we assume (knowing it is still oversimplistic) that the hydrogen and oxygen atoms are spheres, and then correct this a little by assuming that the spheres can interpenetrate each other, somehow. To see this for yourself, make three spheres out of clay, or better yet, tufts of cotton. Two as the spheres should be smaller than the third. The small spheres represent hydrogen atoms and the other larger sphere represents an oxygen atom. Mash the clay, or work the cotton, spheres to form the kind of shape shown in Figure 4–1.

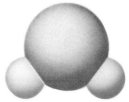

Figure 4–1.   Representation of a water molecule.

As another example, let us take ammonia, which has the molecular formula NH_3. (Formula means "how many atoms of each," as you have probably already guessed.) The model that best fits the facts puts the nitrogen atom in the center, with the three hydrogen atoms all on the bottom (or side or top if you want to twist your head enough):

$$\underset{H \quad H \quad H}{\overset{N}{\diagup | \diagdown}}$$

Here, the flat page diagram is misleading because the H, N, H angle for all possible combinations is about 107°, according to the accepted model; so a clay or cotton sphere model like Figure 4–2 is preferable.

Figure 4–2.   Representation of an ammonia molecule.

As long as we are on the subject of ammonia, it is interesting to note that the most realistic model that we have at present for an ammonia molecule says that it turns itself inside out several times per second. The nitrogen atom passes through the plane that locates the three hydrogen atoms, right through the middle of the three, and out on the other side. Effectively, the ammonia molecule is now upside down or inside out, compared to Figure 4–2. Then, in

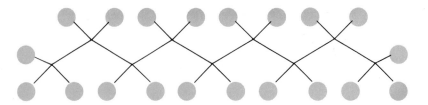

Figure 4–3. A nonane molecule, schematic.

the next instant, the nitrogen atom passes up, through the middle of the hydrogen atoms to the top again, and so on.

For a third example, consider normal nonane, a constituent of gasoline. The formula is C_9H_{20}; we can sketch it like this:

$$\begin{array}{c}
\text{H} \quad \text{H} \quad \text{H} \quad \text{H} \quad \text{H} \quad \text{H} \quad \text{H} \quad \text{H} \quad \text{H} \\
| \quad | \quad | \quad | \quad | \quad | \quad | \quad | \quad | \\
\text{H}-\text{C}-\text{C}-\text{C}-\text{C}-\text{C}-\text{C}-\text{C}-\text{C}-\text{C}-\text{H} \\
| \quad | \quad | \quad | \quad | \quad | \quad | \quad | \quad | \\
\text{H} \quad \text{H} \quad \text{H} \quad \text{H} \quad \text{H} \quad \text{H} \quad \text{H} \quad \text{H}
\end{array}$$

Although useful, this structure is misleading, because all the bond angles are about 109.5°. A clearer picture of the shape of a normal nonane molecule is the schematic diagram shown in Figure 4–3. In this schematic diagram, the hydrogen atoms are represented by dots at the ends of the lines and the carbon atoms are represented by intersections of lines. The line segments represent the pairs of electrons, as usual.

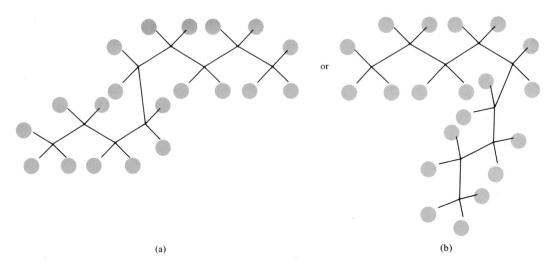

or

(a)   (b)

Figure 4–4. Nonane molecules in two other configurations.

The normal nonane molecule is believed to be in constant motion at ordinary temperatures, wiggly, something like a nervous snake. Thus, at one instant, it might look like Figure 4–4a and in the next instant be contorted still differently, as in Figure 4–4b, and so on.

THE RELATION BETWEEN MOLECULAR STRUCTURE AND MOLECULAR FUNCTION
Morphine and how it may work

The formula of morphine is $C_{17}H_{19}NO_3$; this molecule is shown in Figure 4–5 as it might be imagined to fit a contoured shape of a portion of a much much larger molecule which is part of a biological cell. The imaginary cell molecule fragment is shown as a phantom, the morphine molecule is shown boldly; line segments symbolize shared pairs of electrons, hydrogen atoms are symbolized by H's, oxygen atoms by O's, the nitrogen atom by an N, and, except for one carbon atom symbolized by a C, the other 16 carbon atoms are symbolized by line intersections. The oval shaped symbol, inside a hexagon of carbon atoms on the right, symbolizes six pairs of slightly confused electrons going around in circles.

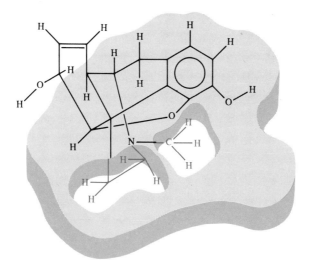

Figure 4–5. Morphine, schematic, showing one possible fit of the morphine molecule to a portion of a cell molecule.

It is to be particularly noted that this sketch shows only one of several possible shapes of a morphine molecule. We could even imagine that, before being attached to a specially shaped part of a cell molecule, the morphine molecule has some other shape, and that, in attaching to the special shape of the portion of the cell molecule, it is deformed to some extent to fit the cell molecule better. This would be like a key to the door of a house being used to unlock the door of an automobile, by using a deformable key. (See Figure 4–5.)

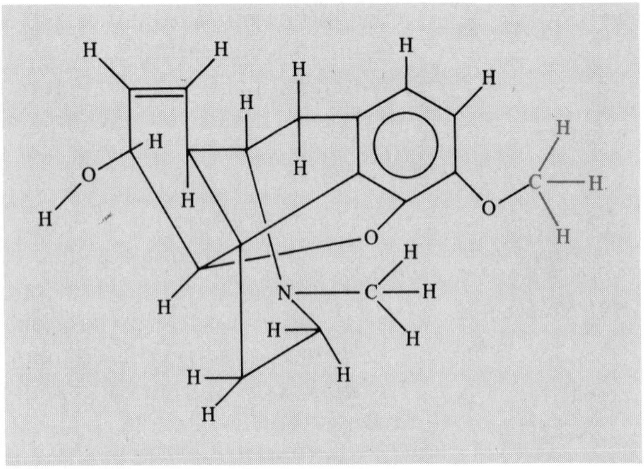

Figure 4–6. Codeine.

Molecules of similar structure

If our idea is to be found useful, it would predict that other molecules, with a similar shape or not too dissimilar shape, could also have the same biological effect that morphine has. Figure 4–6 shows codeine; note the similarity. Codeine is about one tenth as potent as morphine, yet the difference is a "substitution" of a single hydrogen atom by one carbon and three hydrogen atoms. Figure 4–7 shows desomorphine, far more potent than morphine; yet it differs from morphine in having one less oxygen atom and two more hydrogen atoms.

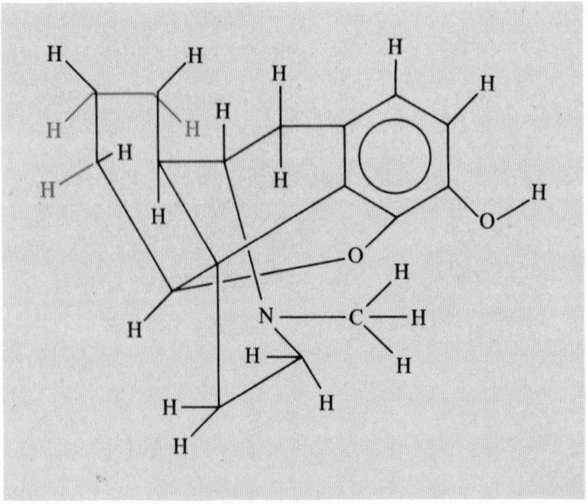

Figure 4–7. Desomorphine.

Notice especially that a very small change in the atomic constitution of a molecule can produce a large change in the physiological properties. Perhaps this is explained by a requirement for a special lock and key kind of fit, so that a small change produces a much poorer, or much better, fit. A poetic way to "explain" the large effect of a small change may also be helpful. Everyone knows from their own experience that the small change of smiling instead of frowning produces a very large difference in the way others react to our presence.

To continue, Figure 4–8 shows demerol, an addictive drug that also resembles morphine in some of its physiological effects. In this case, the differences are quite noticeable, along with a trace of similarity.

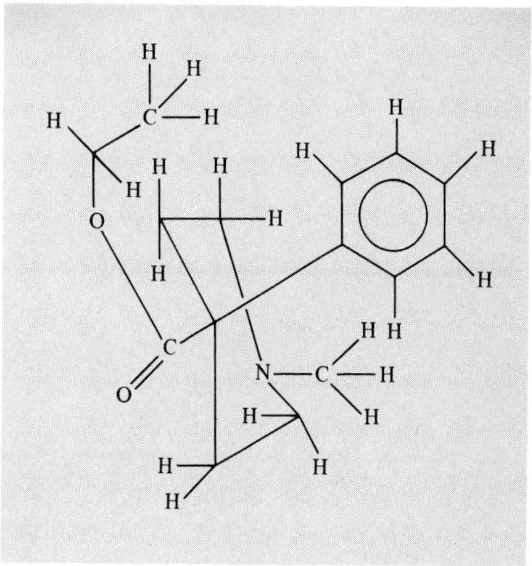

Figure 4–8. Demerol.

Finally, Figure 4–9 shows methadone, also an addictive drug. However, it has been used to aid persons who wish to break their addiction to morphine (or heroin). Methadone lasts longer; that is, it requires less frequent doses, and its withdrawal symptoms are much less dramatic than the withdrawal

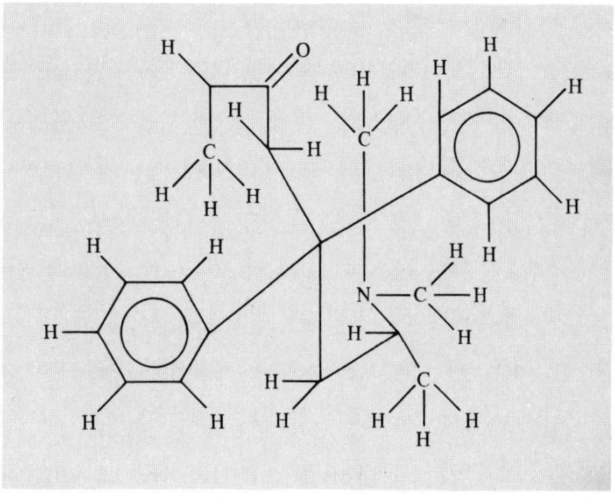

Figure 4–9. Methadone.

symptoms of morphine (or heroin). Again, notice the similarities and the differences compared to morphine.

Heroin itself is obtained by a chemical process in which the morphine molecule is slightly altered. On paper, you can make the change by replacing the "—H" symbol on the left and right "O—H" groupings in the schematic drawing of the morphine molecule with this set of symbols:

$$-\overset{\displaystyle \underset{O}{\|}}{C}-\overset{\displaystyle \overset{H}{|}}{\underset{H}{\underset{|}{C}}}-H$$

Try it, for exercise.

Chemistry as an intuitive art

Our story up to this point surely does not prove that the shape of a drug molecule is closely related to its physiological properties, although it clearly supports such a model. We are not able to cite the facts that would prove the efficacy of shape, in any event. Science is not certain of any of its models, as you already know; only the observable facts are sure. At this point let us digress from our scientific story and focus on a couple of related matters, art and economics.

First, what makes us think that the shape of a morphine molecule is what this book says it is? Slippery answer: We don't really know the shape; a lot of observed facts lead in a very roundabout and indirect way to the shape presented in this book as a reasonable model. Those facts, and their subsequent interpretation in an imaginative way took hours upon hours of the time of hundreds of people. We will get to some of this in following chapters. Here, we have a much more important question: What on earth ever gave anyone the idea that a certain small change, or larger change, in the morphine molecule would produce another, different, drug, like methadone or demerol or one of the others, which might possibly have slightly different properties and thus be more useful to man than morphine itself? Answer: Did you ever paint, or draw a picture, or write a poem? When you finished the first attempt, did you next try a few changes to see if this would improve the result? How did you know, or why did you hope, that a certain change in the sketch or the poem would perhaps make it somehow better? In most of such instances, you, the artist or poet, did not know for sure, you felt it emotionally; you then tried it, to see if it produced an improvement.

Exactly the same process was used for the modifications of the morphine molecule. The chemists who made the changes were artists, molecular artists, if you wish, trying a small change here, a big change there, sometimes succeeding with a modified molecule, much more often failing. (We have only described a few of the successes in our story.) If you asked any one of those chemists while they were making an attempted modification why that particular modification was being attempted, you would get a sketchy or detailed answer (depending upon the personality of the chemist) with strong, detectable, emotionally and artistically based overtomes.

Chemists began with nature itself, since morphine originally comes from poppy seeds. The first problem was to elucidate the molecular shape, very much as a sculptor brings to light the shape of a statue that "exists" inside a block of stone. Once this had been done, the molecule became a sort of canvas or poetic expression ready for further artistic modification. Just as the change of a single word can alter the thrust of a poem, the change of a single atom can alter the properties of a molecule. Who really knows why a poet changed that word, or a chemist substituted another atom? Not the poet, really; nor the chemist.

The costs of molecular modification

Now for the economic aspect. Molecular artistry is expensive in terms of time and materials and sophisticated instruments used to obtain the facts that lead to the indirect conclusion that the molecule has been changed in the manner originally planned. Further, new, modified molecules must be produced in sufficient quantity for testing in nonhuman biological systems before it can safely be used in the first human test cases. The first quantity production processes are usually special and costly. The prehuman test procedures are detailed almost beyond belief. Almost all modified drug molecules turn out to be less effective than the original drug molecule, or worse. This adds up to a very expensive operation.

To recover the investment, a pharmaceutical manufacturer might understandably decide to market a modified drug molecule that, in fact, is not very much better than the original drug or than a similar drug manufactured by a competitor. The result of this practice is that physicians receive information about a very large number of "new" drugs, each one touted to be superior in some way to any other ever produced. Naturally, this is confusing to physicians and pharmacists and gives the public an unpleasant image of the pharmaceutical industry.

Today, advances in our understanding of the model based on the shape and electronic interactions of drug molecules have barely begun to suggest that in the future we will be able to design molecular modifications more efficiently. With each failure and success of a modified drug molecule now, a few more facts are obtained. Ultimately, these observations will lead to a more informed, surer, molecular artistry. At that time it will not be as necessary to attempt to sell such a profusion of varied drugs in order to attempt to recover an investment. It is to be noted that today well over 90% of present prescriptions are for drugs that were not available 15 or 20 years ago. Some very useful drugs have been literally created by the very economic process that has currently produced too wide a variety of choices among drugs.

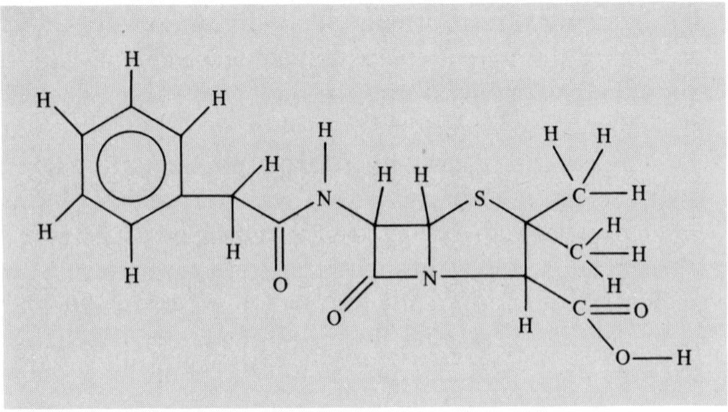

Figure 4-10. Penicillin G.

MECHANISMS BY WHICH DRUGS ACT
The antibiotics interfere with bacterial processes

Another useful drug, to return to our fundamental topic, is penicillin G. The molecule is not literally flat, although it is shown that way in Figure 4-10. (S represents a sulfur atom.) Penicillin G is usually administered to patients in the form of the "sodium salt." This means that the molecule has one of the hydrogen atoms replaced with a sodium atom, symbolized by Na (see Figure 4-11). (The symbol for sodium comes from the German word for this element, "natrium.") Now penicillin G and its sodium salt (Figure 4-11) are not very effective against organisms that are called gram-negative bacteria. A modified molecule, somewhat more effective against these kinds of bacteria, is called ampicillin (Figure 4-12). And carbenicillin (Figure 4-13) is even more effective against gram-negative bacteria.

According to the most likely model each of these drugs does the same nasty thing to a poor bacterium, which is a single celled organism. As we know a

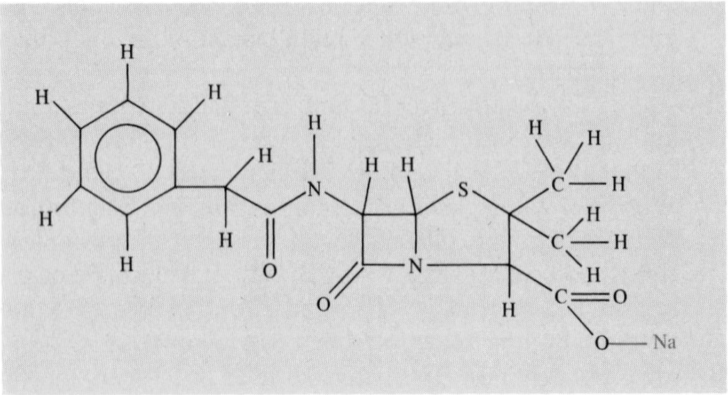

Figure 4-11. Penicillin G, sodium salt.

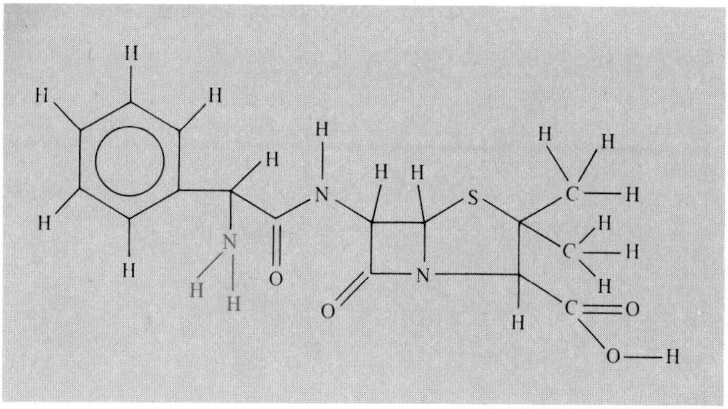

Figure 4–12.   Ampicillin.

biological cell consists of a sack, or wall, inside of which living processes take place. If the wall is broken, the cell will die if the wall is not promptly repaired. What with one thing and another, cell walls are always getting broken and repaired as part of the living process. At the present time, we do not know very much about cell-wall repair, but we do know that it is tricky; a catalyst is required.

A **catalyst** is a substance that enhances a chemical process. For example, when a sugar cube is burned, a chemical process or reaction occurs. Try to burn a sugar cube; light it with a match, if you can, in a saucer or tin can lid. As you will notice, it is difficult to do this. Some help would be useful. Try a catalyst. Rub some tobacco ash from a cigarette, cigar, or pipe thoroughly over all the surface of the sugar cube. Now, try to light it and watch it burn. The tobacco ash is a catalyst for the burning of sugar.

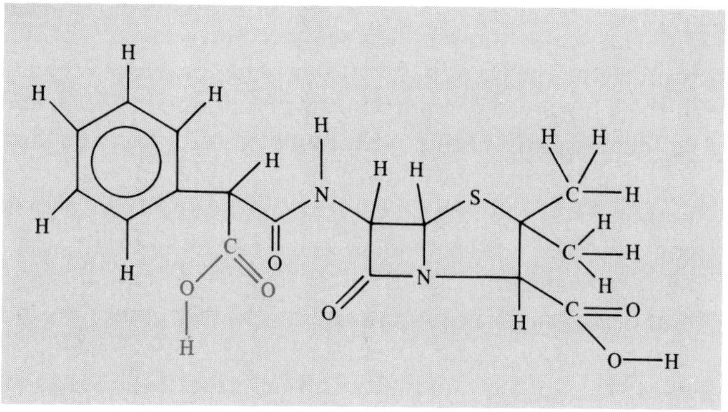

Figure 4–13.   Carbenicillin.

In bacteria there is a catalyst that enhances the chemical process by which the cell walls are repaired. Unfortunately for the bacterium that same catalyst enhances the reaction in which the penicillin G (or any related drug) molecule partially breaks apart to form the modified molecule shown in Figure 4–14. The net result is that, while the poor bacterium's **enzyme** (that's the name for a biological catalyst) is tied up with the penicillin molecule, it is not available for cell-wall building. End of bacterium!

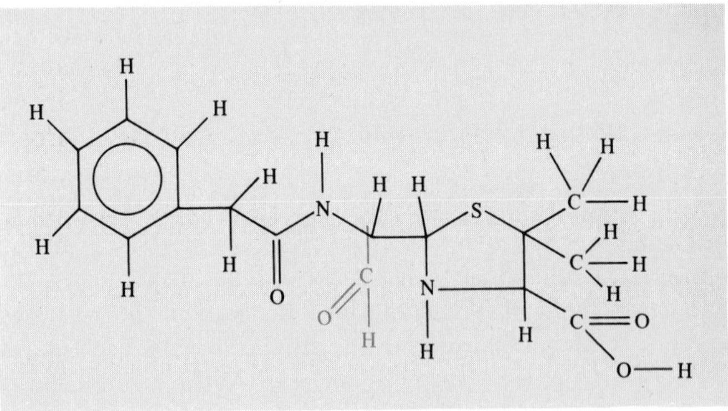

Figure 4–14. Modified molecule, derived from antibiotic drug molecule.

As you have guessed, the enzyme molecule has a specially shaped surface that happens to fit nicely with the shape of the penicillin (or modified) molecule, and this is in part at least (according to the model) why penicillin is effective against a bacterial infection.

Source of antibiotics

Penicillin was discovered accidentally, probably several times, but the significance of those observations passed unnoticed. A large number of related molds, such as those which seem to thrive on decaying oranges and melons, produce penicillin as part of their living process. When these molds are grown in the laboratory in a nutrient broth under controlled conditions, bacteria that would ordinarily thrive in the same broth eventually die. The penicillin, which is in the broth in very low concentration is the cause, of course. Alexander Fleming was the first to take advantage of this effect after he observed the bacteriostatic action of the broth in 1928. Fleming reported these results in a scientific journal in 1929, after further study and observation demonstrated that his initial findings were valid. However, there was not much interest in the report until about ten years later, when crude penicillin powder was used in an effort to cure a dying patient. It arrested the infection in that patient almost immediately, and the treatment was continued until the available supply of penicillin was used up. Unfortunately after the supply was gone the patient relapsed and died. During the early years of World War II, stimulated no doubt by the potential military value of such a potent curative agent, penicillin was first produced on a large scale, in the United States.

Clearly, if one kind of mold produces a bacteriostatic agent, perhaps others would also. Investigators sought varieties of molds all over the world, in samples of soil and in decaying fruit, for example. Today, many of the redundant drugs we use can be traced to these origins, especially if the drug name ends in a suffix such as "mycetin" or "mycin," which is derived from the Greek word for fungus or mold. Two examples among many we might have selected are streptomycin and chloromycetin. The molecule of streptomycin is shown schematically in Figure 4–15. Almost all drugs, even aspirin, have undesirable side effects; streptomycin, for example, produces deafness in some patients after prolonged therapy. To minimize this, an alternative, dihydro-streptomycin is used; with this antibiotic hearing loss is not as likely nor, generally, as severe. Notice the slight difference in the schematic of dihydro-streptomycin presented in Figure 4–16.

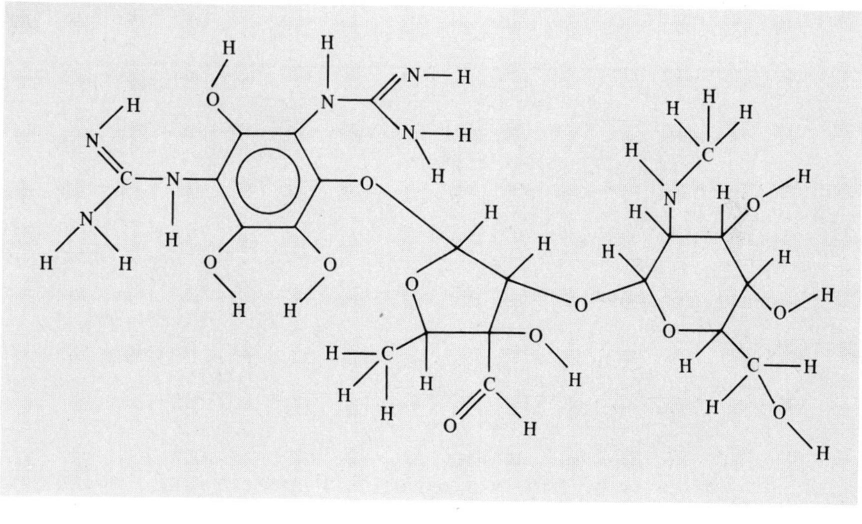

Figure 4–15. Streptomycin.

Figure 4–17 shows chloromycetin. The symbol for a chlorine atom, as you can guess, is Cl. Chloromycetin is notable for an interesting reason. As you looked at the molecular structures for drugs in this chapter, you noticed the presence of nitrogen atoms attached to carbon and to hydrogen atoms only. However, in the structure schematic for chloromycetin, the nitrogen atom is attached to a carbon atom and to two oxygen atoms. It is unusual for a mold to produce a substance in which nitrogen atoms are bound to oxygen atoms.

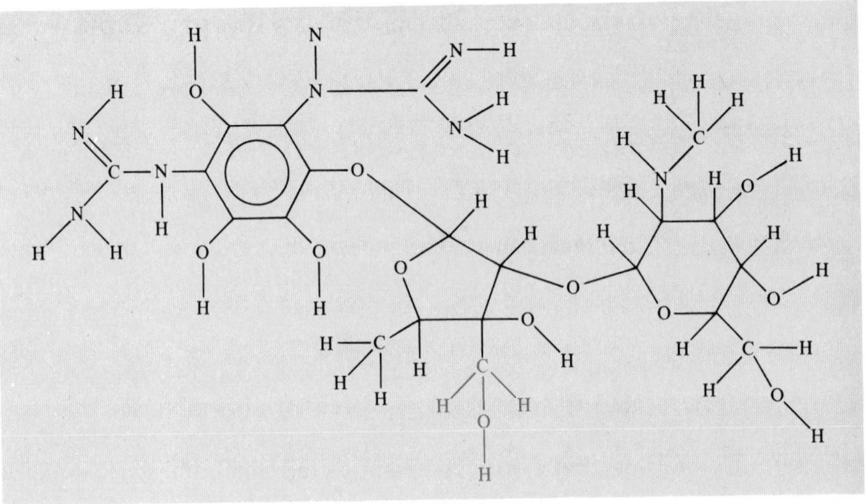

Figure 4–16. Dihydrostreptomycin.

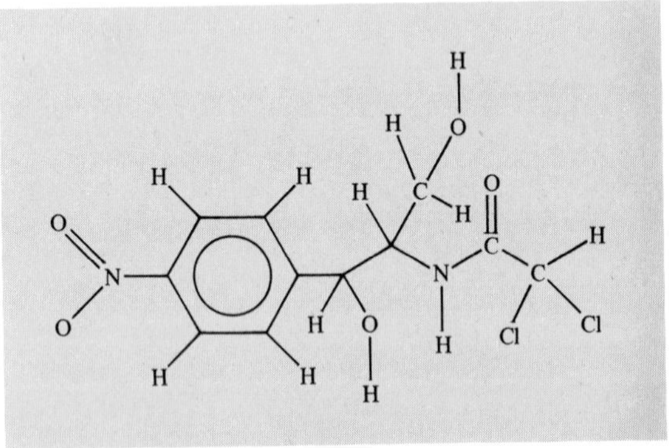

Figure 4–17. Chloromycetin.

Special considerations in the development of drugs

One of the major thrusts of modern drug research consists of a variety of attempts to learn more about the reasons why a particular drug is effective in a particular way. In general, little success has so far been achieved; we have no model which describes the action of aspirin in relieving pain or reducing fever, although several models have been proposed that partially describe the possible action. No one knows why morphine or morphinelike drugs do what they do. We have presented the general features of one of the partially explicatory models in this chapter. We think we know something about the way that penicillin G interferes with bacteria, but we do not know how a penicillin G molecule really fits itself to the enzyme involved, if indeed, this model is correct at all.

Modern drug research is also concerned with other, more immediately practical aspects. For example, it is desirable to get a drug to the location where it can do its work, and not to other places in the body. For gas in the stomach, we wish to use an agent that stays in the stomach, not one that passes through the stomach walls into the bloodstream. For a headache, on the other hand, a drug that remained in the stomach would have little value. To affect the pain centers a drug must reach the bloodstream. Therefore, a detailed knowledge of the permeability of the stomach walls is very useful in developing a new drug which is to be taken by mouth.

As another example, take the useful drug insulin, which cannot be administered by mouth because the insulin molecule is broken up into nondrug entities by the digestive processes. As a result, insulin must be administered by injection. Fortunately for some diabetics at least a sulfa drug can be taken by mouth since it passes into the bloodstream by that route without being destroyed by the digestive processes.

Some drugs act more rapidly than is therapeutically desirable, some more slowly. Modern drug research is concerned with altering the rate of drug action to enhance the desired effect. Thus, the sulfa drug called tolbutamide lowers the blood sugar, and is therefore suitable for some diabetics, but it is a fast acting drug, and soon is gone from the body. A modified drug, chlorpropamide, is much slower acting and therefore better, at least in some instances. Figure 4–18 shows tolbutamide and Figure 4–19 shows chloropropamide; notice the differences.

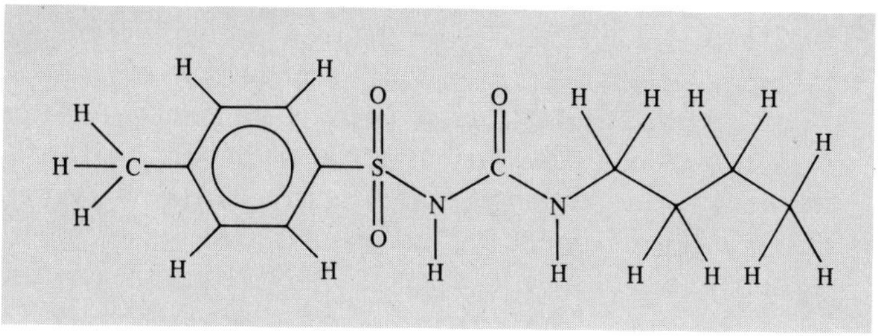

Figure 4–18. Tolbutamide.

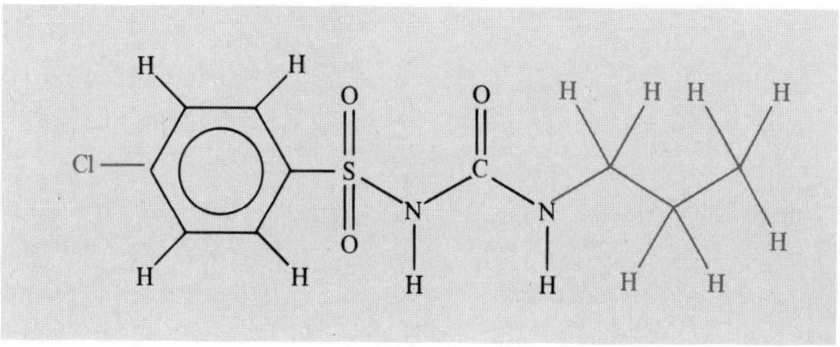

Figure 4–19. Chloropropamide.

Still another concern of modern drug research involves cell permeability at the site of the action. We have already mentioned permeability of the stomach wall; now consider permeability of the wall of a bacterial cell. Once in the bloodstream, where it can be carried to the vicinity of an infecting bacterium, the drug still must act on the organism. In some cases at least it can be effective only if it can get inside the cell, through the cell wall. Contrary to what you may have inferred from other mentions in this chapter, a cell wall is not absolutely impermeable. Various substances constantly pass through it into or out of the cell. It is as though the cell wall is in one sense like a sieve or screen and in another sense like a curtain, which is usually tight but which can be opened to allow a larger particle in or out and then promptly close again. Penicillin is an example of a class of drugs that are effective only when inside the cell itself. Clearly, penicillin would be more effective if, somehow, the molecule could be modified to enable it to sneak inside the cell more rapidly or more efficiently.

A brief look at chemothera-peutics

In this chapter we have considered the subject of chemotherapeutics, or molecular pharmacology. It is appropriate to close with a brief recounting of the story about how it all began.

In prehistoric times, no doubt various enterprising and imaginative individuals established themselves as Medicine Men (and Women). Generally, they prescribed concoctions that were as unpalatable as they were useless. Such things as ground up toad skin mixed with, perhaps, fried cockroaches, and the whole stirred into vulture's blood come to mind. Here and there effective drugs were indeed detected in this manner; perhaps fermented alcoholic beverages were revealed in this way. Various poisons were no doubt discovered by their unfortunate results upon a friendly patient, then later used as part of an offensive technique in tribal warfare. Quinine, one of the most useful specific drugs known today, was used by early man. By and large, however, it was not until a German scientist, Domagk, fed some red dye to sick rats in 1934 that the effective use of specific drugs could be established as a therapeutic regimen.

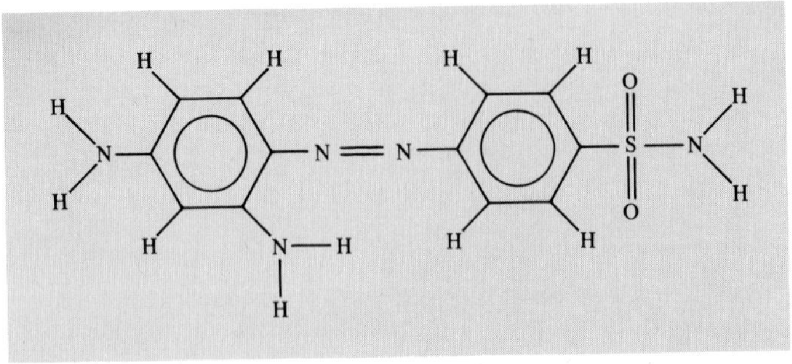

Figure 4–20.   Protonsil rubrum.

Figure 4–20 gives a structural formula for the dye, protonsil rubrum; it is an ordinary appearing red colored powder that can be used to dye fabrics. Protonsil rubrum never did, and never will, harm any bacteria, but, when fed to rats, or humans, the dye is changed into two other molecules (Figure 4–21); the molecule shown in Figure 4–21b is called sulfanilamide. Sulfanilamide is toxic, especially to bacteria such as streptococci, staphylococci, clostridia, and pasteurella. For these bacteria, the substance para-aminobenzoic acid is a nutritional essential. It can be thought of as a vitamin for these bacteria; without it they cannot live. Figure 4–22 shows para-aminobenzoic acid. Notice its similarity to sulfanilamide. According to the model, the bacterium gets mixed up and uses sulfanilamide instead of para-aminobenzoic acid. This is especially true if the patient has been given a large dose of sulfanilamide which thus increases the chances of a sulfanilamide molecule being in the right (wrong for the bacterium) place at the right time, and being mistaken as a vitamin by the unfortunate victim.

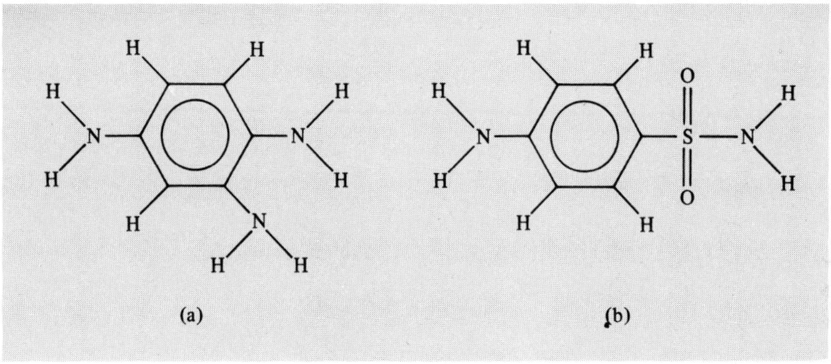

Figure 4–21.   (a) 1,2,4-triaminobenzene and (b) sulfanilamide.

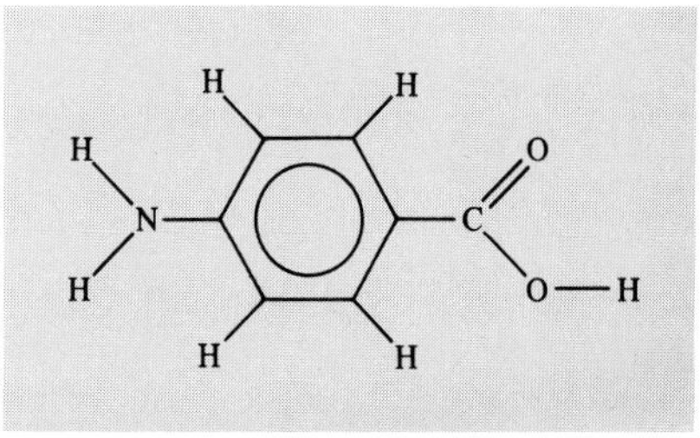

Figure 4–22.   *p*-Aminobenzoic acid.

Unfortunately for the patient, sulfanilamide is not the pleasantest drug to take; it has some undesirable side effects. Following the initial discovery, several thousand variations of sulfanilamide have been prepared and tried. Of these, many were worse than the original drug itself. A few of the acceptable

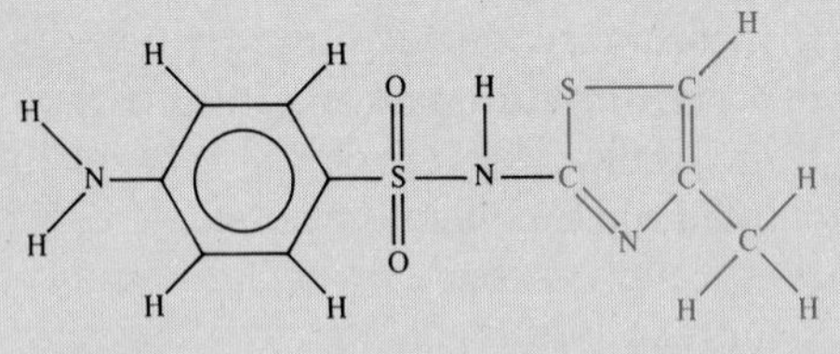

sulfathiazole

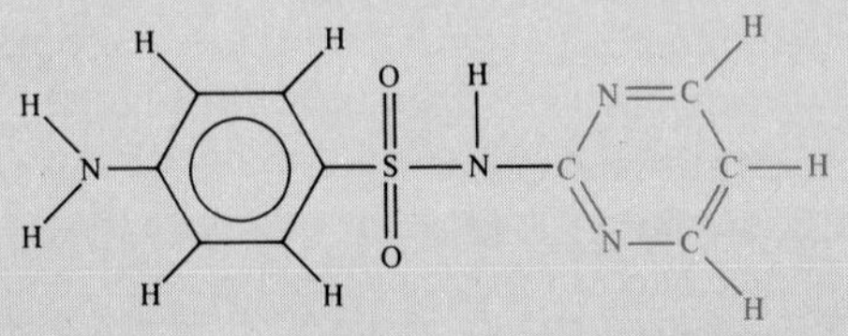

sulfadiazine

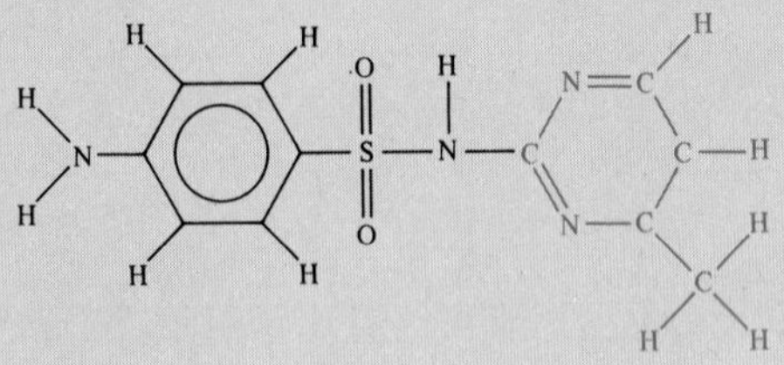

sulfamerazine

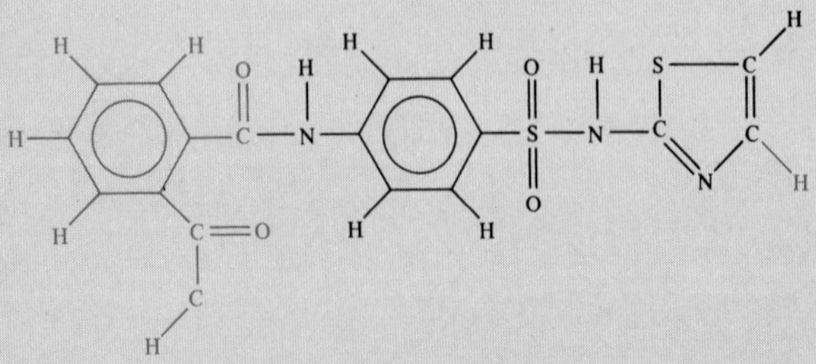

sulfathilidine

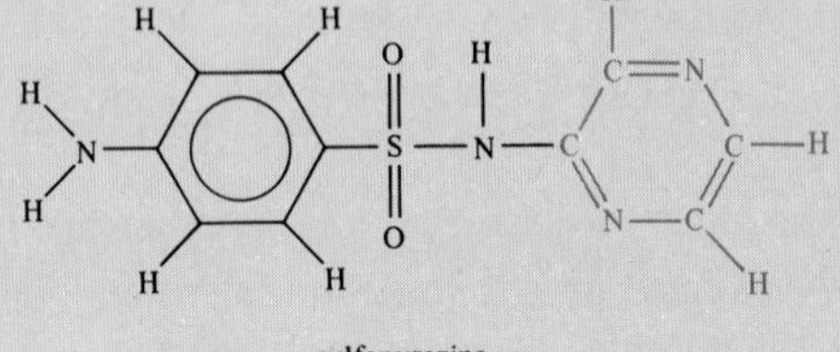

sulfapyrazine

Figure 4–23.
Five drugs.

variations in the fundamental molecular structure are given schematically in Figure 4–23. In general, these variations were introduced to enhance cell wall permeability in the bacterium, to prevent passage of a different variant through the stomach and intestinal walls so that it would stay in the gastrointestinal tract to eliminate infection there, to regulate the rate of action, and so on.

In action, each of these drugs is presumed to break apart with one of the large fragments being sulfanilamide. An interesting side effect was noted in the tests with several of the sulfa drug variations. These were found to be not only bactericidal but to lower the blood sugar content in laboratory animals. Unfortunately, these particular variations had other side effects that made them unsuitable for human use. Eventually, however the drug tolbutamide was synthesized and found to be suitable for human patients with diabetes, at least in some cases. Later developments produced chlorpropamide, which is also used to treat the same disorder. It is possible that some other variation of tolbutamide can eventually be contrived that will be useful for all forms of diabetes.

In the long range view of historical development, with less than 50 years elapsed since molecular pharmacology began to be a science, future progress can be anticipated at least equal to that achieved thus far. This chapter has emphasized drugs for the reason that most people are more interested in this topic than in chemistry. But note the chemistry that you have now learned. We can sum it in a few words: Molecules are composed of atoms, those atoms are bonded one to another in a structurally ordered manner. The next chapter takes up this theme once more.

FOR FURTHER INFORMATION

The Merck Index of Chemicals and Drugs, 9th ed. 1976. Merck and Co., Rahway, N.J. This publication is revised at irregular intervals, published in successive editions. Any edition is an interesting and encyclopedic listing of all sorts of drugs, fully described.

R. F. Gould (ed.): Molecular modification in drug design. *American Chemical Society Advances in Chemistry*, Series 45, Washington, D.C., 1964. A collection of papers presented at a symposium on drug design in 1963. Some of the papers are a bit technical, but others can be read and understood, given the background presented in this chapter.

N. B. Eddy and E. L. May: The search for a better analgesic. *Science*, **181**:407 (1973). The details of recent work on the search for solutions to morphine dependence. Also see an anonymous article in *Chemical and Engineering News*, July 3, 1972, p. 14.

S. H. Snyder: Opiate receptors and internal opiates. *Sci. Amer.*, **236**:44 (1977). Morphinelike substances called enkephalins are effective in relieving pain. This article describes how they might be bound to specific receptor sites in the brain and spinal cord and thus produce their effect.

D. Perlman: Antibiotics—a status report with prognostications. *Chemtech*, **1**:540 (1971). Among other things, this article lists antibiotics that are commercially available in other countries but not in the United States.

I. J. T. Davies: *The Clinical Significance of the Essential Biological Metals.* Wm. Heinemann, London, 1972. An interesting book on a topic that is related to the matters discussed in this chapter, and Chapters 9 and 11.

A. Albert: *Selective Toxicity*, 3rd ed. John Wiley, New York, 1965. A book about substances that injure undesirable cells without too much harm to other cells. The treatment is on the full side; you may end up learning more than you really wanted to know about molecular pharmacology from this book.

A. S. V. Burgen and J. F. Mitchell: *Gaddum's Pharmacology*, 7th ed. Oxford University Press, London, 1972. Something like the *Merck Index*, but very British, and flatly presented. It is best if your biological type vocabulary is ready for action.

G. T. Stewart: *The Penicillin Group of Drugs.* Elsevier, New York, 1965. The title sounds a bit dry; however this is delightfully written, almost a biography of the penicillin drugs and the men and women who carried out the initial work. Recommended.

A. Korolkovas: *Essentials of Molecular Pharmacology.* Wiley–Interscience, New York, 1970. Almost the bible on the introductory aspects of this science. Technical, of course, but readable with attention. Also see W. C. Holland et al.: *Introduction to Molecular Pharmacology.* Macmillan, New York, 1964; F. W. Schueler: *Chemobiodynamics and Drug Design.* McGraw-Hill, New York, 1960; and L. S. Goodman and A. Gilman (eds.): *The Pharmacological Basis of Therapeutics*, 3rd ed. Macmillan, New York, 1965.

J. R. Lambert: The shapes of organic molecules. *Sci. Amer.*, **222**:58 (January, 1970). Not about drugs, but about molecular shapes. Easy to read compared to some of the other references cited here. Also see the Amateur Scientist pages in the *Scientific American* for February, 1973, p. 110, to get some ideas about how to make better looking "molecules" than you can make with clay or cotton balls.

All reputable drug manufacturers publish ephemeral notices describing the properties and uses of their products. These are difficult to obtain but worth the effort if the topic of this chapter has generated more than a casual interest in you. Try your friendly druggist, or your physician, or your physician's office attendant, or your drug salesman neighbor.

D. A. Buyske: *Drugs From Nature. Chemtech*, **5**:361 (1975). A listing of the sources of the important antibiotics current at the time of publication, probable sources of future antibiotics, accompanied by a brief discussion of the role of these drugs and their functions. Also see J. Z. Majtenyi: Antibiotics—drugs from the soil. *Chemistry*, **48**:6 (January, 1975) and 15 (March, 1975), a closely related two-part article.

G. D. Ruggieri: Drugs from the sea. *Science*, **194**:491 (1976). Marine plants and animals contain substances that have druglike properties, with molecular structures unlike any of the drugs used today. These new and different structures could form the basis of new and different, and more effective, drugs once we know more about the details.

M. Gates: *Analgesic drugs. Sci. Amer.*, **215**:131 (November, 1966). On the effects of analgesics such as opium and morphine and the shapes of these molecules.

C. Hansch: Drug research or the luck of the draw. *J. Chem. Educ.*, **51**:360 (1974). How a little serendipity was involved, along with the concept that the shape of a molecule is important. The development of such drugs as antifebrin, phenacetin, and allpurinol, plus the use of computers in modern searches for shapely drug molecules are described.

C. Hansch and J. Fukunaga: Designing biologically active materials. *Chemtech*, **7**:120 (1977). Dr. Hansch is known for his work in relating the structure and shape of a molecule to its physiological effect. Here is an article by the master and an associate on this topic, a sort of how-to-do-it manual, somewhat on the mathematical side.

J. B. Dence: Conformational analysis, or how some molecules wiggle. *Chemistry*, **43**:6 (June, 1970). On both the wiggling and the shapes of molecules that are simpler than those discussed in this chapter.

For schematic pictures of several narcotic antagonists, see the anonymous article in *Chem. Eng. News*, **14** (July 3, 1972).

R. R. Levine: *Pharmacology: Drug Action and Reaction.* Little, Brown, Boston, 1973. An authoritative reference, definitely on the difficult side unless you already know the subject well, but definitely worth dipping into, here and there, for interesting nuggets of information on how a drug acts on living tissue.

R. G. Jones: Antibiotics of the penicillin and cephalosporin family. *Amer. Sci.*, **58**:404 (1970). This is a historically oriented discussion, illustrated with structural formulas of several antibiotic compounds, on the modification of molecules in order to enhance the properties of naturally occurring drugs. Readable and interesting.

H. M. Sobell: How actinomycin binds to DNA. *Sci. Amer.*, **231**:82 (August, 1974). Illustrated, showing how the shape of an anticancer drug molecule is presumed to be responsible for its therapeutic action.

Anon.: *A Prognosis for America*, Pharmaceutical Manufacturers Association, Washington, D.C. 1977. A report of ongoing research projects in the search for new and better drugs. A little biased, perhaps, but informative.

chapter 5

Bonds and Structure

The content of this chapter is largely abstract, an excursion into a chemical myth, or model. To keep some grasp on reality, therefore, a discursive outline of the material may be helpful. We will begin with ancient Greeks and soda straws and examine some of the interesting properties of the Platonic solids. In that discussion you should particularly notice the emphasis upon the tetrahedron (and octahedron); such solid shapes turn out to be essential to an understanding of our chemical model. Up to this point the major chemical topic has been primarily related to molecules. In this chapter we will continue this emphasis, but we will also introduce nonmolecular models. Not all substances are composed of molecules as their fundamental component—salt, baking soda, steel, and copper, are some examples of nonmolecular solids. However, sugar is composed of molecules, and a single diamond can be thought of as a single, very large molecule, incredibly larger, say, than a molecule of sugar or of penicillin G. Finally, in a short appendix to this chapter, we can look at the energy of electrons in atoms once again, this time to pick up a bit of vocabulary that is not important for this book but would be useful should you wish to read almost any other chemistry book with understanding.

THE SHAPES OF SOLIDS

So, to the Greek beginnings. Everyone knows what a **cube** is; it is a three dimensional solid object, with squares on all six sides. A **square** is a regular polygon; a cube is a regular polyhedron, also called a regular solid. A **polygon** is any flat figure with straight lines for all the edges. A regular polygon is any flat figure with straight lines of equal length for all the edges, and with equal angles where the lines join. A **regular solid** is any solid shape with all the sides composed of identical regular polygons. The simplest regular polygon is a triangle with equal length sides, an **equilateral triangle**. Next in complexity is the square, then the **pentagon** with five equal length sides, and equal internal angles. Next, the hexagon; and so on as far as you want to go. Speaking exactly, there is an infinite number of conceivable regular polygons.

The regular solids

Speaking exactly, there can only be five conceivable regular solids. Why the limitation? We can see why with a handful of soda straws, all of equal length. You will also need some pins to hold the straws together at the ends, or use pieces of clay, or a needle and thread if you wish.

   The tetrahedron. We will construct our first regular solid from the simplest regular polygon, equilateral triangles. Make one equilateral triangle from three straws, as in Figure 5–1a. Fasten the ends to each other. Next, with two more straws, using one of the straws on the side of the first equilateral

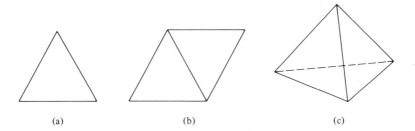

(a)   (b)   (c)

Figure 5–1. Steps in the construction of a tetrahedron from equilateral triangles.

triangle, make a second equilateral triangle, as in Figure 5–1b. Fasten these together, then lift the tip of the second triangle so that the tip is above the first triangle. Finally, make a third equilateral triangle, with one more straw, and the two other straws already in place (the sides of the first and second triangles you have previously made), as in Figure 5–1c.

Notice that when this third triangle is constructed, you have also made a fourth triangle, and you have a solid figure. This regular solid is called a **tetrahedron**. It is the simplest regular solid, composed of equilateral triangles with three triangles meeting at each apex. Before continuing, it is worth examining for its esthetic qualities. Turn it over, so that a different triangle is on the bottom. Notice that it still appears the same. Hold one of the straw edges of your tetrahedron vertically, notice that the opposite edge is now horizontal. Try it with another edge held vertically; always, the opposite edge is at right angles, horizontally disposed.

Regular solids have mirror images. Make two more tetrahedra, each from six straws. (Incidentally, for someone who does not know about tetrahedra, it is amusing to challenge them to construct a figure with four triangles from six straws, or toothpicks, under the restriction that the straws, or toothpicks, can only touch each other at their ends.) With some self-sticking tape, label each apex of each tetrahedron with the letters *A*, *B*, *C*, *D*, exactly as shown in Figure 5–2. Next, put tetrahedron I on top of II, so that the letters correspond on the pairs of apexes, as shown in Figure 5–3. Try to do the same thing with tetrahedrons II and III. No matter how you twist and turn them, it is impossible to get all the pairs of apexes for II and III to correspond. Yet, if you will hold either II or III in front of a mirror, you will notice that the mirror image of II is superposable onto III. Or differently put, II and III (or I and III for that matter) are mirror images of each other. Since II and III are mirror images of each other and are not superposable, this pair is given a special name, **enantiomer**. That is, II and III constitute an enantiomeric pair; they are enantiomers of each other. To get personal, your right and left hand are

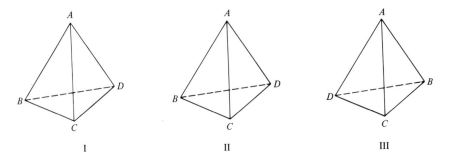

Figure 5–2.   Three tetrahedra, not all identical.

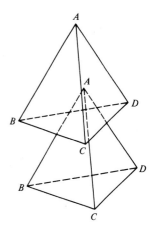

Figure 5–3. Tetrahedra, test for identity.

enantiomeric; one is the mirror image of the other and you cannot superpose one upon the other. To show this easily, recall what happens when you try to put a right hand glove on your left hand; yet the image of your right hand glove in a mirror would fit your left hand easily.

The octahedron. As we now know, a tetrahedron is a regular solid, with three triangles meeting at each apex. Next, construct the next simplest regular solid, with four equilateral triangles meeting at each apex. It will require 12 straws and will look like Figure 5–4 when you are finished. This regular solid is called an **octahedron**. Again, notice its esthetic qualities; for example, by turning it this way and that, it presents the same appearance. You might wish to find out if an octahedron has the property of enantiomerism.

The icosahedron. Next, with five equilateral triangles joined at each apex, an icosahedron (Figure 5–5) results. Thirty straws are required and twelve pins, or pieces of clay, or equivalent fastening devices.

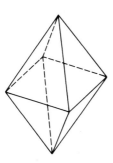

Figure 5–4 Octahedron.

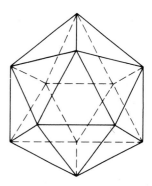

Figure 5–5. Icosahedron.

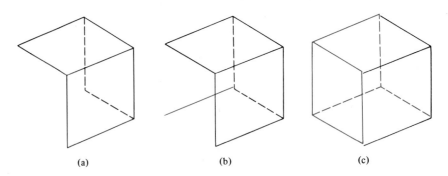

(a)   (b)   (c)

Figure 5–6.   Steps in the construction of a cube from squares.

Next, try six equilateral triangles. We get a flat figure; not a solid. Seven equilateral triangles produce a puckered configuration, and so does eight, nine, ten, and so on. Thus far, that is, using triangles only, we can construct only three regular solids.

The cube. The next attempt will use the next to the simplest regular polygon, the square. First, make a square from four straws. On one edge of the square, with three more straws, construct a second square. Then with two more straws, and using one edge of each of the two squares already constructed, make a third square. It will look like Figure 5–6a by now. Then, with only three more straws (as in Figure 5–6c) make three more squares. The resulting regular solid is a **cube**. It is rather floppy, as you can see (like the octahedron). A square has six sides, twelve edges, and eight corners, or apexes. By selecting the proper four (of the eight) apexes you can produce a structure that has an enantiomer. Try it, using two cubes.

In a cube, three squares meet at an apex. Four squares meeting at an apex produces a flat figure. Five or more squares cannot meet at an apex. This gives us four regular solids, three from triangles, one from squares.

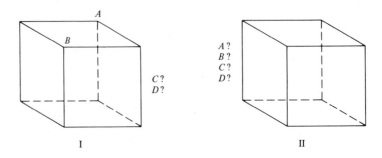

I   II

Figure 5–7.   Enantiomeric cubes puzzle.

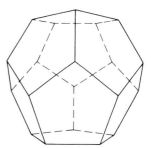

Figure 5–8.   Dodecahedron.

The dodecahedron. We try next with the next most complex regular polygon, a pentagon. Make a pentagon from five straws. Then, with four more and one of the straws on a side of the first pentagon, make a second pentagon. Then, with three more straws and two of the straws already used, one from the side of each of the first two pentagons, make a third pentagon. Continue with more straws; it will require a total of 30 (counting the ones already used) straws and 20 pins or equivalent fasteners at the apexes. The completed figure looks like Figure 5–8.

Again, notice the esthetic qualities of the construction. This regular solid, our fifth, is called a **dodecahedron**. The dodecahedron has three pentagons meeting at each apex. Next, try four pentagons joined at one apex. The structure is puckered, as it is also for five, or more pentagons meeting at an apex.

Only five regular solids are possible. We next use the hexagon as our basic regular polygon for constructing a sixth regular solid. Make a hexagon from six straws. Then with five more, and one of the sides of the first hexagon, construct a second hexagon. Then, with four more straws and one straw from a side of each of the two hexagons already built, make a third hexagon. The structure is flat; more hexagons will also produce a bigger, more extensive, flat arrangement. It is not possible to construct a solid when three hexagons meet at an apex. Four hexagons at the same apex will be puckered, as will five, six, or a higher number. We are stuck with only five regular solids so far.

Try to make a solid with a heptagon, seven straws. You will find that it is impossible to get three heptagons to meet at a common apex and still retain their regular shape. The same applies to an octagon, eight sides, and to a nonagon and to all regular polygons with still more sides. There are only five three-dimensional figures that can be constructed from regular polygons: Three from equilateral triangles, the tetrahedron, octahedron, and icosahedron. One from squares, the cube. And one from pentagons, the dodecahedron.

THE SHAPES OF MATTER

The existence of five, and only five, regular solids was known to the Pythagoreans, and perhaps in even earlier times. Plato, who lived about 100 years later, wrote about these five solids (Pythagoras and his followers pretty much kept it secret) and today they are sometimes called the Platonic solids. At about that same time, early ideas about the nature of the physical universe used a model that said that all matter was composed of four elements: earth, air, fire, and water. This model of four elements was incorporated with the five regular solids. Thus, fire was likened to a tetrahedron, which has the sharpest edges of the five solids. The cube was presumed to represent Earth, perhaps because the cube has a kind of squareness, or solidity, about it. The octahedron was related to air, which is itself closely related to the Earth, as, perhaps, the octahedron is to the cube. This left the icosahedron for water, at that time, since dodecahedrons had not yet been discovered. Later, when this fifth regular solid was identified, it was assigned to the fifth element, an ether, or quintessence.

Modern models of matter are also related to these same concepts. For example, the flat figure obtained by joining three hexagons at each apex produces the kind of repeating figure, or tesselation shown in Figure 5–9.

Today, this is believed to represent the structure of graphite, a form of carbon. One flat hexagonal tesselation would be expected to slide easily over another similar layer. This model would "explain" the lubricating properties of graphite. Molecules of water, in ice, are said to be arranged as in a dodecahedron according to one model (which has not yet been completely accepted by all those who have studied water intensively). The cube forms the basis of some crystalline structures in modern models. You can see this for yourself by closely examining crystals of salt. Sprinkle a few crystals into the palm of your hand; notice the cubical shape of many of those salt crystals. The element, boron (a component of borax) forms crystals with boron atoms at each apex of an icosahedron. The octahedron shape is directly related to the mechanism by which green-leaved plants are able to absorb energy from the Sun. The octahedron is also involved in the mechanism by which oxygen is transported from the lungs, via the hemoglobin in our red blood cells, to other parts of the body where it is needed. And the tetrahedron also is a useful model, as we shall soon see.

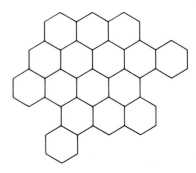

Figure 5–9. Tesselation (hexagonal).

A FIRST LOOK AT THE BONDING OF MATTER

However, at this point we will change the subject and consider the implications of what might be called the harmony, or balance, of opposites. It is a common human experience to notice all sorts of oppositions. Parents generally seek to achieve and maintain the welfare of their children. Children, equally properly, seek to achieve and maintain their individuality and independence as they grow older. The resulting opposition, somehow, usually results in a kind of balance or harmonious relationship (with occasional disruptions). Two friends will often hold opposing views on important matters. In seeking to enhance the friendly relationship there are arguments; yet, somehow each person grows and matures partly as a result of what each learns through such vigorous disagreements, and usually they manage to maintain the harmonious balance of friendship.

Attraction and repulsion

Now, to get to some chemistry, we can ask the question: What holds two atoms together in the form of a molecule? Or, for more complex molecules, such as water, H_2O: What holds the three atoms together? The answer in a nutshell is a balance of opposites, a balance of repulsions and attractions. Another analogy will help us to see how a combination of repulsion and attraction can produce a balanced, steady, condition.

Consider the situation faced by a small boy who has been thoroughly washed and cleaned and finely dressed in his best clothes by his mother, with the promise of a cookie in the near future, but who has, alas, been too tempted by a nearby puddle not far from his home. Now in a bedraggled condition, he hesitates to present himself to his mother, repelled as it were by the thought of her displeasure. On the other hand, he is still attracted towards home by the thought of the promised cookie. In this situation the typical small boy will move about in an approximately circular and wayward path, say, 25 ft (more or less) from his home. The cookie attraction tends to prevent further straying; the repelling anticipation of a scolding tends to prevent a closer approach.

This analogy describes a general principle: *Whenever two, or more, objects are both attracted and repelled, the system of objects tends to attain a condition of balance, a distance of separation which maximizes the attraction and minimizes the repulsion as much as possible.*

Consider a water molecule. It is composed of two hydrogen atoms and one oxygen atom. According to our best model, each atom has a positively charged center, surrounded by negatively charged electrons. Positive charges repel other positive charges and negative charges repel other negative charges. Opposite charges attract. (Why this might be so, our model does not say.) So with these three atoms, two hydrogen and one oxygen we have a system involving attractive and repulsive forces. We predict that a balance will be achieved with the atoms separated a certain distance from each other because of repulsions, yet also attracted because they do not tend to stray any further apart.

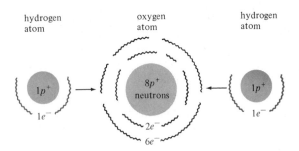

hydrogen
atom

oxygen
atom

hydrogen
atom

Figure 5–10.   Two hydrogen atoms bonded to an oxygen atom (very diagrammatic).

*The covalent
bond in water*

This kind of atom to atom bonding is called a **covalent bond**, or sometimes a **polar covalent bond**. Instead of going into all the details about balance and harmony and attracting and repelling, the covalent bond is usually said to be a shared electron pair bond. The argument goes as follows:

An oxygen atom has (according to the model) eight electrons batting about, with a positively charged nucleus (eight protons and some neutrons) in the center. Two of those eight electrons have a lowest possible energy and tend to spend most of their time fairly near the nucleus. The other six have a bit more energy and buzz about further away from the nucleus. Each of the two hydrogen atoms has one electron, and these two, then, join in the fun and games with the six outer, or **valence** electrons of the oxygen atom, dragging their positively charged, single-proton nucleus along with them. At this point, a *diagrammatic* picture might look something like Figure 5–10.

For reasons that are not well understood, electrons have a tendency to form pairs, and four pairs of electrons seem to be, according to the best model, a particularly favored condition. The two electrons from the two hydrogens plus the six more energetic electrons of the oxygen can pair up, to form four pairs of electrons.

Now, as we know, electrons have negative charges, and therefore would tend to repel each other. Why they then instead form pairs is a good question, but at least we can say that a pair of electrons would tend to repel another pair of electrons. So, for example, if that oxygen atom had only two pairs of electrons (instead of four), we might imagine that one pair would be on one side of the nucleus and one on the opposite side, as far apart from each other as they could get (Figure 5–11). If the oxygen atom had three pairs of electrons, they

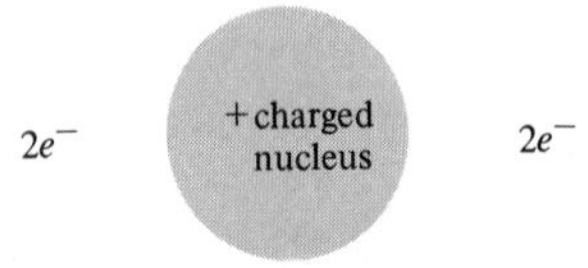

$2e^-$

+charged
nucleus

$2e^-$

Figure 5–11. Disposition of two sets of negative charges around a positive charge.

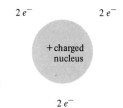

Figure 5–12. Disposition of three sets of negative charges around a positive charge.

would stay as far away from each other as possible (on the average, anyway), as in Figure 5–12.

Now, according to our model, the oxygen atom has four pairs of electrons around it, and you might think that they would arrange themselves as in Figure 5–13a, on the average. This would be correct, if the oxygen atom were flat; however, it is not. In three dimensions, the four pairs of electrons can get further away from each other by hanging around the four apex positions of a tetrahedron, as in Figure 5–13b.

We now have the following situation: The nuclei of the two hydrogen atoms are attracted, each, to one pair of electrons. They are therefore expected to be fairly close to a pair, except that, the closer they get, the more they are repelled by the positive charge of the oxygen atom nucleus; therefore, they don't get too close to the electron pair. The oxygen nucleus, similarly, is attracted to all those electrons around it, which includes the two (one each) from the hydrogen atoms. However, it is repelled by the positive charges of the nucleii of the hydrogen atoms. We end up with a balance; we can say that a molecule of water looks something like Figure 5–14, with a pair of electrons each between each of the hydrogen nucleii and the central oxygen nucleus. Notice that the tetrahedral arrangement (on the average) of the four pairs of

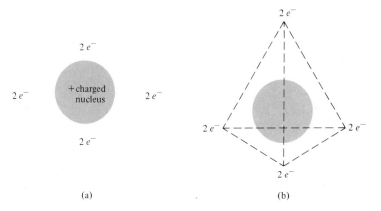

(a)

(b)

Figure 5–13. Disposition of four sets of negative charges around a positive charge: (a) if restricted to a plane; (b) if permitted to occupy three dimensions.

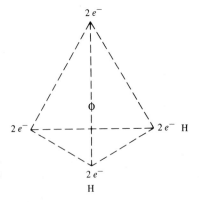

Figure 5–14.   **Two nonbonding pairs of electrons and two bonding pairs, involving hydrogen atoms, around the central positive charge from an oxygen atom; a water molecule, schematic.**

electrons around the oxygen causes the H—O—H angle to be bent; it is not a straight line. The other two, nonbonding, electron pairs are repelling each other and the two bonding electron pairs (see Figure 5–14).

If we only consider the bonding electron pairs, the pairs between the nuclei and draw them as short straight lines, we have this diagram, or picture, of a water molecule:

A special kind of bonding— hydrogen bonding

Finally, to complete this introduction to water, the oxygen atom has a stronger attraction for the shared electron pairs than the hydrogen atom. That is, the negative charges of the electrons are not distributed equally among all three atoms. The oxygen atom is a bit more negatively charged, the hydrogen atoms are a bit more positively charged, compared to the hypothetical condition for equal sharing of all the charges.

This leads to a very important result. Most other molecules that are about the same size and weight as a water molecule are gases under ordinary conditions. Water is a liquid. The slightly negative (oxygen) end of a water molecule attracts either one of the two slightly positive hydrogen atoms of a different water molecule, and the slightly negative oxygen in that second water molecule attracts the slightly positive hydrogen of a third water molecule, and so on. This builds up a kind of network of water molecules and prevents water from being a 100% gas under ordinary conditions. The force of attraction between one water molecule and another is called a **hydrogen bond** (Figure 5–15).

If we wish to get liquid water to evaporate, the hydrogen bonds between one water molecule and another must be broken, to make lighter, single, water molecules, which then assume a gaseous condition. We can break these bonds only by adding energy to the liquid water. Just as energy is required, even to

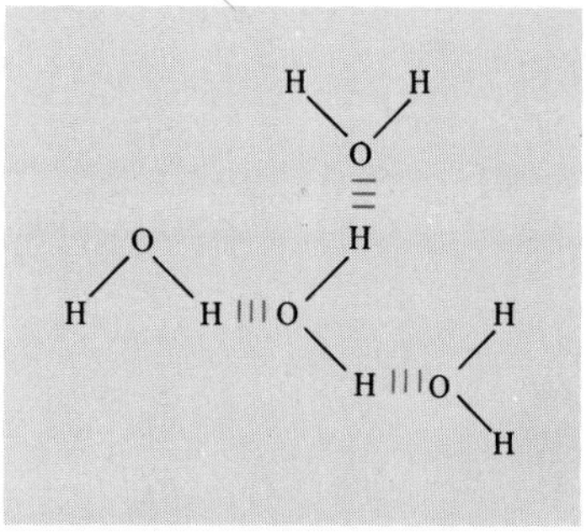

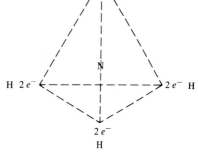

Figure 5–15. A hydrogen bond, symbolized as three short transverse lines, inhibits the free movement of one water molecule with respect to another.

break a piece of butter, or cheese, apart, energy is required to break hydrogen bonds. Typically, we might gasify water by heating it in a pan on the stove. Or, a puddle of water will evaporate in the open air, faster if it is in the sun, where more energy is given to it, than in the shade.

Other substances with covalent bonds

For review, and also to move our story along, we will consider ammonia, NH_3, next. This molecular substance is a gas with an odor; you can smell it, as already noted, by carefully sniffing an opened bottle of so-called ammonia water, used in household cleaning.

There are five valence, or outer, electrons associated with a nitrogen atom. (For each nitrogen atom there are seven electrons, total. Two with a lowest energy assignment, and five with higher energy assignments.) These five electrons plus three more, one from each hydrogen atom in ammonia, form four pairs of electrons around the nitrogen. As we expect, according to the model, the pairs arrange themselves at the apexes of a tetrahedron, with a hydrogen nucleus associated with three of the four pairs (Figure 5–16).

Our third example is methane, CH_4, a gas under ordinary conditions. Methane is a major component in some fuel gases. There are four valence

Figure 5–16. One nonbonding pair of electrons and three bonding pairs, involving hydrogen atoms, around the central positive charge from a nitrogen atom; an ammonia molecule, schematic.

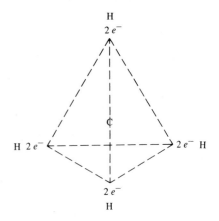

Figure 5–17. Four bonding pairs of electrons, involving hydrogen atoms, around the central positive charge from a carbon atom; a methane molecule, schematic.

electrons associated with a carbon atom. With four more, one each from four hydrogen atoms, we have a total of eight electrons that we can "assign" to the carbon atom. Again, we expect a tetrahedral arrangement for these eight electrons, which form four pairs of electrons, each with an assigned hydrogen nucleus hanging around, nearby (Figure 5–17).

Bonding angles   Let us get back to geometry for another look, using methane as a starting point. We can imagine a hydrogen at each apex of a tetrahedron made of six straws and a carbon atom inside the tetrahedron at the center. What is the size of the angle with the carbon atom at the center and two hydrogens at each end? (Any two hydrogens may be used, since they are geometrically equivalent.) For a perfect tetrahedron, using a little trigonometry, this angle turns out to be 109°28'; or almost 109.5°. Data obtained from laboratory studies of methane are consistent with this conclusion from the simple geometric model shown in Figure 5–18.

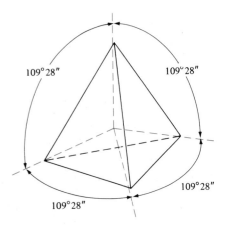

Figure 5–18. Tetrahedral angles, apex to center to apex (two of the six are not identified).

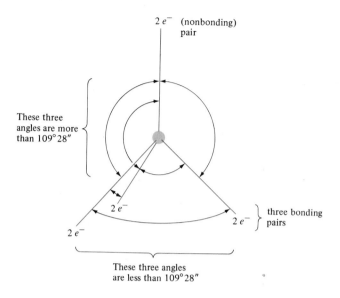

Figure 5–19. **Warped tetrahedral angles for one nonbonding and three bonding pairs of electrons.**

For ammonia the situation is different. The pair of nonbonding electrons can be imagined to be more repulsive than any of the three bonding electron pairs because that nonbonding pair is attracted only to the positive charged nitrogen nucleus. The three bonding pairs of electrons are attracted by both the nitrogen nucleus and their associated hydrogen nucleus. Hence, we can imagine that the bonding pairs are pushed a little closer together, because they don't repel as effectively, by the nonbonding pair, which repels very effectively (see Figure 5–19). This would tend to make the H—N—H angles less than 109.5°. Laboratory data yield an interpreted value of about 107°, consistent with the prediction we have drawn from our model.

The situation is similar for water. With two nonbonding pairs associated with the oxygen and two bonding pairs, we would expect an even greater repelling force upon the two bonding pairs; thus, the H—O—H angle would be predicted to be smaller than the corresponding angle in ammonia. Laboratory data from studies with water yield an angle of about 105°

The reality of the impalpable   It is important to pause in discussions like this, now and then, and reflect upon what has transpired. Atoms and molecules may not exist in the same sense that a tree, or a dog, or this page in this book exists. For a page in this book, existence is palpable; you can feel the page, see it, you can crinkle it a bit, and hear it. So far no one has heard, or seen, or felt, or tasted a molecule or an atom.

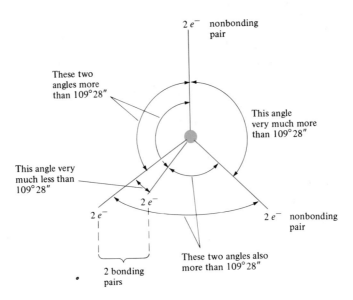

Figure 5–20.   Warped tetrahedral angles for two nonbonding and two bonding pairs of electrons.

Yet no one acquainted with all of the details of palpability of real objects doubts the existence of molecules, atoms, electrons, nucleii, and all the rest. Because they are impalpable, we are faced with a kind of paradox. They are the products of human imagination, generated out of what we can be sure, of, paper crinkling and noise, a dog running, our nose itching, candles burning, a colored neon sign, and a host of other incontrovertible facts. In this case, reflect upon where our imagination has taken us. Not only do we say that there are molecules, we can even say a few things that make sense about their shape. We can apply this imagined detail of shape to make useful conclusions about the actions of drugs, as we have seen in the preceding chapter. Human imagination is rather remarkable; it produces poems, works of art, and statements about molecular detail. It delights, instructs, and entertains.

THE PERIODIC TABLE

We have been told that carbon atoms have four valence electrons, nitrogen atoms five, and oxygen atoms six valence, or outer, electrons. There are more than 100 known elements. Does the number of outer electrons (the ones with the most energy) in an atom increase up to some number larger than 100? Answer: No. After many attempts to imagine a variety of models, with intense discussions, some polemic arguments, and a few damaged scientific reputations, a remarkable model that explained many puzzling points in the behavior of the elements was developed. This model, called the periodic table, provided hints, predictions of then-undiscovered properties, that helped chemists and physicists decide that the number of valence electrons was limited.

There are four more elements, along with carbon, that have four valence electrons per atom: silicon, germanium, tin, and lead. There are four besides nitrogen with five valence electrons: phosphorus, arsenic, antimony, and bismuth. There are also four besides oxygen that have six valence electrons: sulfur, selenium, tellurium, and polonium. What about the other elements? Hydrogen has one valence electron; so do lithium, sodium potassium, rubidium, cesium, and Francium. If you count them, that's 22 named elements, so far. With more than 80 elements not yet considered, this can get complicated. It will be neater and easier to tabulate this information. The table used for this purpose is the periodic table shown in Figure 5–21.

Since we have been emphasizing imagination as a theme, it is interesting to note here that the periodic table was first conceived by (historically attributed to, that is) a Russian chemist, Dmitri Mendeleev, the youngest in a family of 17 children, over 100 years ago when he was 37 years old. Mendeleev knew nothing of valence electrons. He used the relative weights of the atoms of

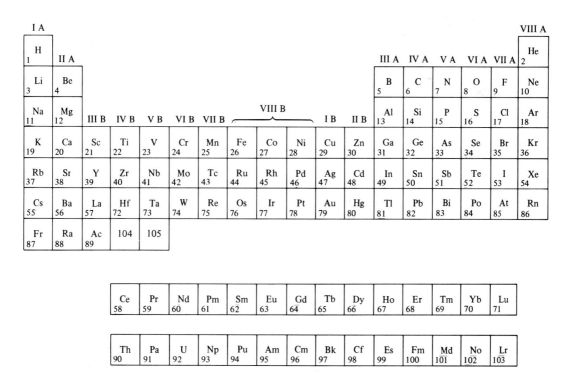

Figure 5–21. The periodic table of the elements.

the different elements, with some notable exceptions, to arrange these elements according to their increasing atomic weights. Mendeleev also based his tabulation, that is, the rows and columns in his periodic table, according to the similarity of certain properties among the elements. He did not imagine that the similarities were caused by having the same number of valance electrons. In fact, he asserted in his published work that all this "modern" fuss about charges in atoms was balderdash, a figment of the imagination that could not possibly ever lead anyone anywhere.

Today, if we were making the periodic table for the first time, we would put the elements with the same number of valence electrons in the same columns of our table. Examine Figure 5-21 carefully. Do not be concerned about your lack of familiarity with many of the symbols. (Each one represents a different element.) Do not be concerned about memorizing numbers of valence electrons. Do notice, however, the number associated with each elemental symbol; these are called **atomic numbers**, and they are the same as the numbers Mendeleev gave the elements except that he called them ordinal (counting) numbers.

Note also that the columns are headed with Roman numerals and the letter A or B. Roman numeral IA signifies one valence electron for the element in that column. The IIA signifies two valence electrons, and so on. The headings IB, IIB, IIIB, and so on, signify a more complex energy assignment for the electrons with highest energies in those atoms. The elements copper (Cu), silver (Ag), and gold (Au) in column IB each have one valence electron for sure, but sometimes they can use two or three electrons as valence electrons. The elements in column IIB, zinc (Zn), cadmium (Cd), and mercury (Hg), have two valence electrons, except sometimes mercury uses only one. All the rest, columns IIIB, IVB, and so on, up to and including VIIIB, generally have two, or more, valence electrons. Finally, the two rows of elements, numbers 58 to 71, inclusive, and 90 to 103, inclusive, should be all put in the table in the same places, where element 57, lanthanum (La) is (for 58–71), and where element 89, actinium (Ac), is (90–103).

In most cases, the symbols for the elements correspond to the English, German, and French names, or to one of them, with a few symbols related to the Latin names for elements that were known in ancient times. Thus, C for carbon or carbone; K for kalium (potassium, in English); Cu for cuprum (copper). The origin of the names given to more recently discovered elements is an interesting story. Cobalt is derived from kobolos, meaning goblin; gallium signifies France, hafnium is related to Copenhagen; thallium from a green, budding twig; vanadium from a Greek goddess with fickle tendencies; niobium and tantalum from Greek mythology, daughter and father.

The atomic number   According to our model for atomic detail, the atomic number is the number of protons, the number of positive charges, in the nucleus of an atom. Example:

fluorine, atomic number 9. In the nucleus of each and every fluorine atom there are nine protons (there are some neutrons also, but we'll get to that in Chapter 6). Fluorine is in column VIIA, or group VIIA, so we know that it has seven valence, highest energy, electrons. It also has two other electrons, at lower energy levels, in each atom. A fluorine atom has nine electrons, nine negative charges, outside the nucleus, batting about somewhere more or less mostly near the nucleus, but not too close generally. The atomic number specifies the number of electrons associated with a neutral atom; **neutral** signifies that the number of electrons outside the nucleus is equal to the number of protons in the nucleus. However, an atom need not be neutral; it could have an extra electron or electrons; it could also lose one or more electrons.

ELECTRO-NEGATIVITY AND IONIZATION POTENTIAL

Some atoms have a greater tendency to acquire extra electrons, or a lesser tendency to lose electrons (practically the same thing), than other atoms. The tendency to acquire extra electrons is called **electronegativity**. The tendency to lose electrons is associated with the phrase **ionization potential**. The higher the electronegativity of an element, the greater the tendency of its atoms to acquire extra electrons. The highest electronegativity value, 4.0, is assigned to fluorine. The lowest possible electronegativity value, 0, is not assigned to any element. The atoms of all elements have some tendency to pick up an extra electron. Ionization potentials are measured on a different scale than electronegativities, with numbers in the hundreds. Since fluorine is very electronegative, we would expect it, correctly, to have a high ionization potential. As it happens, helium and neon in group VIIIA (column VIIIA) have higher ionization potentials for losing one electron than does fluorine; yet helium and neon have very low electronegativity.

The covalent bond

It is time to put this discussion into context. The topics identified in the title of this chapter are bonds between and among atoms and the structure of the combination thereby achieved. So far, we have identified a covalent, or sharing of a pair of electrons, bond. These few words are used to describe the bond between atoms that results from a balanced opposite charge attraction and similar charge repulsion, identified with respect to the positively charged nucleii and negatively charged electrons of atoms. From the geometrical discussion, the shapes of the covalently bonded atoms in molecules was related to electron pair repulsions. The preceding chapter dealt with shapes of molecules from another point of view, and we now see that the shapes of those complex molecules are determined by electron pair repulsions associated with each of the atoms in those molecules.

All molecules consist of covalently bonded atoms, with electron pair repulsion (according to our model) determining bond angles and, therefore, the shape of the molecule as a whole. But covalent bonds are formed only between

Figure 5–22.   The first row of eight elements.

Li	Be	B	C	N	O	F	Ne
3	4	5	6	7	8	9	10

atoms when both have a reasonably high ionization potential and both have a reasonably high electronegativity. Other kinds of bonds are involved for other combinations of ionization potential and electronegativity values of the constituent elements. As we shall see, there are two other kinds of bonds, but to get at that, it is useful to take a closer look at ionization potential and electronegativity.

The periodic table as a guide to ionization potential

Consider the elements in the first row of eight in the periodic table. From left to right these are lithium (Li), beryllium (Be), boron (B), carbon, nitrogen, oxygen, fluorine (F), and neon (Ne). They are listed here (Figure 5–22) in a contiguous fashion, along with their atomic numbers, the number of positive charges in each nucleus. Consider ionization potential first. From which of these might it be easiest to take one electron away? Or, which would require the least amount of energy to knock off one electron? The answer is clear, from the one that holds any of its electrons the least tightly. That would be the atom with the smallest positive charge, lithium. The ionization potential of lithium is the smallest of the eight. Neon, on the other hand, over on the right with ten protons in the nucleus of its atoms, would present the most difficult electron removal problem; it would require more energy, a harder kick or something, to remove one electron from neon than from any of the other seven. We would expect the ionization potentials of the six elements in between to be intermediate and gradually increasing from beryllium to fluorine. Our model is a bit oversimplified, since measurements of the ionization potentials of these elements in the laboratory show that the ionization potentials of boron and oxygen are "too low," but not very much so, compared to our prediction.

In general, for any horizontal row in the periodic table the ionization potential increases from left to right, one element compared to the next (with a few exceptions here and there which we will ignore).

The periodic table as a guide to electronegativity

What about electronegativity? Which element in the first row of eight would have   least tendency to pick up another, extra, electron? Easy, the one with the lea   positive attraction for the extra, negatively charged electron. That would be lithium. And so it goes; beryllium with one more positive charge in its nucleus has a bit higher electronegativity. Boron a higher electronegativity than beryllium, and carbon more than boron. The electronegativity of nitrogen is still greater, oxygen more yet, and fluorine still more.

The electronegativity of neon, however, is close to zero. What happened? Answer: Remember we mentioned before that four pairs of electrons are a specially favored situation. Well neon already has eight valence, or

Figure 5–23.   Next comes sodium, Na, number 11.

Na
11

highest energy electrons (that is, four pairs). It is as though the eight outer electrons in a neon atom "close out," or shield, the ten positive charges in the nucleus from effectively attracting an extra electron.

A sodium atom has one more electron than neon. It is this electron that is the outer electron and places sodium in group IA. The other ten electrons in a sodium atom constitute an inner (less energy) collection of eight, and a really way inner (still less energy) collection of two electrons. These two "closing out collections" of **completed shells of electrons** shield the full effect of eleven positively charged protons from being available near the (more or less) edge of the atom. So, the electronegativity of sodium is low; there is not much tendency to pick up an extra electron. The ionization potential of sodium is also low, it is fairly easy to take away that single outer electron.

The columns of the periodic table

Compare lithium and sodium. Which has the least electronegativity? Neither one has much. But sodium has the least. A sodium atom is bigger than a lithium atom; after all it has eight more electrons, and electrons account for almost 100% of the size of any atom. So, in attracting an extra electron, the sodium is at a disadvantage. The extra electron would have to remain at a greater distance from the positively attracting nucleus on the edge of the sodium atom than it would if attracted by the positive charge in the smaller lithium atom.

If we continue, downward now, in group IA, potassium (K) has less electronegativity than sodium, and a lower ionization potential, also. The reasons are comparable to those we suggested for lithium and sodium, and the tendency continues. Of all the elements in group IA, francium (Fr), number 87, has the lowest ionization potential and the least electronegativity. As it

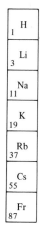

Figure 5–24.   Group IA.

happens, however, francium is an artificial element; it does not occur naturally except in very very small trace amounts. So, the honor of being the element with the least electronegativity, and the lowest ionization potential, is generally accorded to cesium (Cs), number 55.

In general, as we go from top to bottom in any column, that is, any group of the periodic table, the electronegativity and the ionization potential both decrease. For all practical purposes (omitting group VIIIA, since those six elements are not important for our discussion) and omitting francium, the elements with the higher ionization potentials and electronegativities are found in the upper right corner of the periodic table; the elements with the lower ionization potentials and least electronegativities are located in the bottom left corner of the periodic table.

ANOTHER LOOK AT BONDING

As we already know, atoms of two or more elements combine, somehow, to form compounds: water, demerol, alcohol, nonane, salt, soda, ammonia, sugar, candle wax, chlorophyll, and so on for millions of compounds. Some compounds are composed of several different atoms of different elements; some of fewer numbers of different elements. To keep this as simple as possible, we will consider compounds with no more than two different elements, such as water with hydrogen and oxygen, ammonia with nitrogen and hydrogen, carbon dioxide with carbon and oxygen, nonane with carbon and hydrogen, and so on. For the two different elements, then, there are the nine arrangements of extremes with respect to ionization potential and electronegativity shown in Table 5–1.

TABLE 5–1.   Nine Extremes

The Ionization Potential of the Two Elements	The Electronegativity of the Two Elements	Comments
Both high	One high, one low	No such combination possible*
	Both high	Will form covalent or polar covalent bonds
	Both low	No such combination possible
Both low	One high, one low	No such combination possible
	Both high	No such combination possible
	Both low	Will form "metallic" bonds
One high, one low	One high, one low	Will form "ionic" bonds
	Both high	No such combination possible
	Both low	No such combination possible

* With a few exceptions, no elements are known to have a high ionization potential and a low electronegativity, or the converse, so the combination of an atom of such an element with any other elemental atom is "not possible." The major exceptions are the so-called noble gases, the elements in group VIIIA, which do have high ionization potentials and very low electronegativities.

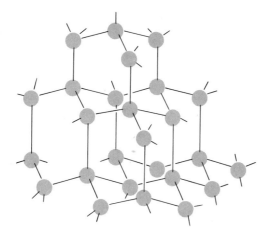

Figure 5–25. The diamond lattice. In this figure, the dark circles represent carbon atoms, the lines represent shared electron pairs. The figure itself is a small fragment of a diamond. Notice that (except for the carbon atoms on the edges) each carbon atom is bonded to four other carbon atoms. These four are at the apexes of a tetrahedron which has the fifth, central, carbon atom at the center of that tetrahedron.

Covalent bonds again

We will conclude this chapter with a brief look at examples of the three types of bonding. First, a few more covalently bonded substances: Carbon atoms will share electrons with other carbon atoms, as we know. When this is carried to extremes, one carbon sharing with four others, and each of those with three more others, and so on, the result is called diamond. Notice that each carbon atom has four valence electrons; if it shares one each of these with four other carbon atoms that are similarly sharing their four valence electrons, we generate a structure of tetrahedra arranged in a three dimensional network. (See Figure 5–25.)

These shared electron pair bonds are sometimes called **pure covalent bonds**, to distinguish them from polar covalent bonds. A polar covalent bond, you will remember from our discussion of the bonding in water, is a shared electron pair bond for which one atom is more and the other less electronegative. As we now know, oxygen is more electronegative than hydrogen, so the bond between oxygen and hydrogen has a negative end, or pole, attributed to the oxygen position, and two positive ends, or poles, attributed to the hydrogen positions. In diamond, the electron pairs are shared between neighboring carbon atoms, each of which have the same electronegativity (of course), so the bond is purely covalent, equal sharing, no positive or negative poles.

Quartz is composed of silicon, Si, and oxygen atoms joined in a network similar to that for diamond. Conceptually, we may substitute a silicon atom for each carbon atom in the diamond network, and then, between each two silicons, we insert an oxygen atom. This describes the structure of quartz. Notice that the bonds between silicon and oxygen are polar covalent. If we were to count the silicon and oxygen atoms, we would find twice as many oxygen atoms as silicon atoms. The formula for the quartz molecule would then be

$$\text{Si}_{\text{big number}}\text{O}_{\text{twice as big number}}$$

and the actual value of the numbers would depend upon how big a piece of quartz we had. In such cases, the formula is simplified to be SiO_2. This formula seems to describe a regular molecule with one silicon and two oxygen atoms; there is no way to tell that instead it signifies an atom ratio counting just by looking at it; you have to know that silicon dioxide, or quartz, is a network polar covalent substance.

Quartz is a major constituent of granite, a common form of rock. Most sands consist of small grains of quartz, broken from larger rocks by freezing and erosion processes. It is interesting to examine a single grain of quartz sand very closely. Generally, a grain of quartz sand will be semitransparent, not clear as is pure quartz, as a result of various impurities in the sand. Glass is closely related to quartz. Although glass is amorphous, it can be thought of as a quartzlike but irregular network of silicon and oxygen atoms in which some of the silicon atoms are replaced with calcium atoms and, in the open spaces here and there in the network are some extra sodiums and oxygens trapped, not unlike a bird in a cage. So called soft glass is made by heating sand, calcium carbonate (limestone), and sodium carbonate (washing soda) together until the mixture fuses and softens into a very viscous, very hot, liquid. This thick liquid can be formed into desired shapes and when cooled retains that shape. Borosilicate glass, or heat resistant glass, is prepared by heating to fusion a mixture of sand, borax, alumina, soda, and potassium carbonate. The resulting glass is similar to soft glass in its structure except that, in this case, atoms of boron and aluminum replace some of the silicons, and the sodium, potassium and other oxygens are trapped.

Covalent bonds in organic molecules

Figure 5–26 is a diagrammatic illustration of a sugar molecule. The line intersections represent carbon atoms, as does the symbol, C. The lines represent shared electron pairs (or where those electrons are most of the time perhaps) between atoms; O symbolizes oxygen atoms, and H symbolizes hydrogen atoms. This particular sugar is called sucrose, the kind you put in coffee to sweeten the taste. There are several other kinds of sugar. These can be visualized as similar to the sucrose molecule, except for interchanging one or more of the O—H groups with an H atom; put one on top and the other on the bottom, instead of the other way around. Such changes generate a different molecule, still a sugar but not as sweet. Or, still other sugars have only one "ring" instead of two as sucrose has.

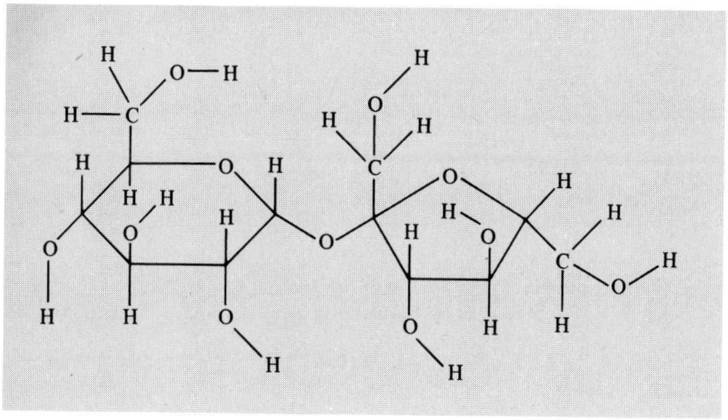

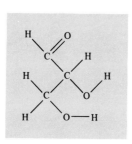

Figure 5–26. Sugar (sucrose).

Speaking a bit loosely, starch can be described as similar to sugar, except that, instead of two joined rings, there are hundreds or thousands of rings joined in a long chain. Cellulose is quite similar to starch, except that the way the rings are joined to form very long chains is ever so slightly different.

The simplest sugar is glyceral (Figure 5–27) it has three carbon atoms per molecule. Glycerine, another three-carbon sugar, has a slightly sweet taste. It is used in candies to enhance the soft, or moist, character of the candy. Glycerine tends to hold water molecules near it by hydrogen bonds (Figure 5–28), as we might have guessed, since there are three O—H groups on a glycerine molecule, each of which has a polar covalent bond.

Figure 5–27. Clyceral.

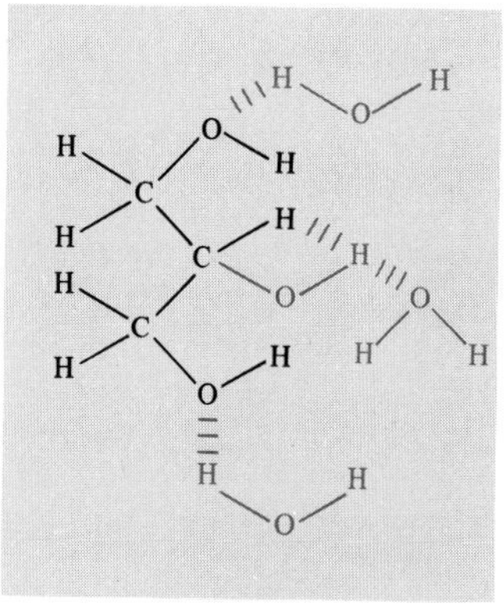

Figure 5–28. Glycerine with three water molecules hydrogen bonded.

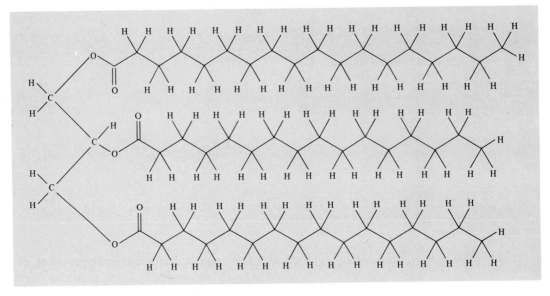

Figure 5–29.   Typical fat molecule.

All fats "contain" glycerine (and are examples of polar covalently bonded molecules). Figure 5–29 shows diagrammatically a typical fat molecule; notice the three-carbon glycerine "base" depicted on the left.

Finally, proteins are polar covalently bonded molecules with thousands of atoms per molecule. Figure 5–30 is a diagrammatic sketch of a very small

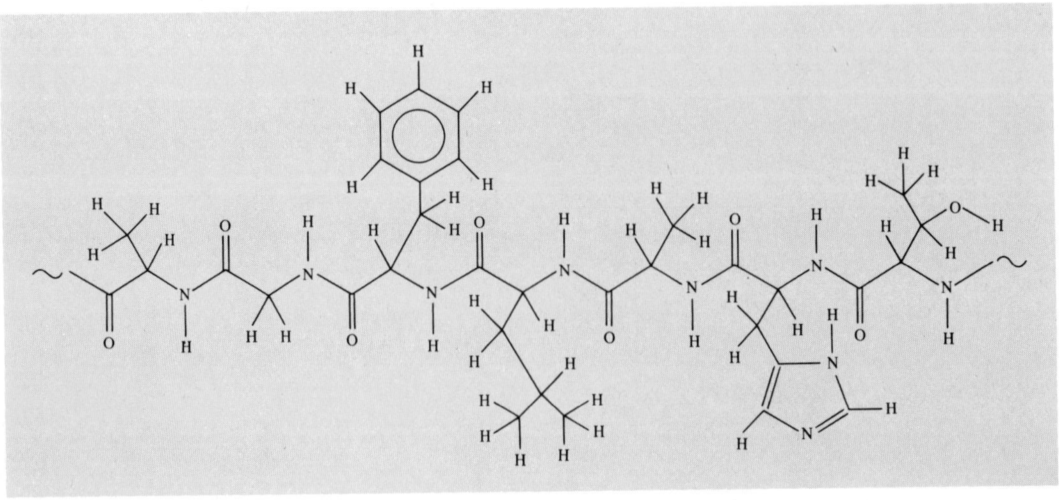

Figure 5–30.   Portion of a protein molecule.

portion of a typical protein molecule. Notice the oxygen atoms and the nitrogen atoms; compared to the hydrogen atoms to which they are attached, the oxygen and nitrogen atoms can be expected to have a slight negative charge and the hydrogen atoms a corresponding slight positive charge. In a real protein molecule, such as those in the muscles of your body, different parts of the same protein molecule are attached together by the hydrogen bonds which arise from this kind of detail. Further, adjacent different protein molecules are attached to each other here and there by hydrogen bonds. To a considerable extent, the structure and movement of our bodies depends upon the flexibility of long protein molecules and on the way they are coiled and configured. And this, in turn, depends upon the formation of hydrogen bonds, in many instances, within the same molecule or between molecules.

Metallic bonds

Of the more than 100 elements, about 15 are not metals. These are the elements with high electronegativity. The others, more than 85, are generally defined as metals. These are shown in the segment of the periodic table given in Figure 5–31. (The elements with symbols in colored areas are on the border-line, definable as either metals or nonmetals depending upon which definition or degree of electronegativity or other kind of criteria you might choose to

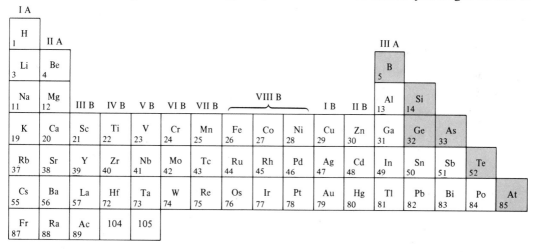

Figure 5–31. The metallic elements.

apply.) Notice that a metal generally has a low ionization potential; if you provide even a reasonably small amount of energy, the atom will give up at least one electron. Notice that a metal has a low electronegativity. If some other atom gave up an electron, that metal atom would not be very interested in accepting the electron.

Iron is a metal. What holds the iron atoms together in, say, a nail? Think of a large number of iron atoms, each of which has given up an electron (or maybe some have given up a couple). Now all those iron atoms have a positive charge because each has a higher positive charge in its nucleus than the number of negatively charged electrons outside the nuclear region. Such an iron nail would fall apart; the positively charged particles, called positive **ions**, would repel each other.

The loose electrons, however, are not likely to get very far from those positively charged iron ions; they are attracted by the positive charges. They act a little like a crowd of small boys turned loose in a store stocked with free ice cream cones. The boys would, at first at least, run about, from a strawberry cone to a chocolate to a vanilla to a butter pecan, and so on, pausing momentarily at this cone then that (until finally there would be one big mess). The loose electrons in an iron nail also move about, from this positive iron ion to the next, and so on, in constant motion, pausing momentarily only, then moving on.

It is these freely moving negatively charged electrons that hold together the positive iron ions by a balanced attraction and repulsion. That is, all the positive iron ions repel each other and so do all the negatively charged freely moving electrons, but there is an attraction between the positive iron ions (which are pretty much fixed in place because they are so heavy) and the wildly moving negatively charged electrons (in part freely moving because they are so light).

The model is the same for any metal, copper in a penny, gold in a ring, zinc (and other metals) in the "white metal" fittings used in automobile door handles, the mixture of iron, chromium, and nickel in stainless steel, the aluminum in a pan; all metals are held together by this same kind of metallic bond.

Technically, the metallic bond is described as a balanced repulsion-attraction involving delocalized electrons and structurally fixed positively charged metal ions. This model is consistent with several facts that we have already observed for metals. Why is a metal shiny? Delocalized electrons, it turns out, are one of the best reflectors of light you can think of (if you are mathematically inclined and like to think about such things). Metals conduct an electric current readily; how so? We will look at the electrical aspects of chemistry in chapter 7, so here it suffices to say that an electric current is simply a bunch of electrons all going more or less on the average in the same direction, like a bunch of air molecules moving through the trees in a forest (called "there's a wind blowing in that forest"). Delocalized electrons are free to move in any direction; when we get them mostly all moving in the same direction, in between or through the structured positive metal ions, this is an electric current being "conducted" through the piece of metal.

The structure of crystals

Let us turn our attention now to the positive ions that are pretty much fixed in position. For this you will need a cigar box or other similar sized fairly sturdy, shallow, box (no lid needed) and a large number of marbles.

Hexagonal close packed crystal structure. Put one layer of marbles in the bottom of the box, jiggling the box a bit to get the marbles settled in nicely. That layer will look like Figure 5–32 (ignore the letters in the figure for now). Those marbles represent one layer of a metal crystal, with each marble representing one positively charged metal ion. Next, put a second layer of marbles on top of the first. There are two ways to do this. Either put the second layer on the interstices marked y or z. There is not enough room to put marbles on both the y and the z marked places.

Next, put in a third layer. There are two ways to do this. The third layer of marbles can have each marble exactly above the corresponding marble in the first layer. Or, each marble can be exactly above the y or z marked spaces, whichever one you did *not* choose to use for the second layer.

The succeeding layers are applied similarly, to build up a simulated crystal. The first layer is called the A layer, the second is the B layer. If your third layer has marbles directly above the marbles in the first layer, it is another A layer; your fourth layer would then have marbles (if you do it properly) directly above the marbles in the B layer, and it too would be a B layer. And so on. You have built an ABABABA · · · crystal structure, which is usually called a **hexagonal close packed** crystal structure.

Face centered cubic crystal structure. The other crystal structure that you might have chosen is called a **face centered cubic** crystal structure. For it, the first A and B layers are the same as before. But for a face centered cubic arrangement, you must place the third, C, layer so that the marbles are not directly above any marbles in the A layer. Then, the fourth layer will have marbles that are directly above the marbles in the A layer, so the fourth layer is

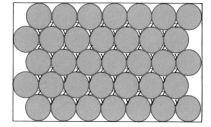

Figure 5–32. First layer, metal crystal, hexagonal or face centered cubic arrangement.

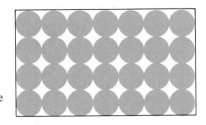

Figure 5–33. First layer, metal crystal, simple cubic arrangement.

an A layer, too. The fifth layer will be a B layer, the sixth will be a C layer, and so on. You have built an ABCABCACAB · · · crystal structure. This structure is known as a face centered cubic because, if you had those marbles glued together just right, you would now be able to see that a slanting sectioning of the ABCABC · · · structure can also be described as being composed of cubes, as well as hexagons. There is no way fully to understand this explanation without actually building the simulated structure from marbles or other spheres in a suitably sized box.

Why metals can be deformed. Most metals have either the hexagonal close packed or the face centered cubic crystal structure. You might notice that one layer of marbles will slide over the underlying layer rather easily, with only a little push (except for the sides of the box getting in the way). This is similar to what happens when a piece of real metal is smashed, or drawn into a wire, or otherwise deformed to a different shape. In the microscopic sized crystals in a real metal, layers of ions slide over one another as the new shape is achieved.

Simple cubic and body centered cubic crystal structure. One more marble exercise. This time some delicate maneuvering is required, put a layer of marbles in the box as illustrated in Figure 5–33. Then, and this is very tricky, put a second layer right on top of the first, marble directly above marble. Look at any four marbles in the second layer which form a square. Directly under these four are four more. The eight marbles describe a cube. This crystal structure is called a **simple cubic** structure; it is of no concern to us in a discussion of metals. But, notice the open space in the center of that cube; you could almost put another marble in that open space, in the center of the body of that cube. If you did this for every cube in the two layers you now have in the cigar box, and if the sides of the box would stretch enough for you to do it, you would have the beginning of a simulated **body centered cubic** structure. Just add more layers similarly to develop a fuller simulation. To do this actually would require some glue and a lot of patience.

The effect of crystal structure on properties

Some metals have a **body centered cubic** crystalline structure; chromium, molybdenum, and tungsten, group VIB, and vanadium, niobium, and tantalum, group VB, are examples. All of these metals are fairly hard; they do not deform easily. Gold, on the other hand, is an easily deformable metal. Gold has a face centered cubic structure. If you have built a simulated marble model of a face centered cubic and of a body centered cubic structure, you would notice, after a thorough examination, that a face centered cubic structure has many more ways for a layer to slide over another layer than a body centered cubic structure. This is consistent with the relative deformability (malleability) of gold compared to chromium.

In fact, gold is too soft to be satisfactory as a precious metal. A ring made of pure gold, for example, would be bent out of shape merely by putting it on your finger. How can gold be hardened? It is not possible to alter the crystal structure (except by cooling or heating) so instead we melt the gold and mix in with it a little molten copper. When pure, the copper also tends to form a soft, malleable, face centered cubic structure, but copper ions are smaller than gold ions. Therefore they interfere with the regularity of the crystal structure. It would be as though you used two different sized marbles in the cigar box and did the best you could to construct a face centered cubic arrangement. The mixture of two different sized marbles in that arrangement would tend to prevent one layer from slipping over another as easily as if all the marbles were the same size.

Mixtures of metals are called **alloys**. In general, an alloy is harder than either, or at least one, of the two pure components. Brass is an alloy principally composed of copper and zinc. The nickel coin we use is an alloy of copper and nickel. Sterling silver is an alloy of silver and copper. (Old so-called silver coins were made of sterling silver, not pure silver.) Dental amalgam is an alloy of mercury, a metal that is a liquid under ordinary conditions, and silver; soft and pasty when first mixed, this alloy becomes harder a few hours later as the liquid mercury penetrates the solid silver crystal structure.

In general, the alloys just described can be thought of as "substitutional"; the ions in the crystal structure are replaced, or substituted, with ions of other metals that are larger, or smaller, than the original ions. Another way to harden a metal is by filling some of the interstices, the tiny spaces in between adjoining ions. Thus, in a simulated face centered cubic structure, or one of the other two simulated structures in that cigar box, we could put a few very small marbles, here and there, in the empty spaces between adjoining marbles. If those tiny marbles were big enough practically to fill that space, they would tend to lock one layer to the next, and make it more difficult for one layer to slide over another.

This is done with iron by a process called **case hardening**. The gears in an automobile transmission (the discs with teeth in them around the circumference) must be very tough, not hard, in order to withstand the forces imposed when an automobile is driven. The teeth themselves, or at least their surfaces, must be hard in order to resist wear; if the surface is soft, the gears will wear out in a few thousand miles as the teeth of one gear rub against the mating teeth of another gear. To solve this, a gear is treated before use by heating it in surroundings containing nitrogen. The solid substance potassium cyanide, KCN, is often used. During the heating process, nitrogen atoms from the potassium cyanide work their way into the surface of the gear teeth and eventually lodge in the interstices of the iron crystals, thus locking one layer to the next. After such treatment, the surfaces of the gear teeth are hard and durable against severe wear.

Faults in crystal structure

Before going further, a possible misunderstanding should be cleared up. As presented, you now have the impression that a metal crystal is a regular, ordered, perfectly arranged structure of ionic spheres. This would be correct if there were such a thing as an ideal metal crystal. In practice, no metal crystal is perfect. We really could not expect it to be; even the smallest of metal crystals contains zillions of metal ions and electrons. Some of those ions are bound to be out of place, or missing, or an extra one happens to be squeezed in where it doesn't really belong, or an impurity metal ion, or several of them here and there, get into the crystal.

Ionic bonds

You will recall that in covalent bonds an atom in a molecule was attached to another specific atom, and, if three or more atoms were present in that molecule, the third atom was bonded to the second at a particular angle with respect to the first and second, and so on, with specific angles always involved. We could say *this* atom is bonded to *that* one, through a sharing of electron pairs. On the other hand, for the metallic bond, we could still talk about shared electrons, even if they were delocalized. Any one metal ion is equally bonded to all its neighbors, on all sides and above and below.

An ionic bond involves ions (of course). It does *not* include electron sharing as an important aspect; in this way it differs from covalent and metallic bonding. It is like metallic bonding in that any given ion is bonded to each and all of its oppositely charged neighbors, not to any specific one. All three types of bonding involve a balance of repulsion and attraction, but for each of these the forces are exerted in a slightly different way. And finally, before getting to details, it is important to note that there is no perfect example of any of the three types of bonds. In any real substance, there is a blend of bond types, although usually one type will predominate. There is not much ionic bonding nor covalent bonding in an iron nail. Sugar is mostly covalently bonded, with a trace of ionic bonding. Salt, our first example of an ionic bond, is very slightly polar covalent; in our discussion, however, we will consider it as though it were perfectly ionic.

To start find sodium (Na) in the periodic table in group IA, and chlorine (Cl) in group VIIA. Notice that from their positions in the periodic table, we can say chlorine is very electronegative and sodium has a low ionization potential. We might expect that if a bunch of sodium atoms ever got close enough to an equal number of chlorine atoms there would be a one-sided trade;

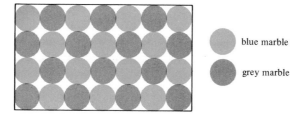

Figure 5–34.   First layer, sodium chloride crystal.

each sodium atom would give up its single valence electron and each chlorine atom would take on one more electron, making the sodiums into positively charged ions and the chlorines into negatively charged ions.

Then, we might imagine, these oppositely charged ions would arrange themselves into a harmonious, balanced, configuration involving attractive, positive to negative and vice versa, and repulsive, positive to positive and negative to negative, forces. Our imagination is correct, and in addition when this happens a great deal of heat and light is emitted. The final result is salt, sodium chloride; NaCl is the simplest formula, representing

$$Na_{big\ number}Cl_{same\ big\ number}$$

The crystal of sodium chloride can be visualized with several colored marbles, say some grey and the rest blue. Arrange these to form a layer in the bottom of a cigar box in a square array, like Figure 5–34, with colored marbles alternating, like a chess board. For the second layer, repeat the first layer, except put a grey marble on top of a blue and a blue on top of a grey marble. The third layer alternates the colors similarly, and so on. The resulting structure simulates the crystal structure of sodium chloride. The grey marbles represent the sodium ions, positively charged, and the blue marbles represent the chloride (not chlorine) ions, negatively charged. Actually, it would be better if the sodium ion marbles were about half the diameter of the chloride ion marbles; this would more closely represent the actual sizes of the two ions in salt.

As you can see from the periodic table, comparing sodium *atoms* and chlorine *atoms*, sodium atoms are larger. But the loss of an electron makes the sodium ion smaller, and the gain of an electron makes the chloride ion larger.

Ionic crystals   The crystal structure of salt is face centered cubic. This way: Look at the top layer in your box and notice only the grey marbles. They form a square array. Two layers lower there is the same array. Any four grey marbles that outline a square at its corners also correspond to four others, two layers below that also outline the corners of a square. All eight outline the corners of a cube, with six faces. In the center of each face of that grey marble cube you can see another grey marble. As far as the grey marbles are concerned, 14 marbles, eight at the corners plus six (one in each face) define a face centered cubic arrangement. Further, the same is true for the blue marbles. In fact, if you look closely (it helps to have a transparent sided cigar box) you will see that the grey marble cubes interpenetrate the blue marble cubes.

Strictly speaking, salt is a crystalline structure consisting of an interpenetrating lattice of face centered cubic sodium ions and of face centered cubic chloride ions. Each positively charged sodium ion is surrounded by six negatively charged chloride ions (except at the surface of the crystal) and by twelve positively charged sodium ions (except at the surface). The same can be said, conversely, for each chloride ion. The crystal is held together by a balance of attractive and repulsive forces. No single sodium ion can be said to be bonded to any single chloride ion, and conversely. The whole crystal is the unit, really. Although we say that the formula of sodium chloride is NaCl, and this looks as if it might represent a molecule of sodium chloride, it does not. Unless you know that sodium chloride is bonded ionically, not covalently (not very much covalently, that is), there is no way to tell merely by looking that NaCl is, or is not, a molecular formula.

There are other examples of ionic compounds: baking soda, sodium hydrogen carbonate, $NaHCO_3$; limestone, calcium carbonate, $CaCO_3$; gypsum, or plaster, hydrated calcium sulfate, $CaSO_4 \cdot \times H_2O$; milk of magnesia, magnesium oxide, MgO (the white solid that is mixed with water to make a thick liquid); soap, sodium oleate is one example, $NaO_2C_{18}H_{33}$; lye, potassium hydroxide, KOH; and thousands upon thousands of others.

Some ionic compounds have crystal structures similar to that for salt. There are several other crystal structures possible in principle. In general, the particular crystal structure applicable to a particular ionic compound depends upon the relative sizes of the positive and negative ions. For sodium hydrogen carbonate, for example, the sodium ion is very much smaller than the hydrogen carbonate ion. It is interesting to note the components of the hydrogen carbonate ion, to look inside this negatively charged ion so to speak. The carbon, the three oxygens, and the hydrogen are polar covalently bonded and the whole structure has a single negative charge, because it has one extra electron, captured from some sodium atom, or acquired in some other way. So, baking soda, sodium hydrogen carbonate, is an ionic compound consisting of an ordered crystalline structure of sodium ions and hydrogen carbonate ions. The sodium ions are sodium atoms that have lost one electron each. The hydrogen carbonate ions are combinations of covalently bonded atoms (one hydrogen, one carbon, and three oxygens) that have one extra electron.

If you walk into almost any chemistry laboratory, you will see a large number of chemical compounds, many of which are ionic compounds. You could guess from this that ionic compounds are important to chemists. But outside the laboratory, ionic compounds are comparatively rare. It is the covalent bonded compounds and the metallic bonded metals and alloys that are more numerous.

COMMENT

At this point you have a choice, depending upon how much this chapter has turned you on. Either read the appendix to this chapter or go directly to the suggested topics for discussion at the end of the appendix. The next chapter continues with a further look at our theory of atomic composition, this time at the atomic nucleus itself.

APPENDIX TO CHAPTER 5

s, p, d, f, AND ALL THAT

According to our model, atoms are made up of a positively charged nucleus, quite small in size, centered in a cloud of electrons that surround the nucleus and occupy a space much larger than the small nucleus. According to the model, it is not possible to pinpoint exactly where the electrons are at any instant. We can only say more or less where the electrons are likely to be most of the time, as long as we specify a reasonably large volume of space as the place where they probably are. However, it is possible to state the energy of those electrons, almost exactly. It turns out that the different energy specifications correspond to different places in space where the electrons probably are most of the time. From here on, it gets complicated.

If you want to know what the energy of an electron is in some particular atom, you first must discover an important piece of information about the whole atom: whether it is in the lowest possible energy state or in some higher energy condition. Once you know the energy state of the whole atom, you can talk about the energy of a particular electron in that atom. The answer would turn out to be a long mathematical equation that would be useful only in special cases. A more generally useful answer is a kind of summarizing, or abstracted, answer for an atom in its lowest possible energy state.

Quantum numbers

So, we ask the question: How do you describe the energy of an electron in an atom? Answer: You describe the energy by stating four quantum numbers. Of course the next question is, what is a quantum number? A quantum number is a number, like 1, or 2, or 3, and so on, as far as you want to go, no fractions allowed, and the two special fractions $+\frac{1}{2}$ and $-\frac{1}{2}$.

Let us try a silly example to illustrate the idea; we will use a male college student who is taking this course. How would you describe the academic condition of that male student? You might do it by stating four things:

1. How long ago he had anything to eat.
2. If he remembers that he has not begun work on that term paper, due this Tuesday.

3. His opinion of the probable difficulty of the next exam.
4. The difference between the grade he would like to get and the grade he thinks he will get.

Knowing these, we could then reach some reasonably good conclusions about the academic condition of that student.

It is the same for an electron. To describe the energy of an electron in an atom, you need to know:

1. The value of the principal quantum number; whether it is 1, or 2, or 3, or 4, or any integer, 5, 6, and so on upward.
2. The value of the orbital quantum number. This can be any number, including 0, or 1, or 2, and so on, provided that it is *not* equal to or greater than the value of the principal quantum number and is positive or zero, not negative.

Example: If the value of the principal quantum number is 3, then the value of the orbital quantum number could be 0, or 1, or 2; no other choices are possible. If the value of the principal quantum number is 1 (the lowest possible value it can have), the orbital quantum number can only be 0.

3. The value of the magnetic quantum number. This can be any value, positive or negative, including zero, but limited by the value of the orbital quantum number.

Example: If the value of the principal quantum number is 3 and the value of the orbital quantum number is 2, then the value of the magnetic quantum number could be any integer, including 0, between -2 and $+2$, inclusive. These would be $-2, -1, 0, +1,$ or $+2$. If the value of the orbital quantum number were 1, the magnetic quantum number could be $-1, 0,$ or $+1$. If the orbital quantum number were 0, the magnetic quantum number can only be 0.

4. The spin quantum number is independent of all this; it can have a value of either $+\frac{1}{2}$ or $-\frac{1}{2}$.

How quantum numbers are used

So, the question is, how do we specify the value of the energy of such and such an electron in an atom? According to what we have learned thus far, the answer would be to say any number (any positive integer) for the principal quantum number; and then from that, within the limits described, give other numbers for the orbital and magnetic quantum numbers, and then say either $+\frac{1}{2}$ or $-\frac{1}{2}$ for the spin quantum number. This is the correct response, thus far; but it needs a little refinement to be closer to the model used. We specify two more rules:

5. No two electrons in the same atom can have all four quantum numbers the same. Three or two can be the same, or one, but not all four.

6. When you have a choice, pick the lowest quantum number. (This rule has some special exceptions; we will take these up later, as rule 8, p. 121.)

Now we can try an example. A hydrogen atom has one electron; when that atom is in its lowest energy state, how would you describe the energy of its electron.

- First you would pick the lowest possible principal quantum number. This would be the number 1.
- Automatically, this makes the choice for the orbital quantum number limited, to 0.
- And this limits the magnetic quantum number, also, to 0.
- Finally, we select the lowest, $-\frac{1}{2}$, for the spin quantum number.

So, the final answer is, in order of quantum numbers, principal, orbital, magnetic, spin: 1, 0, 0, $-\frac{1}{2}$.

How quantum numbers are expressed

We need a counting rule. Actually, we don't need a counting rule but without it, our way of specifying electron energies will not be easily understood by others. In order to communicate anything to anybody, both must accept and use the same rules of communication. The counting rule is

7. Orbital quantum numbers are expressed by letters, not numbers.

This is shown in Table 5–2.

TABLE 5–2. **Orbital Quantum Numbers Counting Rule Table**

Orbital Quantum Number	Corresponding Letter
0	s
1	p
2	d
3	f
4	g
5	h
6	i
7, 8, 9 · · ·	j, k, l, · · ·

Using the table, the energy of that single electron in a hydrogen atom that is in its lowest energy state is given by this set of four symbols: 1, s, 0, $-\frac{1}{2}$.

Quantum numbers for the lighter elements

A helium atom has two electrons; in its lowest energy state, specify the energy of each electron. *Answer:* For the first electron, pick the lowest possible principal quantum number, 1. As before this makes the others fall into line. The complete specification is 1, s, 0, $-\frac{1}{2}$. For the second electron, start all over. Pick the lowest possible principal quantum number, 1. As before this makes the others fall into line, with one change. The orbital quantum number is still s and the magnetic quantum number is still 0. But the spin quantum number cannot be $-\frac{1}{2}$ because if it were, all four would be the same as for the first electron. So, pick the other spin quantum number, $+\frac{1}{2}$. For the second electron the answer is 1, s, 0, $+\frac{1}{2}$.

Try another example, lithium in its lowest energy state, with three electrons in an atom. For the first and the second we get the familiar results. First electron, 1, s, 0, $-\frac{1}{2}$; second electron, 1, s, 0, $+\frac{1}{2}$. But for the third electron we cannot pick 1 for the principal quantum number. If we do the third electron would not have at least one of its four quantum numbers different. (Try it.) So, pick the next lowest possible principal quantum number, 2. The lowest possible orbital quantum number is still s (0, that is); and this makes the magnetic number choice 0 also, and we pick $-\frac{1}{2}$ for the spin. This gives for the third electron, 2, s, 0, $-\frac{1}{2}$.

Table 5–3 shows the quantum numbers for boron (four electrons) and beryllium (five electrons) in their lowest energy states.

TABLE 5–3. **The Quantum Numbers for Boron and Beryllium**

Element	Electron	Principal	Orbital	Magnetic	Spin
Boron	First	1	s	0	$-\frac{1}{2}$
	Second	1	s	0	$+\frac{1}{2}$
	Third	2	s	0	$-\frac{1}{2}$
	Fourth	2	s	0	$+\frac{1}{2}$
Beryllium	First	1	s	0	$-\frac{1}{2}$
	Second	1	s	0	$+\frac{1}{2}$
	Third	2	s	0	$-\frac{1}{2}$
	Fourth	2	s	0	$+\frac{1}{2}$
	Fifth	2	p	-1	$-\frac{1}{2}$

Notice that for the fifth electron in beryllium we could not select s (that is, 0) for the orbital quantum number. If we did, then the four quantum numbers for the fifth electron would have to be all the same as for one of the other electrons. So we selected the next lowest possible orbital quantum number, p (which represents 1). From this, by rule 3 and rule 6, the magnetic quantum number is -1.

Electron configurations

We have been building up the sets of quantum numbers, generally called the **electron configuration**, atom by atom from the simplest atom, hydrogen with one electron, then helium with two electrons, and so on. We are now ready for carbon with six electrons. However, to treat carbon correctly it is necessary to introduce our last rule:

8. (The exceptions to rule 6, p. 119.) Make the magnetic quantum number different whenever you can for electrons that have the same orbital quantum number as required by rule 6. Or: rule 6 (p. 119) applies to the other three quantum numbers always, when you have a choice of more than one magnetic quantum number, make it different if you can; do not make it the lowest possible if you have already picked the lowest possible preceding electron. Note that this rule only applies to electrons that have the orbital quantum numbers, p, d, f, and so on; it does not apply for orbital quantum numbers s.

With all that in mind here is the listing for the six electrons of carbon:

First	1	s	0	$-\frac{1}{2}$
Second	1	s	0	$+\frac{1}{2}$
Third	2	s	0	$-\frac{1}{2}$
Fourth	2	s	0	$-\frac{1}{2}$
Fifth	2	p	-1	$-\frac{1}{2}$
Sixth	2	p	0	$-\frac{1}{2}$

Without rule 8, the specification for the sixth electron would have been $2, p, -1, +\frac{1}{2}$. Under the provisions of rule 8, however that would produce a violation, since the orbital quantum number is p, and the magnetic quantum numbers for the fifth and sixth electrons would both be -1. To satisfy rule 8 we must use 0, the next lowest magnetic quantum number.

Nitrogen has seven electrons. Of these the first six are the same in quantum number specification as for carbon. (You have doubtless noticed this kind of repetition in the preceding examples.) For the seventh electron in nitrogen:

Seventh	2	p	$+1$	$-\frac{1}{2}$

For oxygen, the first seven electrons are as described for nitrogen; the final, eighth electron is

Eighth	2	p	-1	$+\frac{1}{2}$

Notice that rule 8 (above) is not effective. We cannot make the magnetic quantum number different for the eighth electron and still keep the orbital quantum number the same, with the value of p (or 1). That is, we have "used up" all the values of the magnetic quantum number available for a p value given to the orbital quantum number; these are (of course) -1, 0, and $+1$. So, to maintain rule 6 (p. 119), which is still in effect, we must pick a -1 for the magnetic quantum number for the eighth electron. This then requires a different spin quantum number. (Otherwise the fifth and eighth electrons would have all four quantum numbers the same, which violates rule 5. And rule 5 is *never* violated.)

If all this sounds complicated, it is. A simple sloppy statement of rule 8 (p. 121) is: For electrons with the same orbital quantum number, make the magnetic quantum numbers different but don't worry about it if the other rules prevent this. Try to apply rule 8 when you can, if the only other rule you violate is rule 6 (p. 119); but don't violate any other rules.

On to fluorine with nine electrons. We have the electron configuration for the preceding eight by looking at oxygen. For the last electron in fluorine:

| Ninth | 2 | p | 0 | $+\frac{1}{2}$ |

And for neon, with ten, the tenth electron is described:

| Tenth | 2 | p | $+1$ | $+\frac{1}{2}$ |

We could keep on going since there are more than 100 known elements and, in chemistry books written for students majoring in chemistry, the information does continue to flow. Here, we will stop the detailed examination and conclude with some generalizations.

Electron configurations of heavier elements

First, the electron configuration, which is the specification of the quantum numbers that describe the energies, of the electrons in successive atomic numbered elements does follow a pattern, as you have indeed noticed. The pattern gets more complicated at about element number 21, scandium, and it really gets complicated at element number 58, cerium. Still, it is a pattern or a model that correctly predicts ionization potentials and electronegativities and observable properties, such as color, melting points, bonding to other elements, and so on.

Second, the specification of the four quantum numbers only partially identifies the energy of an electron. To identify that energy fully we need a bit more information; the number of protons in the nucleus, for example, is also needed. Thus, the four quantum numbers for the fifth electron, say, in boron and in aluminum (number 13) are the same:

$$2, \quad p, \quad -1, \quad -\frac{1}{2}$$

But boron has five protons in the nucleus and aluminum has 13. The actual energy of that fifth electron in aluminum is much lower than for the fifth electron in boron because of the additional eight protons in the aluminum nucleus. The same comments apply to any other electrons that are in different atoms which have differing numbers of protons in their nucleii but happen to have the same set of four quantum numbers. (To avoid confusion, this is not a violation of any rules since we are talking about different atoms, not about two electrons in the same atom.)

The locations of the electrons

Finally, we said earlier in this appendix that, when the energy is specified by using quantum numbers, we can also say something about the space where an electron is likely to be found most of the time. Here we go.

Any electron that has an orbital quantum number, *s*, is probably in a spherical region with the nucleus of the atom defining the center of that sphere. The bigger the principal quantum number that "goes with" that *s*, the bigger the sphere. For example, 2*s* electron probably occupies a smaller sphere than, say, a 4*s* or a 5*s* or 6*s* electron in the same atom. The larger the nuclear positive charge, the smaller the sphere of probable location. A 2*s* electron in oxygen is likely to be found in a spherical space smaller than that in which a 2*s* electron a beryllium atom is found. The positive nuclear charge of an oxygen atom is 8 whereas beryllium is 4.

Actually, when all this was first developed as a model, the counting number *s* signified "sharp" not spherical. The reasons don't concern us, but *p* came from "principal," *d* from "diffuse," and *f* from "fine." For us, *s* suggests spherical, *p* suggests "double pear-shaped," *d* suggests "double double pear shaped," *f* suggests "fancy shaped," *g* suggests "golly, look at that shape, willya!" and so on.

Any electron that has an orbital quantum number, *p*, is probably located in a region shaped like two pears joined at the stems (very short stems) with the nucleus located right where the stem ends touch, and each pear sticking out straight, either right and left, or one pointing at you and one away from you, or one up and one down. The bigger the principal quantum number or the smaller the positive charge on the nucleus of the atom, the larger the pears.

Any electron that has an orbital quantum number, *d*, is probably located in a region shaped like double double (that is, four) pears, all touching at the stem ends with the nucleus in the middle of them all. One arrangement would have one pear pointing north, one east, one west, one south. Another is one pointing up, one to the right, one down, one to the left. There are three more ways (five all together) for those double double pear shaped regions to be located, but two is enough to give the idea.

Any electron that has an orbital quantum number, *f*, is in a fancy shaped region, probably most of the time, anyway. A word description of these seven regions is difficult to compose and likely impossible for anyone but the writer to understand. As you have guessed, the value of the principal quantum number and of the nuclear charge determine the size of the region. This is a good place to stop, mentioning only that the regions identified with orbital quantum numbers *g*, *h*, *i*, and so on, have even more complicated shapes, with their relative sizes determined by the principal quantum numbers and the nuclear charges involved.

SUGGESTIONS FOR DISCUSSION

1. **(a)** Prepare a detailed topical outline for this chapter, identifying a few major topics as essential or very important and similarly identifying a few minor topics.

   (b) From the remaining major and minor topics (not so identified) select perhaps two or three and justify your opinion to the effect that these are not essential or important.

2. If possible, find a classmate who has come to a different decision about one of the topics you chose in your response to question 1ʰ; that is your classmate sees this as an essential or important topic. Discuss this disagreement with that classmate and justify either your decision to retain your original opinion (that the topic is not essential or important) or your change of opinion after that discussion.

3. In your own words demonstrate briefly that there are five and only five regular solids in a three-dimensional universe.

4. What does the concept of an harmonious balance of opposites mean to you?

5. Distinguish between and among these types of bonds: pure covalent, polar covalent, metallic, ionic. In your answer show how these types are similar, or at least closely related, and how they differ.

6. In hemoglobin, an iron atom has six pairs of electrons associated with it in a high energy level. What geometric figure, probably, is associated with this number of electron pairs? Justify your answer.

7. Which, as you see it, is a more important use of human imagination: to write a poem or to devise a model which tries to explain why water is a liquid, not a gas? Why do you think so?

8. If possible, do a little extra reading on the historical development of the periodic table; then suggest a possible historical outcome if Mendeleev's mother had had only 16 children.

9. Distinguish between electronegativity and ionization potential. Show how they are related and in what way or ways they differ. Also, pick several pairs of elements and identify the members of each pair that have the greater electronegativity, or the greater ionization potential, or both.

10. Using examples, show that it is correct to say that some compounds are composed of molecules and that some are not.

11. In words or with sketches distinguish between and among these structures: hexagonal close packed, face centered cubic, body centered cubic.

12. If you read it with sufficient diligence and attention, state in one sentence the essence of the appendix to this chatper.

chapter 6

The Nucleus

Four wall handball is a game in which, loosely, the object is to bounce a ball off of one or more walls and the floor of a room, or court, in such a way as to make it difficult for an opponent to return the ball to you. It is now necessary to imagine a handball game in which the object is to keep the ball bouncing continuously, returning it courteously, though vigorously, to an opponent. Further, instead of one ball and two players, imagine several balls, some purple and some nile green colored, and a number of invisible players.

In such a handball court we would see the differently colored balls bouncing about wildly and continuously. In this imaginary game the balls are legally allowed to bounce off the ceiling as well as the four walls and the floor; this adds to the complexity of the motion of the balls. And, instead of a court with fixed shape, imagine that the court has a changeable shape, generally curved. At times it has the shape of a sphere, then the shape of an egg, then the shape of a short, fat cigar, then the shape of a sphere with bulges in it, and so on. All the while, the balls continue to bounce around this way and that, inside the peculiar handball court.

THE NUCLEAR HANDBALL COURT

This analogy illustrates the major features of the nucleus of an atom. The purple balls represent positively charged protons and the nile green balls represent neutral (no charge) neutrons, moving about within the nucleus. If we look more closely at the analogy, we see more details. Somehow, as they move around, the balls seem to be attracted to one another, as though an object (a meson) was exchanged between two balls now and then. We would notice that a few of the balls seemed each to have about the same energy and others each had a slightly higher energy, and still others each had still more energy.

We might also observe that in this imaginary model there seemed to be more nile green balls than there ought to be. In this case, sooner or later, one of the nile green balls would change color and become purple; at the same time, outside the handball court we would see a tiny, negatively charged, blue ball, a **beta (β) particle** suddenly appear and then move away from the handball court at high speed. Simultaneously with the other two events we would sense, but not observe, that an invisible ball appeared outside the handball court, on the other side, and moved away at high speed. This invisible particle is called a **neutrino**.

In a different handball game, there might be more purple balls than there "should" be. In such a situation one of two things might happen. A purple ball might change color to nile green and a blue ball, this time positively charged, would appear outside the handball court and move away at high speed along with an invisible neutrino moving off rapidly over on the other side of the handball court. Or, a sluggish blue ball from outside might come to the edge of the handball court and pause there for a moment. Then, quick as a wink, that ball would disappear and, at the same instant inside the handball court, a purple ball would change color to nile green.

A fourth phenomenon could be observed in still another handball game. This time, we would see two nile green balls and two purple balls collect together to form a loose unit of four balls all together. This collection, an **alpha (α) particle**, would then pass through the wall of the court, appear outside, and move away at high speed. When this happened, the whole handball court would shudder, and move promptly, but not very fast, in the opposite direction.

In each of the four instances we have described, we would find that the total weight of the balls involved, green, purple, blue, and invisible, inside and outside of the handball court would be greater if we added them up before the event and less if we added them after the event occurred. We would find that the greater the loss in weight, the higher the speeds at which the particles moved off. Finally, with almost every event, we would have observed a flash of light, a **gamma (γ) photon**, emitted from the handball court; sometimes there would be more than one γ photon.

To complete the picture, we could find many other imaginary handball games in which no unusual events occurred. The colored balls would simply continue to bounce around inside a court with continuously changing shape forever.

A MODEL OF THE NUCLEUS
Protons

Now for a translation. The currently accepted model or theory of the nucleus says that a nucleus contains protons. We can think of a proton (not very accurately) as a ball-like tiny particle that weighs a lot more than you would think, considering its small size, and has a positive charge. All protons are the same, you cannot tell one proton from another.

The nucleus of a hydrogen atom has one proton. All other nucleii have two or more protons corresponding to their numbered position in the periodic table.

Neutrons

Generally, a nucleus also contains neutrons. The number of neutrons inside a nucleus can vary. Thus, for example, usually there are no neutrons in a hydrogen nucleus. However, in a few we find one neutron and in still fewer we find two neutrons. All of the aluminum atoms found in nature have 13 protons and 14 neutrons. However, aluminum nucleii have been made artificially with 13, 15, 16, or other number of neutrons. All atoms of tin have 50 protons in their nucleii, but the number of neutrons in tin atoms in nature varies widely. We find that some of these have 62, 64, 65, 66, 67, 68, 69, 70, 72, or even 74 neutrons.

Isotopes

The three different kinds of hydrogen atoms, with zero, one, or two neutrons in their nucleii are called **isotopes** of hydrogen. The single natural variety of aluminum atoms and the others that can be made with less or more than 14 neutrons in their nucleii are called isotopes of aluminum. As you can see, there are ten isotopes of tin that occur in nature.

Some nucleii have a small amount of total energy, some have a large amount. Light weight nucleii with low energy have a few more neutrons than protons. Heavier nucleii with low energy have several more neutrons than protons. Low energy nucleii are said to be stable. For example, a stable oxygen nucleus has 8 protons and 8, 9, or 10 neutrons. A stable arsenic nucleus with 33 protons has 42 neutrons, and a stable gold nucleus with 79 protons has 118 neutrons. Any other nucleii for these three elements, any other isotopes of these three, that is, will have different numbers of neutrons and will be a high energy, or unstable, nucleus.

ENERGY AND THE NUCLEUS

Now we come to a fundamental statement: *All nucleii with high energy tend to lose that energy.* Although there are other ways than the four described by analogy in the handball court description, these are the four most common ways for a nucleus to lose energy. Whenever a nucleus loses energy, it does so by losing total mass. The mass of the nucleus before a particle is emitted is always greater than the total mass of what is left plus the mass of whatever departed. The loss of mass appears as energy, either as energy of motion (a fast moving particle, departing, for example), or as light energy (a γ photon or two), or both energy of motion and a photon or two. The phenomenon of energy loss in this way is called **radioactivity**.

Radioactivity

A radioactive substance is simply a bunch of atoms whose nucleii have a lot of energy, and first one nucleus, then another, then another, releases its energy. If we have very many of these nucleii in one place, that pile of stuff would appear to you and me as though it were practically continuously emitting energy; it would be radiantly active. From this the word, radioactivity, has been derived.

If a nucleus has "too few" neutrons, or "too many" protons (same thing), it tends to lose its energy by emitting an alpha (α) particle, the four-ball, two proton–two neutron, combination, and a γ photon or two. Or, it tends to lose its energy by "capturing" an electron, which was part of the atom outside the nucleus, that got "too close" to the nucleus. When an electron is captured by a nucleus, a proton inside the nucleus changes into a neutron. Or, third choice, instead of capturing an electron, a proton inside the nucleus will change to a neutron, anyway, and a positron, like an electron but positively charged, appears outside the nucleus moving fast; simultaneously a neutrino, with no charge or mass, also appears on the opposite side of the same nucleus, moving fast. Since a neutrino has no charge nor mass, it is difficult to detect, incidentally.

If a nucleus has "too few" protons, or "too many" neutrons, it tends to lose its energy by a different process. Inside the nucleus a neutron changes into a proton, and outside the nucleus an electron appears, moving rapidly; while on the other side a neutrino appears, moving rapidly. The electron in this case is often called a beta (β) particle.

In general, for all these instances, with "too many" or "too few" protons or neutrons initially, some of the energy appears as one or more γ photons. These are simply particles, or quanta, of energy similar to visible light, except that this kind of photon has a great deal more energy than a photon that stimulates the sense of sight.

The half-life

We will need another analogy to illuminate the idea of half-life. We will use a group of anarchists who want to start a riot. To do this, each anarchist has a bomb with a fuse. Some fuses are long, some are short, some in between. Let us suppose that we have 12,000 of these people, all gathered together in the same place and that they have all lit their bomb fuses at the same time. So, first one bomb goes off, then another, then another, and so on. After, say, 5 min, 6000 bombs have gone off. These were, of course, the shorter fused bombs. The longer fused bombs have not yet gone off. If the lengths of the bomb fuses are truly randomly determined, then in the next 5 min half of the bombs that are left will go off, 3000 bombs, that is. It takes a little thinking to see this, but it is an important concept. Now, if the fuse lengths were random, how many bombs would go off in the third 5 min? Correct! Half of those left would go off, that is 1500 bombs, and so on.

The high energy nucleus is like those bombs with random length fuses. Eventually, since the nucleus has a high energy (that is, because its fuse is lit) it will go to a lower energy state. But when it will do this is randomly determined; it might decay to a lower energy now, or not decay to a lower energy for a million years, or some time in between. Whenever we have a large number of nucleii, each in a high energy state, there will be at some time in the future, half of them left. And in that same time interval from then, half of those left, and so on.

Students who think about this model often ask: "OK, what happens when you get down to one anarchist, or one nucleus?" The answer is not fair. What we do is avoid the issue by saying that random processes, which is what we are considering here, cannot be applied to small numbers of objects. So, when we get down to 100 or maybe 50 or maybe 10, the model becomes useless.

One extension of our anarachist analogy is helpful. Our first group of 12,000 anarchists had bombs with fast-burning fuses. Their half-life was 5 min, the time periods in which their number successively decreased by half. We could imagine another group of more cautious anarchists, with randomly distributed fuse lengths, but with fuses that burned slower. For them, the overall result would be the same, except that it would take, say, 15 min (each consecutive interval) for their number to decrease by half. Their half-life would be 15 min, of course.

Each different nuclear isotope that has extra energy has its own, unique half-life. For example, $^{82}_{35}$Br nucleii have a half-life of 34 hr. The lower left subscript specifies the number of protons in the nucleus; the upper left superscript specifies the total number of protons and neutrons ($35 + 47$ in this case). Comparably, $^{80}_{35}$Br nucleii have a half-life of 18 min. The nucleii of $^{228}_{88}$Ra have a half-life of 6.7 years, whereas $^{226}_{88}$Ra nucleii have a half-life of 1620 years. Some nucleii have half-lives that are quite short, for example $^{214}_{84}$Po has a half-life of 1.6×10^{-4} sec. The model currently accepted for the nucleus does not predict how long or short a half-life will be, except in a general way. So, to get the values of the half-lives cited here, laboratory measurements are necessary. As you can guess, a half-life of less than a second is tricky to measure, and a half-life of thousands of years cannot be measured directly, since man has known of radioactivity only since about 1900, less than 100 years ago.

Summary

An overview would be helpful at this point. There are about 100 different elements, nucleii with different numbers of protons. Each different element has several isotopes, although a few elements, such as aluminum and gold have only one kind of nucleus as they exist in nature. But most of the others have two or more low energy isotopes that exist naturally. However, all of the elements with 84 or more protons in their nucleii have only high energy, radioactive, nucleii, and a half-dozen or so of the elements with lower numbers of protons in their nucleii also have naturally radioactive nucleii.

NATURAL RADIATIONS AND THEIR EFFECTS

One of the elements with less than 84 protons, which has a radioactive isotope is potassium. Potassium chloride exists as a trace impurity in salt. Most of those natural potassium nucleii are low energy and stable, not radioactive. But about 0.01% are the $^{40}_{19}K$ isotope; these nucleii are radioactive, emitting β particles (and neutrinos) at high speed, and γ photons, with a half-life of more than 10^8 years. The salt in the toast you had for breakfast, on a hamburger for lunch, or from another food source contained as a trace impurity enough potassium chloride with a fraction of a percent of the $^{40}_{19}K$ isotope to bombard your insides with zillions of β particles since breakfast, or lunch, time.

A β particle, a fast moving electron, can be thought of in this context as a tiny bullet that damages any living tissue it passes through. Broadly, as it passes by a covalently bonded molecule in one of the cells in your body, it hits that molecule or comes close enough to it to knock off one or more of the electrons. When that happens, the molecule can no longer function as intended in that living cell of your body. But, as we know, there are about 1×10^{10} molecules in a typical cell, and the loss of only one molecule probably would not be significant. It depends upon chance whether a β particle will disrupt one, or none, or several, or a few, molecules in a cell. Obviously since man has used salt for seasoning for thousands of years, the damage in this case is either slight or nonexistent. Some scientists who have examined the matter in detail conclude that the effect is beneficial. They argue that, if a cell is killed when only a few of its molecules have been damaged by a passing beta particle, the cell was probably weak anyway and should be eliminated, especially since the body is able to generate a replacement as a natural process.

Cosmic rays

This kind of argument is buttressed further by considering the phenomenon of cosmic rays. No fully satisfactory model has yet been conceived for the origin of cosmic rays. Somewhere out in space, or from several places out in space, protons are formed (probably by stripping electrons from hydrogen atoms, which are plentiful in the universe) that are, somehow, accelerated to very high velocities. These protons travel randomly through space. Because they are all over the place, the Earth is always in the path of many of them. Those fast moving protons hit the gas molecules that make up our atmosphere, knocking electrons off in the collisions. The positively charged ions formed by these collisions then move at very high velocities. Another result of those collisions is that γ photons are produced. The overall effect is much like firing an explosive bullet into an ant hill; you would observe parts of ants flying off in all directions.

Some of the fragments produced by cosmic ray collisions travel through our atmosphere, downward to the ground. Generally, they are themselves moving fast, with enough energy to knock electrons off of any molecules that happen to be too close to their path. While reading this (unless you are in a cave, and even there you are not fully protected) your body has been hit by several million fragments, most of them moving fast enough to penetrate partially or completely through your body disrupting molecules in your body cells as they go. Apparently, the effect is not fatal nor even very harmful at the dosage to which we are all exposed.

On the other hand, exposure to a very severe dose of such radiation can cause permanent damage and even fatalities. If enough cells of the body are incapcitated at a rate greater than the rate at which the body can replace those damaged cells, the ultimate result is fatal. If the rate of damage is not quite that great, the result is a permanent disability. Since the rate at which the body can make replacements is high, only a severe dose of radiation is to be avoided. Cosmic rays, the consumption of salt, and other similar daily events, do not present that kind of challenge to living beings.

There is one exception to this, however. The somatic cells, the cells involved in reproduction, can be damaged in the same way as other cells. In this case, although the disruption may not be fatal, it can cause an alteration in the progeny conceived after that. Sometimes, the new life is unable to survive or is defective in some way; very rarely the mutant thus formed is superior in some way, depending upon the kind of disruption to the somatic cells. It has been suggested that this effect is responsible for the generation of evolutional change, but detailed mathematical calculations show that it cannot cause more than a small fraction of the known changes since life began on the Earth; other effects must therefore be responsible for most of our evolutionary development.

Let us review a bit at this point. Cosmic rays have a very high energy, they interact with molecules near their path as they travel. Each time a cosmic ray particle (a proton or a secondary fragment) interacts with a molecule, knocking off an electron or two or more or perhaps knocking the molecule apart, that particle loses a bit of its energy. From then on, it travels a bit slower, although still very fast. The next molecule it interacts with is also knocked about, but the particle is slowed down a little more. Eventually, it moves so slowly, after zillions of collisions, that it is no longer effective. Now, in our air, there are comparatively few molecules, so a cosmic ray particle can travel a great distance through the air. If it hits us, or a tree, or a pile of dirt, it encounters a lot more molecules; either gets slowed down considerably by the time is passes through and comes out the other side, or it gets stopped somewhere inside. So, for protection from cosmic rays, you could walk around with a big, thick cement hat, or you could live in a cave. Either way (cave is better) you would reduce your exposure to cosmic rays—if you think it would be worthwhile!

Alpha (α)
particles

Measurements show that α particles have a lot of energy although not nearly as much, typically, as an average cosmic ray particle. But, α particles are much heavier than cosmic ray particles. As a result, an α particle loses more energy per collision than a cosmic ray particle.

Why so? To see this, set up a lot of billiard balls on a billiard table, distributing them randomly. Then, project a ping pong ball into the billiard balls at high speed. You would notice that with each collision, the ping pong ball only slowed down a very little; it would continue to bounce along, eventually hitting several billiard balls before it slowed to a stop. Next, try the same thing but with a cue ball. It too bounces around but rather soon comes to a stop after hitting only a few billiard balls. The ping pong ball is so light, compared to a billiard ball, that with each collision it can only lose a tiny bit of its energy. It has to hit several billiard balls before it comes to a complete stop. The cue ball, on the other hand, is about the same weight as a billiard ball. As a result, with each collision it can lose a large portion of its energy. So, it comes to a stop after hitting only a few billiard balls.

It is the same with an α particle. Although it has a lot of energy when it starts out, it only requires a few collisions to stop it cold, relatively speaking. A typical α particle is stopped by traveling through only a couple of inches of air; a piece of paper thinner than the page of this book will stop any α particle. Your skin will stop α particles. We can conclude that α particles are not harmful to man and other living things when they are generated outside of the body. However, note that in any single collision an α particle loses a lot of its energy. Whatever it does hit is really messed up. If, therefore, you ingest a radioactive atom that, at some unknown future time, will emit an α particle from its nucleus inside your body, a lot of damage will result.

For example, $^{226}_{88}$Ra nucleii emit α particles. At one time, compounds of radium were used in the luminous paint on the dials of luminous dial watches to outline the numbers and the hands. This paint was applied with small paint brushes, and, to form the pointed tip, the people who painted the dials would moisten the brush with their tongue and lips. As a result they ingested paint containing radium nucleii that would emit α particles. Now look at the periodic table for radium, number 88. Notice that it is in group IIA, along with calcium, number 20. That is, calcium and radium have similar (not exactly the same) chemical properties. Calcium is used by the body among other things as part of the bone structure. Any ingested radium will be used by the body in much the same way. Your red blood cells are made inside the larger bones in your body. Zillions are made per day, to replace the worn out cells. The α particles from the ingested radium caused an alteration in the cells that made the red blood cells in the bodies of the luminous watch dial painters. Eventually, they became anemic because the red blood cells now being produced in their bodies were defective. Luminous watch dials today are made with a different kind of paint and applied by machine not by hand.

Beta (β) particles and positrons

A β particle is an electron, a negatively charged particle, moving fast, that is generated when a nucleus drops in energy by changing a neutron into a proton. A positron has the same mass as an electron, but it is positively charged and is generated when a proton changes into a neutron inside a nucleus. Both are very light, much lighter than an α particle or cosmic ray particle. Either one has a lot of energy, although not nearly as much, in general, as a cosmic ray particle. So, using the billiard ball–ping pong ball–cue ball analogy we would expect that a β particle or positron has many collisions before it is stopped, but not as many as a cosmic ray particle (even though electrons and positrons are lighter) because a cosmic ray particle has a lot more energy to begin with.

It takes several feet of air to stop a β particle or a positron. Some of the more energetic β particles can pass through a human body and keep on going, interacting with any molecules in their path. A severe dose of β or positron radiation disables or kills.

Gamma (γ) photons

As we know, in almost every instance when a nucleus undergoes a change from a high energy to a lower energy and a particle is emitted or an electron is captured, some of the energy lost by that nucleus appears in the form of one or more γ photons. Now, a γ photon has no mass; it travels at the speed of light (of course, it *is* light, although not detectable by the human eye). A γ photon has a great deal of energy. But, because it has no electrical charge and no mass, it can pass right through a molecule and not lose any of its energy; and the molecule is not the least disturbed. Occasionally however, when conditions are just right (we might think of a really direct hit, perhaps) the γ photon will give some or all of its energy to a molecule in its path. When it does, that molecule is very much disrupted. So, a γ photon could pass through miles and miles of air, or earth, or human bodies, or anything; it is very penetrating. Only occasionally will it cause any damage; however, the damage it does do on those rare occasions is severe. Given a severe enough dose of γ radiation, the results are disabling or fatal.

Thus, γ radiation is perhaps the most dangerous, not because any single γ photon is likely to cause damage to our bodies but because adequate shielding is difficult to obtain. Granted, a lot of γ photons can pass through our body with no harm done, but a few of these will surely interact and cause dramatic damage. The only way to prevent this, if we are exposed to a source of γ radiation, is to use shielding, that is, a lot of innocent molecules and atoms between the source and us. Thick cement, a wall of lead, water, and other substances may be used for shielding; none of these stop γ photons completely; they merely reduce the percentage that gets through to the other side of the shield.

Neutrinos

Neutrinos are emitted in some radioactive decay processes, as we know, when a nucleus loses energy. However, neutrinos are not harmful for the same reason that we do not know very much about them.

This sounds like, "What you don't know won't hurt you," a foolish proverb; but in this case it does apply. Let's be philosophical for a moment. How do we know or what makes us think that β particles, cosmic ray particles, γ photons, and all the rest are harmful? Each of these is known to exist by the observable effect it has. If it had no observable effect, we would not be able to say that it existed. In the case of these particles, the effect they produce, which we can observe, is the removal of electrons from atoms and molecules in their path, or the breaking apart of a molecule if it is hit hard enough in the right way (what ever that means). It is this disruption of molecules that is biologically damaging.

Now, neutrinos are very, very difficult to detect, in fact, it takes enormous effort and very elaborate, sophisticated equipment to observe any effect that can be attributed to the passage of neutrinos through matter. That is, a neutrino can pass clear through the earth, some 8000 miles of solid stuff so to speak, and not bother a thing. Very, very, very, rarely, a neutrino does interact with matter, which is how (apart from models that predict them) that we think they exist. Unless we can detect neutrinos, we cannot learn much about their behavior. But because they do not interact very much, we do know that neutrinos are harmless, although we know very little aside from that. While you have been reading the preceding sentence, several million neutrinos passed through your head; they were generated in the Sun, and only your best friend would notice any effect.

Summary

Before describing some of the practical applications of our knowledge of nucleii, it will be reassuring to review at this point. There are exactly 88 elements that exist naturally. These are the elements numbered from 1 to 92, inclusive, in the periodic table, except for techetium, number 43, promethium, number 61, astatine, number 85, and francium, number 87.* Some of the 88 elements have naturally occurring, high energy, radioactive, isotopes. Thus, all of the isotopes of elements numbered 84 and higher are naturally radioactive. Some of the isotopes of some of the lower numbered elements are also naturally radioactive; we have cited $^{40}_{19}K$ as an example.

By and large, most of the atoms with which we come in contact every day have low energy, stable nucleii, not subject to radioactive decay and not damaging to biological systems because of any radiation emitted. On the other hand, almost everything we have daily contact with does contain traces of radioactivity, small amounts of elements that are naturally radioactive. All pieces of wood, for example, contain exceedingly small amounts of the $^{226}_{88}Ra$ isotope. In addition to these sources of radiation, we are subject to cosmic rays, as we know. Occasionally, these do damage somatic cells in a way that adversely (or favorably) alters heritable traits, but otherwise it seems reasonable to suggest that man and other forms of life have adjusted to this kind of challenge, and perhaps have benefitted from it in the long run.

ATOMIC DECAY—USE AND MISUSE

Now you know more about atomic nucleii and their tendency to change spontaneously from a high energy condition to a lower energy condition, emitting radiation as they do. Such knowledge (in more detail) can be used to make, or force, low energy nucleii into a higher energy state so that their subsequent loss of energy can be employed under semicontrolled conditions. That is, we can now make radioactive species artificially and use them to provide power or for medical or research purposes. Unfortunately all radioactive substances can be dangerous; unless they are controlled very carefully, some men will be undesirably subjected to severe doses of radiation, perhaps accidentally, perhaps purposely, or sometimes as an inadvertent consequence of a planned and itself desirable use of this source of energy. We will consider some of these environmental aspects in a later chapter on energy; here, we will briefly examine some of the other applications of this knowledge.

Radioactive decay

We will begin with an artificial element, $^{238}_{94}Pu$. This isotope can be made in different ways; one way is symbolized by equation (6-1).

$$^{237}_{93}Np + {}^{1}_{0}n \rightarrow {}^{238}_{94}Pu {}^{0}_{-1}\beta \qquad (6\text{-}1)$$

This equation states that when a neptunium nucleus containing 93 protons and a total of 237 protons and neutrons (that would be 93 protons and

---

* These four elements, and elements numbered 93 and higher, are artificial. They are made by adding protons (in effect) to nucleii of naturally occuring elements, thus producing new nucleii with a different number of protons and neutrons from those that occur naturally.

144 neutrons) interacts with a neutron, two products are formed. One is a nucleus of plutonium with 94 protons and 144 neutrons (238 neutrons and protons, total) and a β particle or electron. Notice that the total number of neutrons and protons that we start with, 238 (the 237 in the neptunium nucleus plus the other neutron), is the same as the number that we end with, all 238 in the plutonium nucleus. Notice also that the total positive charge that we start with, 93 protons (the neutron has no charge) is the same as the final positive charge, 94 protons (or 94 units of positive charge) less one negative charge attributed to the electron; or, $94 - 1 = 93$. This same balance, or equality of total number of neutrons and protons, between initial and final condition is always true of nuclear reactions. Also, the net charge is fixed—93 positive charges at the start and at the end in this case.

Probably, what actually happened was that a neutron was absorbed by the neptunium nucleus,

$$^{237}_{93}\text{Np} + ^{1}_{0}n \rightarrow ^{238}_{93}\text{Np}$$

That nucleus then had "two many" neutrons, so one of them changed to a proton and an electron was emitted

$$^{238}_{93}\text{Np} \rightarrow ^{238}_{94}\text{Pu} + ^{0}_{-1}\beta$$

This nucleus of plutonium also has a high energy. Sooner or later, its half-life is 90 years, it emits an α particle and a γ photon.

$$^{238}_{94}\text{Pu} \rightarrow ^{234}_{92}\text{U} + ^{4}_{2}\text{He} + \gamma$$

In this case, the difference in energy between the $^{238}_{94}\text{Pu}$ isotope and the products, $^{234}_{92}\text{U}$ and $^{4}_{2}\text{He}$, is comparatively large. So, the α particle, that is, the helium nucleus, is ejected at high speed, and the gamma photon also has a great deal of energy. One tablespoonful of $^{238}_{93}\text{Pu}$ atoms, for example, emits enough energy to keep a 100-W light bulb continuously lit for several years. Even 1 g of this isotope, to be specific, emits 42,000 watt-hours (Whr) of energy in a 10-year period; 1 g is about 1/30 oz, you will recall.

Power from radioactive decay

Differently put, we can use this isotope of plutonium for a portable, long lived, power source. Plutonium-238 (another way to symbolize $^{238}_{94}\text{Pu}$) powered satellites are now used to monitor the weather on Earth. The same kind of source powers some of the equipment left on the Moon by the Apollo 12 astronauts. Here on Earth, cardiac pacemakers have been made with this isotope of plutonium as the source of power. An operation to replace the power source will be necessary only about every 10 years, rather than every couple of years when a battery is used as the power source. Sooner or later, it is possible that plutonium-238 powered artificial hearts will be designed and used for patients whose own hearts have been irreparably damaged.

Radioactivity and the Earth's molten core

The applications of radioactive phenomena as a source of power that we mentioned in the preceding paragraph are man-made. A different, natural "application" demonstrates the dramatic magnitude of the energy that is sometimes involved.

The substances in the interior of the Earth, just like those on the surface, include radioactive isotopes of various elements. As these decay from a high energy state to a lower energy state, energy is released. Now you must realize that this process started when the Earth was forming and has continued and is continuing right now. But, in the interior of the Earth, the escape of this energy is practically impossible, except very slowly. As a result, the interior of the Earth became warmer and warmer and then hotter and hotter. The evidence we have today indirectly suggests that the core of the Earth now consists mostly of molten iron, so you can tell it is pretty hot.

The heat produced by radioactive decay is the source of the energy we see in volcanoes and in earthquakes. The heat melts some of the solid matter in the crust of the Earth and also causes gases to be formed. The liquids and gases take up more space than the solid matter, and press outward. Sometimes this pressure indirectly causes the solid portion of the Earth to move; this is an earthquake. Sometimes the molten matter finds a weak spot in the solid crust of the Earth and pushes out through it; this is a volcano. It was this source of energy that, years ago, broke up the single large land mass on the Earth, probably then located near the South Pole, and moved it in pieces we call continents to their present positions. This process continues today, slowly but surely. For example, southern California (that is, the southwestern portion of the state) and the Mexican penninsula of Baja California, which adjoins, is breaking off the North American continent.

A research application of radioactivity

Let us get to a less dramatic natural application of radioactive decay. The isotope carbon-14 (that is, $^{14}_{6}C$) emits a β particle; its half-life is 5730 years. Now, somehow, carbon dioxide in the air is used by green plants along with water to form carbohydrates and cellulose. The formation of these useful products, carbohydrates and cellulose, takes place in the green leaves of the plants under the influence of photons from the Sun. The process is called carbohydrate (or cellulose) **photosynthesis**. If we could learn how this is accomplished in a plant, we could hope to duplicate the process artificially, possibly even under controls that would permit a more efficient utilization of the Sun's energy. To begin to learn this, we would ask questions about the details of the chemical interaction between and among carbon dioxide, water, the green chlorophyll in plant leaves, and the sunlight.

If, for example, we could follow the fate of a carbon atom as it breaks away from the two oxygen atoms in carbon dioxide and eventually becomes bonded to other carbon atoms, hydrogen atoms, and a few oxygen atoms to form a carbohydrate molecule, many of the necessary questions would be answered. By using carbon-14, we can follow this process. A carbon atom in a molecule, bonded to other atoms, behaves almost exactly the same no matter what its isotopic composition. The only thing that really counts to make an atom a carbon atom is the posession of six protons in the nucleus; the number of neutrons do not matter.

However a carbon dioxide molecule that contains a $_6^{14}C$ carbon atom does have one difference from an ordinary carbon dioxide molecule. We can tell where it was, by watching for the β particle it emits when it eventually loses energy.

$$_6^{14}C \rightarrow \; _7^{14}N + _{-1}^{0}\beta$$

After emission the molecule is no longer a carbon dioxide molecule, of course; but it was until that moment.

If we start with a very large number of carbon dioxide molecules, each one containing a carbon-14 atom, we can follow the fate of the carbon dioxide by tracing the location of the β particle emissions as they happen to occur, one after another. So far, although many of the details of carbohydrate photosynthesis have been discovered, the information is still insufficient for us to build a model that we might use to make carbohydrates artificially by photosynthesis.

Medical uses of radioactive isotopes

A therapeutic use. Another example will illustrate a therapeutic use of radioactivity. Approximately one third of naturally occurring tellurium isotopes are $_{52}^{130}Te$. These nucleii are bombarded with deuterons to produce a radioactive isotope of iodine. Deuterons are nucleii of one of the hydrogen isotopes, $_1^2H$.

$$_{52}^{130}Te + _1^2H \rightarrow \; _{53}^{131}I + _0^1n$$

This isotope of iodine has a half-life of approximately 8 days. Until its discovery in 1938, a different radioactive isotope of iodine, with a half-life of approximately 25 min, was the only one available for therapeutic work; this was $_{53}^{128}I$.

The problem with iodine-128 is that, after the radioactive isotope is produced in a bombardment, or irradiation, process, it must be purified; after that, somehow, it must be incorporated into a molecule or ionic compound suitable for its therapeutic purpose. All this takes time. As a result, although we might start out with a large quantity of radioactive isotope, every 25 min later we have only half as much iodine-128, then one fourth as much, then one eighth as much, and so on. So, by the time the iodine-128 is incorporated into the human patient, there is only a little left, almost hardly worth the effort.

With iodine-131 on the other hand a half-life of 8 days does not restrict the therapeutic value; this half-life allows ample time for prior preparation of the material. Interestingly, this isotope was prepared, or discovered, almost to order. A specialist in nuclear medicine, Dr. J. G. Hamilton, one day by chance asked Professor Glenn Seaborg if he could make a radioactive isotope of iodine with a half-life longer than about 25 min. Asked for an ideal half-life for his work Hamilton replied, "Oh, about a week." Seaborg and his coworker, Professor Jack Livingood, prepared tellurium targets, which they then bombarded with neutrons and with deuterons, in a series of experiments. From these experiments, iodine-131 (among other isotopes) was identified. Later it was produced in quantity for therapeutic use. Since then, iodine-131 has been used to prolong the life of many people (including, interestingly, Professor Seaborg's mother).

Iodine-131 is used in the diagnosis and treatment of thyroid disease (you may know that the thyroid gland produces thyroxin, a hormone, and that this hormone molecule contains iodine atoms), in the diagnosis of liver and kidney disorders, as a means of screening for pulmonary emboli, to locate brain tumors, to measure cardiac output, and to determine blood volume. More than 2 million patients are treated each year with iodine-131.

We can look at two of these uses in slightly greater detail. The thyroid gland will take iodine, any isotope, from the blood and incorporate it into the thyroxin molecules it makes. So, at all times there is some iodine in the thyroid gland. Even when this gland is diseased, it is generally true that only some of the cells of the gland are not functioning properly. If the defective cells are removed, sometimes natural physiological processes replace them with good cells. As long as the defective cells are there, no replacement will be made, so the thyroid gland limps along with greatly decreased efficiency. Fortunately diseased cells are usually more susceptible to disruption than healthy cells. Iodine-131 decays by β and γ emission:

$$^{131}_{53}\text{I} \rightarrow {}^{131}_{54}\text{Xe} + {}^{0}_{-1}\beta + \gamma$$

As we know, the γ photons generally pass out of the body, but the β particles generally mess up the molecules in the cells they pass through. If the dose of iodine-131 is carefully adjusted so that it is insufficient to damage many healthy cells but enough to damage most of the diseased cells, one can expect that such treatment will be beneficial; and usually it is.

Measuring blood volume. Another valuable technique for which iodine-131 is used is determination of blood volume. For this it is best to use an analogy first. Suppose that you have an odd shaped vase and that you wish to know how much water it will hold. The easy way, of course, is to fill it with water and then pour the water out into a measuring cup. (Of course it might be necessary to fill the cup three or four times plus a final partial filling and then add the numbers to get the total volume.) We cannot do this with your blood to see what your blood volume is without getting into difficulties (possibly fatal ones), so we will assume that we cannot do this with the vase, either.

There is a way to measure the volume indirectly. First we would take 1 cup of water, carefully measured, and add a known amount, say, 6 drops of ink,

To one cup
of water

Add six drops
of ink

Stir the
mixture

Pour the mixture
into this vase

Fill the vase to the brim
with water and stir well

Pour about one cupful
back into the cup

Compare color of this cupful with color
of second cup of water to which one drop
of ink has been added, and stirred

Figure 6–1.   Indirect measurement of the volume of a vase.

stirring well (see Figure 6–1). Now put the inky water into the empty vase and fill the vase with ordinary water, again stirring well. Next remove 1 cup of the slightly inky water by pouring it from the vase back into the measuring cup. It will not be as dark colored as it was to start with. Fill another measuring cup with plain water. Now, add 1 drop of ink at a time to the water in the second measuring cup, stir well each time, and each time compare the color with the slightly inky water in the first measuring cup. If, for example, 1 drop of ink in the plain water produces the same color as the inky water poured from the vase, we could tell that the volume of the vase was 6 measuring cups.

So, for blood volume, we would inject a known amount of iodine-131 (probably as dissolved sodium iodide, NaI, in water) into your blood stream. We would let it mix for a while and then extract a small sample of your blood. By measuring the amount of now diluted iodine-131 radioactivity in the extract blood sample, we could calculate your blood volume. For example, if the extracted blood had a radioactivity that was 1/100 as much as the radioactivity of the NaI we injected, we would calculate that your total blood volume was 100 times larger than the volume of the NaI solution we injected initially.

Of course, the calculation would be correct only if the initial solution was thoroughly mixed throughout your blood before we took out the blood sample. To check on this, we would (ouch) take a second blood sample some time later. If its radioactivity were less than the first sample we would know that adequate mixing had not taken place and we would base a new calculation on the radioactivity of the second blood sample (and check *it* by later taking a third sample, of course).

A diagnostic use. The body reacts differently to different poisons. One way to tell whether or not a patient is healthy would be to give the person a dose of some poison and then perform an autopsy on the corpse to see which organs of the body accumulated the poison and how much. If those organs, say the liver and kidney, had a normal reaction to the poison and only accumulated the "proper" amount, we could tell that the patient's liver and kidney were healthy. This method of diagnosis, however, is not practical for obvious reasons.

Another way is to give the patient a little bit of poison, not enough to cause any harm at all, and then operate on the living patient and take a sample of the liver and kidney to see if they did, or did not, accumulate a normal amount of the poison. This method is not as harmful to the patient since they only risk death from the operation rather than from the poison.

A somewhat less drastic procedure has been made possible by the use of radioactive isotopes. This is to give the patient a little bit of poison that is radioactive and then, from the outside of the body, no operation needed, monitor the radiation coming from the kidney or liver. The amount of radiation absorbed by the organs can be determined as normal or abnormal compared to that of a healthy person, and a diagnosis of the patient's difficulties can then be made. The poison used for this procedure is an isotope of technetium. This element, number 43, does not occur in nature; all of its isotopes are relatively short-lived radioactive species. If any of its isotopes were around when the Earth was formed, they have long ago decayed away. To make the isotope technetium-99, we begin with a naturally occurring isotope of molybdenum (number 42), molybdenum-98.

$$\ce{^{98}_{42}Mo} + \ce{^{1}_{0}}n \rightarrow \ce{^{99}_{42}Mo} + \gamma$$

As you have suspected, the molybdenum-99 isotope has "too many" neutrons. One of these changes to a proton and a β particle is emitted from the nucleus, with a half-life of 67 hr.

$$\ce{^{99}_{42}Mo} \rightarrow \ce{^{99}_{42}Tc} + \ce{^{0}_{-1}}\beta + \gamma$$

So, if we have a pure sample of molybdenum-99 to start with, bit by bit it builds up an impurity of technetium-99. (Obviously the amount of molybdenum-98 decreases in quantity all the while.) The trouble with technetium-99 is not that it, too, is radioactive but that it has a half-life of only about 6 hr. As you have guessed, this 6-hr half-life is simply too short to be practical, unless we could make the technetium-99 right at the patient's bedside, so to speak. Unfortunately, neutron irradiation facilities cannot ordinarily be put at a patient's bedside. One neutron irradiation facility that *is* used in this case is in Pleasanton, California, a somewhat remote community if the patient is in a hospital in, say, Huntington, Indiana. However, everything works out nicely.

Molybdenum-98 is irradiated in California, and the product molybdenum-99 is shipped by air to Indiana. There is plenty of time for this because molybdenum-99 decays with a 67-hr half-life, making itself into technetium-99 all the while. At the Indiana hospital the technetium-99 that is now present is extracted and incorporated into the proper molecular or ionic compound and injected into the patient in very small amount. Shortly thereafter radiation detectors are moved over the patient's body and the radiation from the technetium-99 indicates whether or not there is a normal absorption of this small amount of poison. No harm comes to the patient because the devices used can easily detect radiation from an amount of technetium far, far less than the amount which would harm the patient.

Isomeric transition

Technetium-99 decays by a mechanism we have not yet discussed, isomeric transition. Briefly, as formed from molybdenum-99, the nucleii of technetium-99 are in a unique high energy condition. These nucleii lose energy exclusively by emitting a γ photon. Usually, the γ photon does not get very far; it gives up all its energy to one of the electrons outside the nucleus of the atom. That energy is far in excess of the electron's ionization potential. So, a bit of the energy given to the electron by the γ photon is "used" to free the electron from the atom, and the rest is available for energy of motion; the electron comes out of that particular technetium-99 atom like you know what. For all practical purposes, the electron behaves like a β particle (except it was not generated by a nuclear neutron to proton change) and can be detected outside the patient's body. The cellular damage done in the process of leaving by the fast moving electron is minimal. For a statistic, more than 5 million patients are currently treated each year with technetium-99 for diagnostic purposes; the major disorders so diagnosed include kidney, liver, brain, and thyroid malfunctions.

The detection of radiation

It is appropriate at this point to digress from practical applications to a brief discussion of radiation detection. As we know, radiation, a β particle, for example, knocks electrons off of molecules. We have discussed this before with emphasis upon the disrupted molecules that result. Now consider the electrons that are knocked off. If we could collect these electrons on a metal wire, they would tend to travel along that wire from the place where they were collected to some other place on the wire. In fact, if the wire were attached at its other end to the positive terminal of a high voltage source, the electrons would scurry toward that terminal. In their passage along the wire they would constitute a small "burst" of electric current. This current can be amplified electronically and detected. The instrument used for this is called a **Geiger counter**, and sometimes the detection consists of a conversion of the energy in the burst of current into a clicking sound in a loudspeaker or earphone.

In another kind of detector, called a **scintillation detector**, the β particles are allowed to pass through some tougher molecules; these do not lose their electrons readily. Instead, when they interact with a β particle that passes by, they simply absorb some, or all, of the energy of that β particle and are themselves raised to a higher (molecular) energy state. Then, promptly, they release their extra energy as a photon, a tiny flash of light. This flash of light is detected by an electronic system and the event is transformed into a visible or audible signal that can be recorded or observed by a scientist or technician.

There are several other kinds of detectors, only one more will be mentioned: **photography**. A photographic film consists typically of a layer of gelatin spread thinly and evenly on a plastic film or sheet of paper. A photographic negative, or color slide, began as a gelatin film on a transparent piece of thin plastic sheet. A color print or black and white print is identical, except that the gelatin film is put on a paper base. In most cases, extremely small crystals of silver bromide, AgBr, are embedded in the unexposed gelatin film. This film is then exposed to a source of radiation; ordinary light (if you are taking a picture of your dog, for example) or β particles, or γ photons, or other radiation. When a photon or β particle, for example, passes through the gelatin film near enough to a crystal of silver bromide, it transfers some or all of its energy to the crystal. In this case, the crystal keeps the energy even though it is now in a higher energy state; it does not immediately release that extra energy as the molecule in the scintillation detector did. Later on, at leisure, in a photographic developing process it is possible to change the silver bromide crystals chemically into tiny grains of silver metal. The grains are so small that instead of appearing shiny, as silver usually does, they appear almost black. Further, it is easier to change the silver bromide crystals to silver if they have received, and now hold, an extra bit of energy. So, when a film has been exposed, say, to β radiation and is developed, we will see dark spots wherever a β particle passed near to a silver bromide crystal. (The process by which a color slide or print is made is more complicated but it depends upon the details described here for the initial steps in the process.)

Radioactivity as an analytical tool

Neutron activation analysis is another application of radioactivity. Ordinary paint contains small traces of impurities, and the identity and amount of such impurities vary from batch to batch. These impurities can be detected by neutron activation analysis. Suppose that a batch of paint contained a small amount of, say manganese, atomic number 25, and copper, atomic number 29. By irradiating a small piece of dried paint sample with neutrons, the presence of these elements can be detected.

$$^{1}_{0}n + {}^{55}_{25}\text{Mn} \rightarrow {}^{56}_{25}\text{Mn}$$

and

$$^{1}_{0}n + {}^{63}_{29}\text{Cu} \rightarrow {}^{64}_{29}\text{Cu}$$

After irradiation, many of the manganese and copper nucleii will be in the higher energy condition of the isotopes shown on the right of the arrows as products in the two equations. The manganese-56 isotope decays with a half-life of 2.56 hr, emitting a β particle and several γ photons, each with a precisely known energy. This half-life and the particular energies of those γ photons are unique to manganese-56. If we detect that half-life and those photon energies, we can be certain that the sample of paint had manganese in it; or, that a piece of dirt, maybe, which contained manganese contaminated the paint sample.

The copper-64 isotope decays by β emission and γ photon emission, also. The half-life of 12.8 hr and the γ-photon energies involved are unique to copper-64. Again, their detection (assuming careful sample preparation) certifies that copper was present in the paint. In several criminal court cases, samples of paint taken from a suspect's car have shown the same impurities as fragments of paint found embedded in a hit-run victim's clothing. The method is very sensitive, tiny fragments of paint suffice.

This sensitivity is dramatically demonstrated in the case of "Who killed Napoleon?" During the period 1806 to 1814 Napoleon attempted to put an embargo on British commerce with Europe. To some extent he was successful, but in the process he produced a lot of hard feelings, to say the least. (One of the side effects of Napoleon's anti-British policies was the War of 1812 between the United States and Britain.) Napoleon found it necessary to wage war against Russia in 1812 to enforce the embargo and was soundly defeated. Eventually, by early 1814 Napoleon was forced to abdicate all of his power except his sovereignty of the island of Elba, to which he then retired. Ten months later he was able to take advantage of political chaos in France and returned. Not too long after that, he was defeated at the Battle of Waterloo, in 1815; this time permanently. He spent the remaining $5\frac{1}{2}$ years of his life in exile, a prisoner on the island of St. Helena.

Now, shortly before his death he sent a lock of his hair in response to a request from one of his few admirers. During his imprisonment on St. Helena, Napoleon more than once accused his captors of slowly killing him. In those days, arsenic oxide, As_2O_3, commonly called "arsenic" was a favorite poison. It is sweet tasting, and, if fed in small amounts to a victim over a long period of time, causes disabilities of various kinds and hides the fact of poisoning by a display of misleading symptoms. However, when arsenic oxide is fed to a victim over a long period, it tends to accumulate, in small amounts, in the victim's hair; this, of course, was not known then. About 10 years ago, the lock of Napoleon's hair was borrowed from the museum where it was kept and was subjected to neutron activation analysis for arsenic. Interestingly, arsenic was found in the sample, and more was found at the ends near the hair roots than was found at the tip ends. In extenuation of his British captors, we must admit that it is known Napoleon used a dark colored pomade on his hair. In the last century, most dark colored hair pomades contained arsenic oxide to prevent the formation of mold. It is possible that the arsenic found in the lock of hair could have come from an external rather than an internal source. We may never know the truth about Napoleon's death, but this makes an interesting story to illustrate an application of radioactivity to a historical question.

Radioactive dating

Another application of interest to archeologist-historians is less gruesome. As we know, cosmic rays, mostly very high energy protons, have been battering the Earth steadily for billions of years. When they get into the Earth's atmosphere, the protons interact with anything they come close to, and a secondary shower of pretty high energy particles is produced, and so on. Some of the incoming protons simply shatter the nucleus of an atom in the Earth's upper atmosphere, producing neutrons as one of the results. We have not said so before, but outside the nucleus a neutron doesn't last long; the half-life of a whole bunch of free neutrons is about 12 min. Each neutron decays into a proton and an electron and a neutrino. However, if one of those free neutrons happens to hit a nitrogen-14 nucleus before it decays, a new nucleus, carbon-14, is produced, along with a proton.

$$\,^{1}_{0}n + \,^{14}_{7}N \rightarrow \,^{14}_{6}C + \,^{1}_{1}H$$

As we might guess a carbon-14 nucleus has "too many" neutrons. Eventually, it decays by β emission, forming nitrogen-14 again.

$$^{14}_{6}C \rightarrow {}^{14}_{7}N + {}^{0}_{-1}\beta$$

However, carbon-14 has a half-life of 5730 years, so it hangs around, on the average, for quite a while before decaying. The generation of carbon-14 occurs in the upper atmosphere, 10–20 miles up, or more. These carbon-14 atoms drift downward and generally get combined with oxygen, to make carbon dioxide, which behaves chemically like ordinary carbon dioxide, CO_2, except that sooner or later that carbon atom is going to revert back to nitrogen and emit a β particle in its dying gasp. Meanwhile it is entirely possible that the carbon-14 carbon dioxide will get used by a green plant to make cellulose. Now the carbon-14 atom is trapped in a cellulose molecule in the plant.

Because of the huge number of cosmic ray particles and the huge number of nitrogen atoms in the atmosphere, the random process we described produces a very steady average number of carbon-14 atoms. You can take any tree, any blade of grass, any living thing, a piece of your liver, or a piece of fingernail (which was very recently alive) and you will find that one out of every 6×10^{12} carbon atoms is a carbon-14 atom.

If we calculate the rate of decay from the known half-life of carbon-14 and express our results in grams instead of atoms, 1 g of carbon taken from any living thing has enough carbon-14 atoms in it to emit 16 β particles per minute. (This is going on right now in your body, for example.) It has been going on for millions of years in other people's bodies and in trees, grass, and so on. In a living, breathing plant or animal, as fast as the carbon-14 atoms decay, they are replaced by other carbon-14 atoms, generated steadily in the upper atmosphere, and eventually incorporated into living beings. As long as you live, as long as a tree of blade of grass is alive, 1 g of its carbon atoms will emit 16 β particles per minute.

Suppose now that someone cuts that tree down. It is no longer alive; it will no longer replenish its now steadily dwindling supply of carbon-14 atoms. So, 5730 years from now, 1 g of carbon from that tree would emit half of 16, or eight β particles per minute. If the dead tree lasts that long, somebody could come along 5730 years from now, find the old tree, cut off a piece, heat it severely (without letting it burn) until only carbon was left, take 1 g of that carbon, examine it with a radiation detector, and count only eight β particles per minute. They could then say "Ah ha! Somebody cut that tree down 5730 years ago."

This procedure has been very useful to modern investigators. As recently as 5000 years ago there were men on earth; they did not leave any written records, but they did make fires, they did cut down trees, they did make grass sandals. Some Indians, once, made some grass sandals and left them in a cave. Before they got back a volcanic eruption sealed the cave. Very recently, some acheologists opened the cave, found the sandals, and sent some samples off for radiocarbon dating. The sandals were 9600 years old, give or take about 200 years due to experimental error in the dating procedure. There was no other way to determine a reasonable date for these artifacts.

On the other hand, some specialists in Egyptian archeology had good reasons to think well of two models. One model stated that Egyptian farmers grew wheat before the Pharohs, the other model stated that wheat was not grown until after the Pharohs established their reign. There was no good way to settle the argument until somebody found some wheat in an old, old granary, up in the mountains, not too far from the Nile river. The wheat was found by carbon dating to be about 6300 years old, long before the reign of the Pharohs.

NUCLEAR FISSION AND FUSION

To introduce the last practical application, fission and fusion of nucleii, it will help to emphasize one important topic in a brief review. All spontaneous nuclear change depends upon a drop in energy level. What we have initially is at some higher energy level, and what we get later is at some lower energy level. In ordinary everyday processes we see this. No one in his right mind would jump off of a tall building because he knows he would fall. We can say that the reason he falls is "because" initially energy was expended to place him at some higher energy level (he climbed the stairs or rode the elevator to the top of the building). If he jumps, he will spontaneously achieve a lower energy level (squshed mess on the sidewalk at the bottom of the building) and give up the energy he had (break the sidewalk or smash a car). Roll a ball, fast; it eventually slows to a stop, spontaneously losing energy of motion as it rolls.

The fission of uranium

What's the point? In ordinary everyday processes we can easily observe the tendency to go spontaneously from a high energy to a low energy state. We can *see* the ball slow down; we *know* what happens if we jump. When a nucleus spontaneously loses energy, it does not slow down (it wasn't rolling in the first place); it does not fall downward (nobody made it jump off of something). When a nucleus loses energy, the result weighs less; a higher energy level nucleus loses energy by losing mass. The greater the loss of mass, the greater the energy release.

There are three naturally occurring isotopes of uranium, uranium-234, uranium-235, and uranium-238. Of these, uranium-235 nucleii can be fissioned, broken about in half, and the parts will have less mass than the original nucleus. The process can be controlled, as we shall see, and it also occurs naturally.

In the natural process, sooner or later a neutron from a cosmic ray will charge into a uranium-235 nucleus and, an instant later, it will split about in half. In addition to splitting two or three left over neutrons that did not happen to stick with either half will be tossed out. In the process, a lot of energy (for one nucleus) will be released. Now, if there is another uranium-235 nucleus or two close by and if those left over neutrons happened to be aimed (by chance) in their direction, the next thing you know there will be a couple more fissioned nucleii, and more energy released. And so it goes. All we need to get a lot of energy fast is to have enough (it takes a lot of them, about 5 lb or so) uranium-235 nucleii close enough together so that, after the first one fissions, the rest will follow in very rapid succession. This is called an "atomic bomb, old fashioned variety."

Far from being useful, all that energy delivered that fast is very destructive, as you know. Can we slow it down? Yes. First the uranium-235 nucleii must be farther apart. They can be fairly widely dispersed in a pile of bricks. We could call that a nuclear pile; the bricks are made of graphite for reasons we won't go into here. Then, here and there in that pile of bricks we can put neutron absorbers. Cadmium (atomic number 48) atoms make good neutron absorbers. The metal is generally shaped into long rods, which can be stuck into holes in the pile or pulled partly out.

So, to make a controlled nuclear reaction we first construct a pile of graphite bricks, leaving spaces for the cadmium control rods and for the chunks of uranium-235, here and there, all carefully calculated mathematically, of course. Next, put the cadmium rods in place, all the way in, not even partly pulled out. Then, put in the uranium-235. This was done for the first time under an athletic stadium, in 1942 in Chicago. Now, carefully and very slowly, *one* cadmium rod was pulled out a little way and the pile was checked to see whether any energy was generated. Then another rod was pulled, but only part way out. The pile was checked again. Another rod was pulled, but still only part way. Eventually, detectable energy was produced in this very first nuclear reactor.

Today, other designs are generally used instead of a pile of graphite bricks. The energy that is generated is used indirectly to make steam, which drives steam turbines for powering electrical generators. In a few instances, other heat transfer systems are used also, or instead. For example, the first nuclear submarine used hot liquid sodium metal (in well-sealed tanks and pipes) as a heat or energy transfer medium.

Nuclear fusion   A nuclear fusion process is based upon the same principle, mass loss. Thus, if we take any two nucleii of low atomic number and somehow fuse them together to form one nucleus, that new nucleus will weigh less than the two lower numbered nucleii totaled together. Thus, two hydrogen-2 nucleii weigh more than one helium-4 nucleus. This is the essential of the fusion process that is occurring right now in our Sun and in many other stars, as we know, except in the Sun four hydrogen-1 nucleii are fused to form one, lighter weight, helium-4 nucleus.

The fusion of nucleii requires a lot of energy merely to force the two positively charged nucleii close enough together to fuse into one unit. At the temperature of the Sun, this is no problem. Here on Earth it is indeed a problem, and once it is solved on a small scale in the laboratory, it will be an even greater problem to solve it on a large scale for the production of practical amounts of power. At present, indications are that fusion power (when we get it perfected) will present fewer and less serious ecological problems than those associated with fission power.

Ecological aspects

We will consider the ecological aspects of fission-power in more detail in a later chapter, but a few pressing questions can be briefly treated here. What if a nuclear reactor blows up? Could one ever explode? If it did how many people would be killed? Answers: It could be a mess. Possibly, since reactors are made by imperfect human hands. Depends on where it is and when it happens.

A few facts will help to clarify all this. First, nuclear reactors are built to very stringent safety standards. Since it is impossible even to state, much less obtain compliance with, an ideally perfect set of safety standards, it is also impossible to be absolutely certain that a reactor will be truly safe. Further, it can never be ascertained with absolute confidence that the stringent standards that are used have been fully enforced. The conclusion is certain: If enough nuclear reactors are built, one of them, sooner or later, will explode catastrophically. Given the standards that are imposed and the degree of compliance which is rigidly enforced, it is likely, but not certain, that the number of fatalities will be small.

The real question is how much we are willing to risk in order to make a significant gain in solving our energy problems. In a parallel situation, it seems that people are willing to risk the certain loss of many lives in order to enjoy the advantages of travel by automobile. And it is, indeed, all but certain that the loss of life in all the nuclear reactor explosions that will ever occur will be much less than the loss of life that has already been caused by our use of automobiles.

Next, what about radioactive by-products from a nuclear reactor? The halves of the split uranium-235 atoms are nucleii too, at very high energy levels. Mostly, these decay by β or positron emission. Some have very short half-lives; some have longer half-lives. All of these (despite what you may read in published information written by poorly informed but sincere people who often confuse by-products from atom bomb production with by-products from nuclear reactors) probably can be effectively contained without causing harm to mankind. It is true that in the past some very sloppy (and some would say inexcusable) techniques were used that were ineffective to contain these radioactive species. We can hope, at least that this problem will be solved.

SUGGESTIONS FOR DISCUSSION

1. Either constructively criticize at least two of the analogies used in this chapter, being sure to suggest how they might be altered in order to be more useful to you, or prepare your own description, analogical or otherwise, of at least one of the models described in this chapter.

2. In your own words describe the differences and similarities between and among protons, neutrons, and mesons.

3. In short sentences or phrases, show that you know the meaning of these words: radioactivity, isotope, α particle, energy state, β particle, nucleus, atom, positron, half-life, random processes, carbon-14 dating, cosmic rays, artificial element, deuteron, nuclear reactor.

4. Describe the similarities and differences, if any of either, between these pairs: proton and nucleus of a hydrogen atom; nuclear fission and nuclear fusion; radioactive decay and change in mass; β emission and positron emission; detection of radiation by scintillation and by photographic film; volcanoes and earthquakes; iodine-128 and iodine-131.

5. Which is most to be avoided, which next most, and so on, of the following: cosmic rays, α radiation, β radiation, γ radiation, salt in your food, neutrinos? Justify your ordering of choices.

6. Carbon-14 dating was touted in this chapter as a practical application of radioactivity. Is it, in your opinion? Support your opinion with a logical argument.

7. Briefly discuss the desirability, or the opposite, of an artificial heart which can be purchased for about the price of a luxury automobile. (This is likely to be the approximate cost, at least at first.)

chapter 7

Equations, Formulas, and Reactions

This is a chapter about symbols. In the preceding chapters symbols have been mentioned and used with the presentation itself. They have been related as much as possible to other matters and the symbols themselves were presented lightly, as an integral part of a discussion. Sooner or later, for any real discussion in chemistry it is necessary to face up to the use of symbols.

In many respects the challenges we face in this chapter are like those involved in learning another language. As we all know, this requires a conscious effort. Fortunately, as with a new language, there is a systematic, ordered, set of involvements within the concept of chemical symbols, and once past the first shock it gets a little easier. In any event, an involvement with symbols is not unique to chemistry; we see it everywhere.

Consider the use of symbols in religion, for example. A universal symbol for peace is to raise the right arm slightly, with the hand flat, palm out, toward another person. We find this symbol in all human cultures; the stylized pose of some statues of Budda intended to convey a peaceful attitude, or one of fearlessness (closely related to peace), is called *abhaya mudra*; the right arm is slightly raised, the hand is open, palm outward. The meaning of this pose is clear to anyone, no matter what his own religious background might be, whether or not they can pronounce the word-name of the pose and whether or not they know anything about Oriental cultures. Probably, this symbol for peace developed from the much earlier practice of carrying a weapon in the right hand, so that an obviously unarmed, open hand suggested a peaceful intent. The meaning of this symbol is therefore relatively easy to discern.

WRITING CHEMISTRY
Chemical symbols

Some of the symbols used in chemistry are about as straightforward; H for hydrogen, C for carbon, and so on. Almost as obvious are H_2 for hydrogen molecules, Co for cobalt, Cl for chlorine. We will use an arrow, $\rightarrow$, as a symbol, and we have already noted in the preceding chapter that an arrow symbolizes a change of some sort, a kind of "motion" from a starting condition to a different condition later. The point is that chemical symbols make sense; they permit the communication of information in a special way. Here we go.

$$H_2O(l) \rightarrow H_2(g) + O_2(g)$$

In words, liquid water is decomposed into gaseous hydrogen molecules and gaseous oxygen molecules. The letters in parentheses, (l) and (g), signify liquid and gas. If one of the substances had been a solid, we would have used (s). The point is that it is simpler, once you catch on, to write the sentence "Liquid water is decomposed into gaseous hydrogen molecules and gaseous oxygen molecules" in symbols than to write the English sentence. We can consider this use of symbols as a chemical sentence.

Chemical sentences

There is a big difference between an English sentence and a chemical sentence. For example, it is perfectly good, grammatical English to say, "It will rain yesterday." This sentence is nonsense, but it conforms to the rules of grammar. A chemical sentence, on the other hand, is restricted to a symbolic description of *reality*. We cannot write the following sentence and claim that it is a chemical sentence:

$$H_2O(l) \rightarrow C_2H_5OH(l)$$

Liquid water cannot be transformed into liquid alcohol, except miraculously, and this is beyond the scope of our discussion. Although in English we can speak of water changed into wine, we cannot do so chemically. Water can be changed into hydrogen and oxygen, though.

Find a friend with a battery operated transistor radio and persuade him or her to lend you the 9-volt (V) battery from his/her radio. Dissolve about a teaspoon of salt in a glass of water, and immerse the battery in the water completely. You will observe bubbles of gas originating at the positive and at the negative terminals of the battery. Hydrogen molecules are formed from water at the negative terminal, and oxygen molecules at the positive terminal. If you observe closely, you will notice that there seem to be more bubbles, or bigger maybe, of hydrogen than of oxygen. A careful measurement would show half as much oxygen as hydrogen. The correct sentence is

$$H_2O(l) \rightarrow H_2(g) + \tfrac{1}{2} O_2(g)$$

This chemical sentence is more detailed than the first one. It conforms more precisely to reality. It is called a **chemical equation**, or equation, for short. A chemical equation is a symbolic sentence that describes a chemical change or reaction. The terms "chemical change" and "reaction" are synonymous.

CHEMICAL REACTIONS

A reaction, a chemical change, is a process (as mentioned in Chapter 3) in which one or more substances disappear and new substances, not present before, appear. In a reaction the mass of the new substances is exactly the same, as closely as can be measured, as the mass of the substance or substances that disappeared. We can tell that the original substances disappear because we can no longer detect their presence when we look for the properties of those original substances. We can tell that new substances are formed because we now detect new properties, not present before.

With the radio battery and the water, we would be able to measure a loss in weight for the water that was decomposed; we would see new properties, bubbles of gas. And, if we checked carefully, we would detect different properties for each of the two gases. Hydrogen, for example, is lighter than air and oxygen is heavier. A mixture of these two gases in the proportions produced in this reaction is explosive; either gas alone is not.

Incidentally, although it is a little hard on the battery, this is a convenient way to check a partially used radio battery to see if it has any life left. A dead, or almost dead, battery will not decompose water. The salt dissolved in the water is used to promote the rapidity of the decomposition. Without salt (sugar won't work) the water will indeed decompose, but so slowly as to not be noticeable without fancy equipment. (Some water does contain enough dissolved substances as obtained from a water tap to permit a noticeable decomposition, however.)

The energy of chemical reactions

As we know, water will not decompose spontaneously. Energy is required; that's why this procedure is hard on the radio battery. Energy from the battery is required in order to decompose the water. As we know, water molecules consist of an oxygen atom in the center bonded to two hydrogen atoms. To decompose water, these two O—H bonds must be broken; this requires energy. On the other hand, when two oxygen atoms, so formed (from two water molecules) join to make an oxygen molecule, O_2, the formation of the bond releases energy. Similarly, when two hydrogen atoms join to make a hydrogen molecule, H_2, energy is released. In this case, more energy is required to break up the bonds in the water than is released by the formation of the new bonds. The net effect is a requirement for energy from an external source, the battery.

We could say that a bunch of hydrogen molecules and half as many oxygen molecules "contain" more energy than the water from which they were formed. It is convenient to think of this energy as residing in the atom to atom bonds. This manner of considering what happens provides a theoretical basis for defining a chemical reaction. A **chemical reaction** *is a process in which bonds are broken and new bonds* (this atom to a different than before atom) *are formed.*

For almost all reactions, the energy required to break the old bonds is either more, or less, than the energy released by the formation of the new bonds. (Only rarely do we find an example where the energy to break all the old bonds is exactly the same as the energy involved in the formation of all the new bonds.) That is, in general a chemical reaction either releases energy, or absorbs it. Reactions that, overall, require energy are called **endoergic reactions**; those which release energy are called **exoergic reactions**. (Endo is based on a Latin root meaning within, inside of; exo from a Latin root signifying outside; and erg from a Greek root meaning work or energy.) The decomposition of water is an endoergic reaction; the burning of paper is an exoergic reaction.

Equations for chemical reactions

There are still more details symbolized in an equation. In particular, an equation conveys the notion that mass is neither lost nor gained in a reaction. That is, in a reaction no atoms are destroyed, no atoms are created; the total mass does not change. To see this, we need to know that in an equation the symbol $H_2O(l)$ means more than water, it means 1 mole of water molecules (in the liquid state). We already know that 1 mole of anything is a lot, 1 mole of molecules is 6.02×10^{23} molecules. When we check it out, 1 mole of water molecules weighs 18 g, rounded off (actually, a bit more than 18 g).

So, in an equation $H_2O(l)$ means 18 g of liquid water, or—same thing—6.02×10^{23} molecules of water in the liquid state. Similarly, $H_2(g)$ symbolizes 1 mole of hydrogen molecules, which weighs 2 g (rounded off), and $\frac{1}{2} O_2(g)$ symbolizes $\frac{1}{2}$ mole of oxygen molecules, weighing 16 g (rounded off). Both the hydrogen and oxygen molecules are in the gaseous state, of course.

Here is the equation, once again.

$$H_2O(l) \rightarrow H_2(g) + \tfrac{1}{2} O_2(g)$$

The complete English sentence this equation symbolizes is "One mole of water molecules in the liquid state weighing 18 g (and consisting of 2 moles of hydrogen atoms and 1 mole of oxygen atoms) decomposes to form 1 mole of hydrogen molecules in the gaseous state weighing 2 g (consisting of 2 moles of hydrogen atoms) and $\tfrac{1}{2}$ mole of oxygen molecules in the gaseous state weighing 16 g (consisting of 1 mole of oxygen atoms)." You can see why chemists prefer chemical equations to English sentences to express a reaction in symbols.

As we know, the word "mole" is a symbol for a very large number; initially it is troublesome for some students to use the word "mole" in their studies. If you have this difficulty, then remind yourself that "mole" is a kind of "chemical dozen" when you see it. The essential is the idea of counting atoms, not the counting unit. You may be more comfortable with "dozen," at first, but soon you will become quite familiar and friendly to "mole."

Notice the atom counting in the equation. We begin with water, 1 mole of water molecules. In that mole of water molecules there are 2 moles of hydrogen atoms and 1 mole of oxygen atoms. If the water is to be decomposed, then no matter what the products are, they *must* consist of, total, exactly the same atoms: 2 moles of hydrogen atoms and 1 mole of oxygen atoms. Notice that the symbol H_2 signifies 1 mole of hydrogen molecules, each with two atoms, or, a total of 2 moles of hydrogen atoms. Notice that the symbol O_2 signifies 1 mole of oxygen molecules, each with two atoms, or a total of 2 moles of oxygen atoms. In this case, the $\tfrac{1}{2}$ in front of the O_2 symbol conveys the meaning of half as much, or half of 2 moles of oxygen atoms (or a half mole of oxygen molecules, really), which is equivalent to 1 mole of oxygen atoms, as required.

In our very first chemical sentence in this chapter, the $\tfrac{1}{2}$ was omitted in front of the O_2 symbol to keep things simple. That first chemical sentence, strictly, was not an equation since it implied more oxygen molecules (or atoms) than could be observed in reality when the water was decomposed. A chemical equation in which careful attention is paid to the atom count is said to be a **balanced equation**.

Naming chemical reactions

We have thus far belabored the decomposition of water. We could call this reaction a decomposition reaction. There are many other examples of decomposition reactions. This reaction is, also, endoergic. We could call it an endoergic reaction. Some endoergic reactions are also decomposition reactions, but some are not. Some decomposition reactions (most, in fact) are endoergic, but some are not. The water was decomposed by utilizing electrical energy from the battery, so we could call this reaction an electrolysis reaction (lysis comes from a Greek root meaning to break apart). Some electrolysis reactions are decomposition reactions, but some are not. Not all endoergic reactions are electrolysis reactions, and so on. There are perhaps a hundred or more ways to classify chemical reactions, and for most of these the definitions overlap in a confusing way. One of the most useful classifications is oxidation-reduction; the decomposition of water is an oxidation-reduction reaction, for example. To see this clearly, a digression is necessary.

Structural units in chemical reactions

The study of any subject is rendered easier and more manageable by subdivision of the totality into structured units. Thus, a full knowledge of poetry includes information about various kinds of structural units: alliteration, onomatopoeia, dithyramb, verse, line, hexameter, elision, stanza, limerick, anaphora, sonnet, doggerel, and on and on and on. Chemistry is also divisible into structural units that aid study.

An atom is a particle, composed of neutrons, protons, (mesons), and electrons. Elements are composed of atoms, and all the atoms of an element have the same number of protons in their nucleii. An atom has a mass that depends largely upon the number of protons and neutrons in its nucleii. Although we can think of an atom as the smallest particle we could get if we could cut up a piece of an element, say, a chunk of copper more and more finely divided, no one has ever obtained an atom in this way. Rather, an atom is a conceptually derived structural unit; and the same applies to the parts of which it is supposed to be composed. We use the concept of atoms to generate explanations for those things we can indeed see with our own eyes.

We say that atoms combine to form molecules, sharing electrons. Molecule is also a structural unit. Atoms lose or gain electrons to form ions, and the ions "combine" to form crystals, such as in salt. A crystal is another structural unit, not as well defined perhaps, as a molecule. The notion of electron pair sharing, or of charged ions themselves resulting from electron loss or gain, is another structural unit. We say that the formula of water is H_2O, not some other combination of numbers of atoms. The concept that atoms combine in countable small-number ratios is a structural unit. To say that the molecule of water has a unique shape is still another. A structural unit, that is, dithyrambs in poetry or molecules in chemistry, is a way to think about the subject such that the whole totality of the subject becomes more tractable, better or more easily understood.

A dithyramb was an ancient improvised song in honor of the wine god Dionysus, so the original dithyrambs were all Greek. Dithyrambs are rare today, although you can find them in Dryden's works, more or less coincidentally; but these are almost 300 years old by now. The modern calypso, a folk song from Trinidad, probably comes closest to the ancient dithyramb.

OXIDATION NUMBER

Another chemical structural unit, not nearly so complicated as a dithyramb but which will take longer to describe because we will go into more detail about it, is oxidation number. To help keep track of the detail, notice that oxidation numbers are related to numbers of valence electrons in many instances. You will feel better if we point out at once that oxidation numbers need not be memorized; you can work out most of them by using the periodic table for a few seconds. To get started, a little historical background will be useful.

Originally, the formulas of compounds were determined by laboratory analysis. From the laboratory results, which showed factually that salt was 39.34% by weight sodium and 60.66% by weight chlorine, it was determined theoretically that the formula for salt was $NaCl$, not Na_2Cl_5 or Na_3Cl, or some other combination of numbers of sodium and chlorine atoms (ions, strictly). This method is still used to determine the formulas of newly discovered compounds today, but the way we remember the end result, one sodium and one chlorine in the formula itself, $NaCl$, is by using oxidation numbers.

The idea with oxidation numbers is to work things out so that they add up to zero. The oxidation number for sodium is $+1$, and a common oxidation number for chlorine is -1. Since one $+1$ and one -1 add to zero, the formula for sodium chloride is one sodium and one chlorine, $NaCl$. (Whenever chlorine has a -1 oxidation number attributed to it, we use the word "chloride" instead of chlorine.) Similarly, calcium has an oxidation number of $+2$. So, for calcium chloride it is one calcium and two chlorines, $CaCl_2$. That is one $+2$ and two -1's add to zero. If we know, or can look up, all the oxidation numbers, we can figure out the formulas of many different compounds.

The simplest way to summarize the oxidation numbers is to use the periodic table; as you can see by looking at Figure 7–1, the oxidation numbers of the different elements form a fairly regular pattern when displayed in this manner.

We have already noted -1 for chlorine denotes chloride; the -1 oxidation number for fluorine implies the name fluoride, -1 for bromine, bromide, and iodide when -1 is used for iodine. Similarly, a -2 for oxygen implies oxide (another name for water is hydrogen oxide, H_2O; two $+1$'s for hydrogen and a -2 for oxygen add to zero). The same applies to sulfur, -2 implies sulfide, and similarly for selenium, selenide, and tellurium, telluride. To complete the story, -3 for nitrogen is nitride (the compound Mg_3N_2 is magnesium nitride; three $+2$'s and two -3's, $+6$ and -6, add to zero). The same applies to carbon and silicon when they have an oxidation number of -4, carbide and silicide, respectively. (Thus, Ca_2C is calcium carbide; two $+2$'s and one -4.)

Oxidation numbers of polyatomic ions

From the information given in the periodic table and with a few more rules tossed in to help, we could work out all the additional details for the oxidation numbers of polyatomic ions. However it is simpler to give the information as

Periodic Table of the Elements

IA	IIA	IIIB	IVB	VB	VIB	VIIB	VIIIB	VIIIB	VIIIB	IB	IIB	IIIA	IVA	VA	VIA	VIIA	VIIIA
H +1(−1) 1																	He — 2
Li +1 3	Be +2 4											B +3 5	C +2,+4,−4 6	N +3,+5,−3 7	O −2(−1) 8	F −1 9	Ne — 10
Na +1 11	Mg +2 12											Al +3 13	Si −4 14	P +3,+5,−3 15	S +4,+6,−2 16	Cl −1,+5,+7 17	Ar — 18
K +1 19	Ca +2 20	Sc +3 21	Ti +2,+4 22	V +2,+3,+5 23	Cr +5,+6 24	Mn +2,+4,+7 25	Fe +2,+3 26	Co +2,+3 27	Ni +2,+3 28	Cu +1,+2 29	Zn +2 30	Ga +3 31	Ge +2,+4 32	As +3,+5 33	Se +4,+6,−2 34	Br −1,+5,+7 35	Kr (+2,+4) 36
Rb +1 37	Sr +2 38	Y +3 39	Zr +4 40	Nb +3,+5 41	Mo +6 42	Tc +4,+7 43	Ru +3 44	Rh +3 45	Pd +2,+4 46	Ag +1 47	Cd +2 48	In +3 49	Sn +2,+4 50	Sb +3,+5 51	Te +4,+6,−2 52	I −1,+5,+7 53	Xe (+2,+4) 54
Cs +1 55	Ba +2 56	La +3 57	Hf +4 72	Ta +5 73	W +6 74	Re +4,+7 75	Os +4 76	Ir +3,+4 77	Pt +2,+4 78	Au +1,+3 79	Hg +1,+2 80	Tl +1,+3 81	Pb +2,+4 82	Bi +3,+5 83	Po +2,+4 84	At — 85	Rn — 86
Fr +1 87	Ra +2 88	Ac +3 89	104 —	105 —													

Lanthanide series:

Ce +3,+4 58	Pr +3 59	Nd +3 60	Pm +3 61	Sm +2,+3 62	Eu +2,+3 63	Gd +3 64	Tb +3 65	Dy +3 66	Ho +3 67	Er +3 68	Tm +3 69	Yb +3 70	Lu +3 71

Actinide series:

Th +4 90	Pa +4 91	U +3,+4 92	Np +3,+4 93	Pu +3,+4 94	Am +2,+3 95	Cm +3 96	Bk +3 97	Cf +3 98	Es — 99	Fm — 100	Md — 101	No — 102	Lr — 103

Figure 7–1. Periodic table of the elements showing their commonly exhibited oxidation numbers above the elemental symbols. (Oxidation numbers in parentheses are less commonly noted.) In addition to the numbers shown, an element in the uncombined state, or combined only with another atom of that element, is assigned an oxidation number of zero.

matters of fact. Notice that in each case for these polyatomic ions, the oxidation number corresponds to the magnitude of the charge on the ion. The concept of oxidation number was designed to incorporate this feature, and it does so successfully in most cases. In Table 7–1 the formulas for the polyatomic ions are listed along with their English names and oxidation numbers. The oxidation numbers given apply to the whole ion, all the parts taken together.

TABLE 7–1. **The Polyatomic Ions**

Formula of Polyatomic Ion (with electrical charge indicated as a superscript)	Name of Ion	Oxidation Number
NH_4^+	ammonium	+1
H_3O^+	hydronium	+1
OH^-	hydroxide	−1
CN^-	cyanide	−1
$C_2H_3O_2^-$	acetate	−1
ClO^-	hypochlorite	−1
ClO_2^-	chlorite	−1
ClO_3^-	chlorate	−1
ClO_4^-	perchlorate	−1
MnO_4^-	permanganate	−1
NO_2^-	nitrite	−1
NO_3^-	nitrate	−1
HCO_3^-	hydrogen carbonate	−1
HSO_4^-	hydrogen sulfate	−1
O_2^{2-}	peroxide	−2
CO_3^{2-}	carbonate	−2
SO_3^{2-}	sulfite	−2
SO_4^{2-}	sulfate	−2
CrO_4^{2-}	chromate	−2
$Cr_2O_7^{2-}$	dichromate	−2
PO_2^{3-}	hypophosphite	−3
PO_3^{3-}	phosphite	−3
PO_4^{3-}	phosphate	−3
AsO_3^{3-}	arsenite	−3
AsO_4^{3-}	arsenate	−3

Use of oxidation numbers

Proficiency in chemistry demands memorization of the oxidation numbers of the elements and the polyatomic ions. Even for others who do not want to be proficient in chemistry, it is helpful to see how this information is used. For example, to figure out the formula of any combination possible from among those presented, it is first necessary to identify a component with a positive and a component with a negative oxidation number. Thus, we cannot pick sodium, +1, and magnesium, +2. We might pick magnesium, +2, and hydrogen as a (rare but it is OK) −1.

Next using the smallest multiplying factors, set the plus and minus sum to zero. For our example with +2 and −1, magnesium and hydrogen, we need *one* times +2 and *two* times −1 ($1 \times +2$ and 2×-1 add to zero). So, the formula is Mg_1H_2, or simplified, MgH_2. The name is magnesium hydride. (When hydrogen is in the −1 oxidation number condition, hydrogen becomes hydride.)

Try calcium with +2 and sulfate with −2; it is $CaSO_4$ for the formula, calcium sulfate for the name. Try sodium sulfate; Na with +1 and SO_4 with −2 produces Na_2SO_4; $2 \times +1$ and 1×-2, that is.

What about calcium nitrate, with +2 for calcium and −1 for nitrate? We have $1 \times +2$ and 2×-1, which gives the formula $Ca(NO_3)_2$. We put a parentheses pair around the NO_3, and a two outside the parentheses; otherwise, with no parentheses the subscript 3, which belongs on the oxygen, would get mixed up with the subscript 2, which belongs to the whole polyatomic ion. As another example, consider aluminum sulfate, +3 for aluminum and −2 for sulfate. $2 \times +3$ and 3×-2 is the simplest magic combination that adds to zero, so we have $Al_2(SO_4)_3$ for the formula.

Multiple oxidation numbers

You will notice that some of the metals in the periodic table listing have more than one positive oxidation number; iron as +2 or +3 is an example. From this we can conclude that there are two different but possible compounds involving iron and, say, bromine. With a −1 for bromine we can have $FeBr_2$ or $FeBr_3$. Either one is correct, both do exist, and both, as far as our information goes, would be called iron bromide. Needless to say chemists must be able to distinguish between the two when they talk about them, so the oxidation·state is identified in the name. In $FeBr_2$, iron is in the +2 oxidation state, so the compound is called iron(II) bromide, "iron two bromide." For $FeBr_3$, iron(III) bromide is the name.

Try the formula for cobalt(II) arsenate and cobalt(III) arsenate, for practice . . . one of the correct answers is $CoAsO_4$.

OXIDATION-REDUCTION REACTIONS

The concept of oxidation numbers has other utility than figuring out the correct formulas of compounds without needing to go to the laboratory to get percent by weight data [plus perhaps a lot of mathematical hassle before ending up with $Co_3(AsO_4)_2$]. This concept is involved in understanding oxidation-reduction reactions.

We have three terms: oxidation number, oxidation, and reduction. An oxidation number, as we know, is an arbitrary number, zero, positive, or negative that is assigned to an element. Zero is always assigned to an uncombined element and to an element combined only with itself, such as H_2. (Occasionally, zero is assigned to an element in the combined state; carbon has an oxidation number of zero in the compound sucrose, ordinary sugar $C_{12}H_{22}O_{11}$, but this is rare.) Nonzero, that is positive or negative, oxidation numbers are arbitrarily established numbers that enable us to determine the formula of a compound by manipulating the subscripts so as to make the sum of the oxidation numbers equal zero in the compound.

On the other hand, **oxidation** is a process in which an oxidation number increases during a chemical reaction. **Reduction** is the opposite process, a decrease in oxidation number during a chemical reaction. Always, never any exceptions, oxidation and reduction occur together, never one without the other.

Let us go back to our familiar example

$$H_2O \rightarrow H_2 + \tfrac{1}{2}O_2$$

Notice that the oxidation number of oxygen in water is -2 and that it is zero in O_2. The oxidation number of oxygen increased. For hydrogen in water the oxidation number is $+1$ and it is zero in H_2. The oxidation number of hydrogen decreased. That is, in this reaction an oxidation occurred; oxygen was oxidized from -2 to zero. A reduction also occurred; hydrogen was reduced from $+1$ to zero.

Oxidation-reduction reactions are all but ubiquitous. We see them almost everywhere. Combustion reactions, the photosynthesis of starch, cellulose, and sugar in the green leaves of plants, the metabolic and catabolic reactions within our own bodies, corrosion reactions, the chemical reactions in batteries, are all oxidation-reduction reactions.

Here is our example again, this time turned around.

$$\tfrac{1}{2}O_2(g) + H_2(g) \rightarrow H_2O(l)$$

This is an oxidation-reduction reaction, of course; oxygen is reduced (zero to -2) and hydrogen is oxidized (zero to $+1$). This reaction is exoergic, and has been proposed as an important, though partial solution to the energy shortage that will face us as our supply of fossil fuels becomes smaller and more difficult to obtain. In essence, according to the simplest proposal, water can be decomposed into hydrogen (and oxygen) by electrical energy from nuclear reactors. The hydrogen can then be transported in tank cars or by pipeline to locations where fuel is needed and then recombined with oxygen from the air to produce energy. Thus, the energy from a nuclear reactor at some distant point can be as it were moved to another place and made available there. Without doubt, this suggestion is possible. However, there are other, less expensive ways to obtain hydrogen from water (although they are not yet fully developed) instead of electrolysis. We will look at one of these later in this chapter.

For the moment, notice another aspect of the two equations involving water, hydrogen, and oxygen; either one is the *reverse* of the other. We can write them both, as one equation, by using two oppositely pointing arrows.

$$H_2O(l) \rightleftharpoons H_2(g) + \tfrac{1}{2} O_2(g)$$

This set of symbols implies an entirely new concept, one which we have not yet discussed: reversibility.

EQUILIBRIUM   Even without a battery to electrolyze water, it is nevertheless decomposing continuously into hydrogen and oxygen. The rate of this spontaneous decomposition is extraordinarily slow at ordinary temperatures; so, at any given moment above a lake or a bathtub full of water or above a glass of water, the total amount of gaseous hydrogen and oxygen from the decomposition of the water is so small as to be undetectable. Further, if we confine the water in a partially filled, sealed, container with some space above the water and do not allow the hydrogen and oxygen gases to escape, they will recombine to form water again.

There will be an **equilibrium**, a balance, a dynamism—water decomposes into hydrogen and oxygen, very slowly, and, at exactly the same rate, it is reformed. The reactions are **reversible**, we say that a condition of **dynamic equilibrium** exists; the reaction proceeds in opposite directions at equal rates.

At some higher temperature, the same thing is true except that the decomposition is more rapid, and the equal rate of reformation of course also increases, to be equally rapid but in the opposite direction.

Equilibrium can be disturbed, however. Suppose that we wish to obtain hydrogen from the decomposition of water without electrolyzing it or heating it; we simply decide to use its spontaneous, though slow, decomposition at ordinary temperatures. In principle, all we have to do is to put an enclosure around a bathtub (or something) full of water, leaving a space above the water itself. Then, by some trick, we must extract the tiny bit of gaseous hydrogen that is in the air above the water, and put that hydrogen somewhere else. If we do this, the reformation of water cannot occur, but the water will of course continue to decompose, slowly, spontaneously. This particular suggestion is not practical because the rate of decomposition is so very, very slow at ordinary temperatures; however at higher temperatures, when the rate would be more rapid, it could be practical although probably too costly to be economically useful.

USE OF EQUILIBRIUM IN INDUSTRIAL PROCESSES
The need to synthesize nitrates

The concept of removing one product to disturb an equilibrium is used today in a different application. The story illustrates rather clearly how the technological application of science can, and did, alter the history of the world. It also illustrates the point that the results of science can be applied for the benefit or for the detriment of mankind, depending on the choice one wishes to make.

To set the stage, we must look at Europe in the preceding century, some years prior to World War I. Initially, the countries we now know as West Germany, East Germany, Austria, Czechoslovakia, and Hungary were even further subdivided into small principalities. Bit by bit, these petty fiefs agglomerated into larger domains and eventually became a strong politicoeconomic force. This development was marked by wars and intrigue, which also involved the countries we know as France, England, Poland, and Russia from time to time. Further, and still loosely summarizing, the antipathy between and among the central European people and those in the bordering countries, which can be traced back as far as Charlemagne more than a thousand years previous, according to some historians, continued to fester.

At about the turn of the century, it became clear (rightly or wrongly) to the men who held political power in what we will call Imperial Germany that the other European countries must be subjugated and that a war was the appropriate instrument. However, they had a serious problem. Wars are fought with men and explosives. Imperial Germany had the manpower (they thought); it did not have the explosives.

A bit of chemistry will help here. Nitroglycerine is an explosive. In fact, it is a bit too much as an explosive. To be useful as an industrial and mining explosive or as a military explosive, control is essential; one wants the stuff to explode only when the explosion is necessary, not at some other, inopportune, moment. Nitroglycerine will sometimes explode, it seems, if you merely look mean at it; at other times when you want it to explode, it will not. Alfred Nobel accidentally discovered how to control this explosive; it is merely necessary to absorb it into kieselguhr, a porous dried clay. (Nitroglycerine is a thick, oily liquid.) The result is called dynamite, and its explosive characteristics can be controlled most of the time. Other ways have since been found to control nitroglycerine and other explosives.

Equation (7–1) is the oxidation-reduction reaction that describes the explosion of nitroglycerine or dynamite:

$$H_5C_3(NO_3)_3(l) \rightarrow \tfrac{3}{2} N_2(g) + \tfrac{5}{2} H_2O(g) + 3\, CO_2(g) + \tfrac{1}{4} O_2(g) \qquad (7\text{–}1)$$

The reaction is exceptionally exoergic. We begin with a liquid (or a solid if we begin with dynamite) that changes in an incredibly small instant into gases, with a consequent tremendous expansion in volume. The gases, furthermore, are at a high temperature, which makes their volume even larger. What was a liquid is now, an instant later, a gas with a volume 20,000 times greater than the original liquid. Anything that is very near the original liquid (or solid) is necessarily going to be subjected to great violence. The explosive wave of gas has been estimated to move at a speed of about 8000 m/sec (about 5 mi/sec). (As you have guessed, this reaction is not reversible in practice.)

Now, all military explosives contain carbon, hydrogen, nitrogen, and oxygen atoms. The nitrogen and oxygen are always in the form of an NO_3 or NO_2 group. Apparently, it is the presence of either one of these nitrogen oxygen groupings that makes a substance an explosive.

In those days, Imperial Germany had no local source of nitrogen in the form of an NO_3 or NO_2 grouping, and you cannot make explosives without this. The best source in those days was in Chile in the form of sodium nitrate, $NaNO_3$. Friendly diplomatic relations with Chile were enhanced further, and stockpiles of sodium nitrate were built up in Germany by shipment from the Chilean mines.

Meanwhile, the war councils in Germany discussed the difficulties of shipping from the western coast of South America to German ports in the Baltic Sea during wartime against an adversary, England, known to have a strong and vigorous fleet. Either they would need to postpone the war until their stockpile of sodium nitrate was certain to be sufficient, plus some extra as a "safety" factor, or some other way to get the necessary nitrogen and oxygen must be found.

The synthesis of ammonia—the Haber process

At that time chemists knew that nitrogen was present in the air; the air is about 80% nitrogen, actually. One chemist, Fritz Haber, applied the principles of reversibility and solved the problem of Imperial Germany. Hydrogen will react with nitrogen to form ammonia in a reversible (oxidation-reduction, of course) reaction.

$$\tfrac{1}{2} N_2(g) + \tfrac{3}{2} H_2(g) \rightleftharpoons NH_3(g) \qquad (7\text{–}2)$$

The nitrogen was obtained from the air; the hydrogen was obtained from petroleum in a process called steam-reforming, which we will look at later. Under high pressure and reasonable but high temperatures in the presence of an iron oxide catalyst, the reaction proceeds from left to right at a decent rate, and equally, at the same rate, from right to left, of course. However, as the ammonia, NH_3, is formed, most of it is removed. Therefore, the left to right reaction proceeds as before, but the rate of the right to left reaction cannot go as fast because the ammonia is being removed continuously and quickly cooled, so that it does not decompose appreciably into the original nitrogen and hydrogen.

The synthesis of nitrates

The ammonia from the Haber process is reacted with oxygen to form water and the important product, nitrogen monoxide (NO).

$$4\,NH_3(g) + 5\,O_2(g) \rightleftharpoons 6\,H_2O(l) + 4\,NO(g) \qquad (7\text{–}3)$$

The same kind of strategy is used here to promote the production of nitrogen monoxide. If additional oxygen is supplied, nitrogen monoxide will react spontaneously with it to form nitrogen dioxide, NO_2.

$$NO(g) + \tfrac{1}{2} O_2(g) \rightleftharpoons NO_2(g) \qquad (7\text{--}4)$$

About as fast as the nitrogen dioxide is formed, it is sprayed with water droplets, with which it reacts and in which it dissolves, thus preventing most of it from reverting to nitrogen monoxide and oxygen.

$$3\,NO_2(g) + 3\,H_2O(l) \rightleftharpoons 2\,H_3O^+(aq) + 2\,NO_3^-(aq) + NO(g) \qquad (7\text{--}5)$$

The (aq) indicates that these ions, the hydronium ions, H_3O^+, and the nitrate ions, NO_3^-, are dissolved in water.

The nitrogen monoxide product is recycled back into the system to form more nitrogen dioxide by reacting with oxygen. This final reaction is reversible, as you can see from the double arrow symbol. As it happens, when the opposed rates are equal, most of the substances are in the form of the ions, and NO, with very little nitrogen dioxide; the water, of course, is present in excess, since it is used as a spray initially. The tendency of this reaction to go from right to left is of course further hindered by the consumption of the nitrogen monoxide to form more nitrogen dioxide.

This whole business probably sounded about as complicated to the German war council as it does to you, but they trusted Haber and his colleagues, who quickly established factually that they could provide a source of nitrate ions for the manufacture of explosives from local materials. About one year after this, Imperial Germany found an excuse for starting World War I which has continued with short intervals of semipeace, ever since, according to one historical model, at least.

In his personal correspondence and other papers there are statements by Haber that he deliberately set out to find a way to convert the nitrogen in the air into nitrates for the express purpose of aiding the German war plans. Today, this seems reprehensible, or at least difficult to justify; in the context of the circumstances then, it may be possible to conclude that there was some justification, at least according to some of Haber's biographers. In any event, the same chemical reactions can be and have been used for the benefit of mankind.

Nitrogen fixation

Thus, we have two equations to consider. Equation (7–2) symbolizes the production of ammonia from hydrogen and nitrogen, and equation (7–5) symbolizes the production of nitric acid (the solution of those ions) from nitrogen dioxide and water.

$$\tfrac{1}{2} N_2(g) + \tfrac{3}{2} H_2(g) \rightleftharpoons NH_3(g) \qquad (7\text{--}2)$$

$$3\,NO_2(g) + 3\,H_2O(l) \rightleftharpoons NO(g) + 2\,H_3O^+(aq) + 2\,NO_3^-(aq) \qquad (7\text{--}5)$$

Now, if we save some of the ammonia and use the rest to manufacture that solution of nitric acid, we can react ammonia and nitric acid together.

$$NH_3(g) + H_3O^+(aq) + NO_3^-(aq) \rightleftharpoons NH_4^+(aq) + NO_3^-(aq) + H_2O(l)$$

$$(7-6)$$

The product is a solution of ammonium ions, NH_4^+ (not to be confused with ammonia molecules), and nitrate ions in water. By evaporating the water off, we obtain a solid white crystalline product, ammonium nitrate, NH_4NO_3. Ammonium nitrate is a fertilizer with a high nitrogen content, useful in agriculture. Without it and other nitrogen containing fertilizers, all made at least in part from nitric acid, which comes from Haber's contribution to technology (and of course including other fertilizer components which contain phosphorus, potassium, and other elements), the output of the world's agriculture would be far less than the world's needs.

This is the story of **nitrogen fixation** the process by which nitrogen in the air is altered into a fixed or usable form. The Haber process involves a requirement for high pressures and reasonably high temperatures, as well as a formidable arrangement of industrial equipment occupying acres of land. On the other hand leguminous plants, such as clover, alfalfa, soy beans, and others, interact with nitrogen-fixing bacteria which infect the roots of such plants. By means of a series of reactions which are not yet completely understood and are probably much more complicated than the reactions described by equations in this chapter, the bacteria are able to convert nitrogen in the soil (that is, diffused into the soil from the air) into nitrate ion and other nitrogen containing substances, thus fertilizing the soil (in which they are grown naturally) with nitrogen. Eventually, this process, which does not require high pressures, high temperatures, or elaborate equipment (at least not by the bacteria) will be elucidated, and the Haber process may be superceded (see Chapter 11). High pressures, high temperatures, and elaborate equipment are costly; a process which does not involve some or any of these requirements is almost certain to be less costly.

The promise of possible low-cost manufacture of nitrogen containing fertilizers, and therefore of increased profits until economic factors force a lower price, today justifies the expensive research on nitrogen fixing bacteria in attempts to learn their "secret" or to otherwise simulate the results they achieve. Currently, and without much publicity, research in the bacterial fixation of nitrogen is being carried out in at least a dozen different competitive industrial research laboratories and in some university laboratories as well.

The production of hydrogen by steam reformation

As we know, hydrogen can be obtained by the electrolysis of water, but the process uses large amounts of electricity for a comparably small amount of hydrogen (and oxygen). As a result water is electrolyzed to obtain hydrogen only in a few places in the world where electricity is inexpensive. It is generally more economical, although more wasteful of natural resources, to obtain hydrogen in other ways. One of these is called steam reformation. Crude oil is mostly composed of compounds of carbon and hydrogen. One of these is propane, C_3H_8, a gas under ordinary conditions, which will serve as our example (other compounds from crude oil can be and are used in steam reformation). The process is operated at temperatures high enough such that water is a gas, steam.

$$C_3H_8(g) + 3\ H_2O(g) \rightleftharpoons 3\ CO(g) + 7\ H_2(g) \tag{7-7}$$

One of the products, carbon monoxide, CO, reacts with more steam.

$$3\ CO(g) + 3\ H_2O(g) \rightleftharpoons 3\ CO_2(g) + 3\ H_2(g) \tag{7-8}$$

The production of hydrogen and carbon dioxide, CO_2, is favored in this dynamic equilibrium by removing the carbon dioxide from the mixture of gases. This is achieved by treating the whole business at frequent intervals with water. Of all the substances present, only carbon dioxide is soluble in water. Thus it is removed by solution and the continued formation of hydrogen is promoted.

Notice that the hydrogen that is produced comes from the hydrogen in the propane molecules as well as from the hydrogen in the steam (gaseous water) molecules. Water is more plentiful than propane and it is therefore ecologically desirable to use water only as the original source of hydrogen atoms. This is particularly valid since, ultimately, after one or more further cycles of use the hydrogen will recombine with oxygen and form water again.

Hydrogen from water without electrolysis

Several proposals have been made in recent years for processes that use only water as the source of the hydrogen atoms but are likely to be less expensive than electrolysis. One of these is based upon iodine.

Iodine is a good oxidizing agent; it will react with almost anything to increase its oxidation number, while the iodine is itself reduced to a lower oxidation number. Thus, iodine is an excellent antiseptic. If your skin is cut, apply tincture of iodine (iodine dissolved in alcohol); the atoms in any infectious bacteria molecules will be oxidized, and this will kill the bacteria. The trouble is that the iodine is not very particular, it will oxidize atoms in your own tissue molecules also, thus prolonging the time of healing. Our point here, of course, is that iodine will also indirectly oxidize the oxygen in water from -2 to zero. This of course frees the hydrogen, and thus produces the desired result. The whole proposed process involves a series of reactions designed to make the production of hydrogen and oxygen controllable.

The first two reactions are preliminary; as you can see, no hydrogen is produced. The reactants are, besides iodine, lithium ion and hydroxide ion, both in solution, for the first step. In this step, iodine is heated and is a liquid; at ordinary temperatures iodine is a purplish-black solid.

$$3 \, I_2(l) + 6 \, Li^+(aq) + 6 \, OH^-(aq) \rightleftharpoons 6 \, Li^+(aq) + 5 \, I^-(aq) + IO_3^-(aq) + 3 \, H_2O(l)$$
$$(7-9)$$

The products are lithium ion, Li^+ (called a "spectator ion" because it seems to remain unaffected), iodide ion, I^-, and iodate ion, IO_3^-, along with water. Next, alcohol is added to the product mixture and lithium iodate, $LiIO_3$, which is not soluble in alcohol, separates out as a white crystalline solid. (The alcohol is evaporated, then condensed back to a liquid, ready for use again.) The lithium iodate is dissolved in water and a solution of potassium iodide, KI, is added.

$$Li^+(aq) + IO_3^-(aq) + K^+(aq) + I^-(aq) \rightleftharpoons KIO_3(s) + Li^+(aq) + I^-(aq)$$
$$(7-10)$$

At temperatures near the freezing point of water, to which this solution is now cooled, potassium iodate, KIO_3, is only slightly soluble; it separates out, **precipitates** is the fancy word, as a white, crystalline solid.

When heated to about 600°C, potassium iodate decomposes, forming potassium iodide and oxygen.

$$KIO_3(s) \rightleftharpoons KI(s) + \tfrac{3}{2} O_2(g) \qquad (7-11)$$

The oxygen is an almost unnecessary by-product, although uses can be found for it. The idea is really to regenerate potassium iodide for use in the second step, equation (7–10), since the other products in the second step, lithium ions and iodide ions in solution, are the key for getting hydrogen, as we will eventually see. For now, at high temperatures the lithium and iodide ion solution will react with steam to produce gaseous hydrogen iodide, HI, and lithium ions and hydroxide ions in solution.

$$6 \, Li^+(aq) + 6 \, I^-(aq) + 6 \, H_2O(g) \rightleftharpoons 6 \, Li^+(aq) + 6 \, OH^-(aq) + 6 \, HI(g)$$
$$(7-12)$$

(Actually, at the temperatures employed here, it would be correct, but perhaps confusing, to point out that the water of the lithium and iodide ion solution has evaporated and the lithium iodide and lithium hydroxide, symbolically indicated as ions in solution, are really molten liquids.)

In any event, the hot mixture of gaseous hydrogen iodide and steam is cooled, which causes the hydrogen iodide to dissolve, as ions, in the water from the condensed steam. This solution is caused to flow over metallic nickel. (In solution a hydrogen ion, H^+, is symbolized more realistically as a hydronium ion, hydrogen ion "plus" water, H_3O^+.)

$$6\,H_3O^+(aq) + 6\,I^-(aq) + 3\,Ni(s) \rightleftharpoons 3\,Ni^{2+}(aq) + 6\,I^-(aq) + 3\,H_2(g) \tag{7-13}$$

Thus, hydrogen is produced. One more step is necessary to regenerate the iodine for use in the first step (equation 7–9) and the nickel for use in the preceding step (equation 7–13). The water is evaporated from the solution of nickel and iodide ions, and the solid, very dark green crystals are heated to produce nickel metal and iodine.

$$NiI_2(s) \rightleftharpoons Ni(s) + I_2(g) \tag{7-14}$$

The gaseous iodine is cooled slightly to form liquid iodine, ready again for the first step. You will notice that overall, the only substance that is not recycled is water; in the whole cycle, then, water is decomposed to form oxygen and hydrogen. The use of the double arrows throughout indicates a dynamic equilibrium; by removing various products or by simply adjusting the temperature, high or low, each step is controlled to the desired products.

Notice that most of the steps, the first, third, fifth, and sixth reactions, are oxidation-reduction reactions. The first, second, and fifth reactions are exoergic; the other three are endoergic. Since the overall result is the decomposition of water, which is in itself an endoergic reaction, we know that the heat required for the three endoergic reactions is greater than the amount of heat released in the three exoergic reactions.

REACTIONS THAT USE OR PRODUCE ELECTRIC CURRENT
Removal of silver tarnish

After all that we can relax a bit with a homely, practical example: how to get rid of the tarnish on silverware. In the air, as a result both of natural and human activities, there are trace amounts of sulfur compounds. These will react with all sorts of other substances, including silver. The product in the case of silver is silver sulfide, Ag_2S, a black solid that sticks like anything to the silver underneath. There are three ways to get rid of this black tarnish on silver. Scrape it off; this risks scratching the relatively soft silver surface. Dissolve it off; this means that along with the sulfide part, you will also remove the silver and eventually harm the silver object. Besides, the best solvent involves cyanide ion, CN^-, a dangerous poison. You have guessed the third, use an oxidation reduction reaction; this will restore the silver to its original, metallic condition.

As we know if something is to be reduced in an oxidation-reduction reaction, something has to be oxidized. In some of the examples we have used thus far, when a substance is oxidized it is altered quite a bit. In some cases, at least, it becomes useless. In this case we must find a substance that can be oxidized and about which we don't care very much. Our choice is aluminum foil. We will reduce the silver in the silver sulfide from +1 to zero (the oxidation number of metallic silver) and oxidize some aluminum foil from zero in the foil to +3, as a dissolved ion.

$$3 \; Ag_2S(s) + 2 \; Al(s) \rightleftarrows 2 \; Al^{3+}(aq) + 3 \; S^{2-}(aq) + 6 \; Ag(s) \qquad (7-15)$$

Put some hot water in a pan, preferably an enameled pan or a glass bowl. (An aluminum pan may be used, but you then have a slight risk of developing a hole in the aluminum pan. The aluminum atoms in the pan might get oxidized instead of the aluminum atoms in the foil.) Dissolve some baking soda in the hot water; a few spoonfulls is enough. Push a big peice of aluminum foil clear to the bottom of the hot solution; the foil piece should be big enough to wrap the tarnished silver in completely. Be careful that none of the foil sticks up out of the water and work any bubbles out from entrapment in the foil. If you have any bubbles caught in the foil, probably some places on the tarnished silver-ware will remain tarnished and you'll have to do the whole thing over again. Lay the tarnished silverware on top of the foil, under the water, avoiding any entrapment of bubbles. It does no harm if the foil gets punctured here and there. Wrap the rest of the foil around the silverware. Keep the whole thing hot (it works if you don't, but more slowly); inspect the silverware from time to time; remove when the tarnish is gone. Generally, depending upon the degree of tarnish, the process requires from 1 to perhaps 5 hr. The procedure will be unsuccessful on any piece of tarnished silver that does not touch the aluminum foil or does not at least touch another piece of silverware that is touching the foil, and so on; there must be metal to metal contact.

Oxidation and reduction half-reactions

What is going on? To see this, we will digress. You need a silver plated or sterling silver spoon and an iron nail, large size. Clean both since they are going into your mouth. Put the end of the nail (either end) along one side of your tongue, near the back of your mouth. The nail must be big enough so that the other end sticks out of your mouth. Put the spoon similarly along the other side of your tongue, near the back. The other end of the spoon sticks out of your mouth, too, of course. Now, keeping the nail and spoon about where they are, in your mouth, touch the nail and spoon together firmly outside your mouth. The result, which you can taste, is an oxidation-reduction reaction.

On one side of your tongue, iron atoms are oxidized and dissolve in your saliva. The electrons thereby released remain in the nail. If the nail is touching the spoon, those electrons travel over into the spoon.

$$Fe(s) \rightarrow Fe^{2+}(aq) + 2 \; e^- \qquad (7-16)$$

At the other side of your tongue, those electrons leave the spoon and interact with the water in your saliva and some of the hydrogen atoms in that water are reduced, forming hydroxide ions as the other product.

$$2\,e^- + 2\,H_2O(l) \rightarrow H_2(g) + 2\,OH^-(aq) \tag{7-17}$$

On the one side of your tongue, you taste iron ions, Fe^{2+}; on the other side you taste hydroxide ions, OH^-. Not enough gaseous hydrogen is produced to be able to sense the bubbles of this gas.

Equation (7–16) shows only an oxidation, of the iron. Since the other half is not symbolized, this equation is called an oxidation half-equation. Equation (7–17) shows only a reduction; it is a reduction half-equation. The two can be "added," cancelling out the electrons, according to the rules used, to show an ordinary oxidation-reduction equation.

$$Fe(s) + 2\,H_2O(l) \rightarrow Fe^{2+}(aq) + H_2(g) + 2\,OH^-(aq) \tag{7-18}$$

There is a dynamic equilibrium here, as usual. However, it is not symbolized because, at equilibrium, when the opposed rates are equal, there is a great deal more product than original material. That is, if you keep that nail in your mouth long enough, it will corrode away to practically nothing but iron ions.

Batteries

We will get to corrosion later; right now we should notice a fact. There was no taste of iron ions, nor of hydroxide ions, until after the nail and the spoon were brought into contact to allow the electrons to flow from the nail into the spoon. If instead of allowing the two to touch, you had joined the nail to a piece of wire and the other end of the wire to the spoon the electrons would have flowed out of the nail, into the wire, and then into the spoon with the same result in your mouth. The combination, nail, your mouth, spoon constituted a small electrical cell or **battery**. (Strictly, a battery is two or more cells.) All electrical batteries operate on this principle, an oxidation-reduction reaction that can be physically separated into two halves, an oxidation half and a reduction half.

That 9-volt (V) radio battery we used in the first part of this chapter consisted of two separable parts. In one part this oxidation half-reaction took place:

$$Zn(s) \rightarrow Zn^{2+}(aq) + 2\,e^- \tag{7-19}$$

In another part, this reduction half-reaction occurred:

$$2\,e^- + 2\,MnO_2(s) + 2\,H_2O(l) \rightarrow 2\,MnO(OH)(s) + 2\,OH^-(aq) \tag{7-20}$$

The product, $MnO(OH)$, is usually called basic manganese oxide.

These two half-reactions will push electrons out of the negative marked terminal of a cell and take them in at the positive marked terminal with an electrical force of 1.5 V. If six such cells are joined (in series) electrically, the total overall electron push-pull is six times as great or 9 V.

The reactions described here apply to the so-called "dry-cell" flashlight battery, which is rated as 1.5 V, the 9-V transistor radio battery, and several other commercially available cells and batteries.

A different pair of oxidation and reduction half-equations are involved in the physically separate regions of the cells in an automobile storage battery, of a "ni-cad" battery (nickel-cadmium storage battery typically used to power small calculators as "rechargeable" batteries), or of a tiny "mercury cell" used typically in electric watches and in some cameras.

CORROSION

We will conclude this excursion into oxidation-reduction reactions with a quick look at corrosion.

Corrosion of iron

Consider, for example, the steel body of an automobile. The iron (the steel) is protected from oxidation by a coat of paint. If the paint is chipped off, corrosion begins. As we have seen, in the presence of water iron is oxidized, so all we need is a drop of rain water, a splashed drop from a puddle, or any old water.

$$Fe(s) \rightarrow Fe^{2+}(aq) + 2\,e^-$$

The electrons from this process remain in the iron metal, of course. Unless they are removed the oxidation will stop almost as soon as it starts. That is, if further oxidation occurred, the iron would acquire a fairly high negative charge due to all those electrons and therefore attract back to it the positively charged iron ions, Fe^{2+}, reducing them, and dynamic equilibrium would be achieved.

$$Fe(s) \rightleftharpoons Fe^{2+}(aq) + 2\,e^-$$

On the other hand, if the electrons are used up, say by a reduction process, the corrosion of the iron can continue. As it happens, oxygen is dissolved in almost all natural waters. There is probably some dissolved oxygen in that drop of rain water sitting on top of the chipped place in the paint coating.

$$2\,e^- + \tfrac{1}{2}O_2 + H_2O(l) \rightarrow 2\,OH^-$$

Until that oxygen which is dissolved in that drop of water is consumed, the corrosion of the iron will continue. The next time it rains or water is splashed on to the chipped spot, more corrosion. The obvious prevention is either to move to the desert or to reapply a coat of paint over the chipped spot.

Corrosion of tin

The same corrosion mechanism is applicable to tin cans. A tin can is actually mostly steel. Some tin cans have a coating of special varnish on the inside surface that comes in contact with the edible contents; some have the inside surface coated with tin. Almost all "tin cans" are coated with tin on the outside surface. The tin coating is very thin.

Tin is used as a protective coat because it is nontoxic, sticks tightly to steel, is inexpensive to apply, and develops a coating of tin(IV) oxide that adheres to the tin as an impervious coating and prevents further oxidation of the tin. But, if the tin coating is damaged, by cutting the can open, for example, the corrosion of the underlying iron proceeds whenever it is wet with water containing dissolved oxygen.

Corrosion of aluminum

Aluminum cans are used as food and beverage containers in some instances. Aluminum is much like tin; both are readily oxidized, but both form an impervious adherent coating of oxide, aluminum oxide or tin(IV) oxide, that prevents further oxidation or corrosion. As it happens, the aluminum oxide coating on aluminum is soluble in alkaline media, so aluminum cans cannot be used for foods which are immersed in an alkaline water solution. The baking soda you used in conjunction with aluminum foil to remove tarnish from silverware forms an alkaline solution in water. This dissolves the oxide coating and permits the aluminum in the foil to be oxidized while the silver in the silver sulfide is reduced. (We will get to alkaline solutions in more detail later in this chapter.)

Preferential corrosion

Let's get back momentarily to the nail and spoon in your mouth. We learned that in this set-up iron was oxidized. Why is that? Why was it not the silver in the spoon that was oxidized? Answer: It was, but not much.

We have two possibilities:

$$Ag(s) \rightleftharpoons Ag^+(aq) + e^- \qquad (7\text{--}21)$$

$$Fe(s) \rightleftharpoons Fe^{2+}(aq) + 2\,e^- \qquad (7\text{--}22)$$

As it happens, when dynamic equilibrium is established producing equal and opposed rates for the silver–silver ion system, there are very few silver ions (and freed electrons) compared to the much larger concentration of iron ions (and those freed electrons) in the iron–iron ion system. We can express this comparison more simply by saying that iron has a much greater tendency to be oxidized than silver does.

Of all the metals, copper, silver, and gold in group IB of the periodic table have little tendency to be oxidized or corroded. Of the three, copper is the easiest to oxidize and gold the most difficult. This property of gold, along with its relative scarcity in the Earth's crust, is responsible for the high value makind has placed on it. Once you have some gold, it will not corrode; it will "last forever." Only a few other metals are equally or more resistant to oxidation than gold: platinum, iridium, rhodium, and palladium in group VIIIB. Actually, of all the twelve metals in groups VIIIB and IB, only iron is noted for its relative ease of oxidation, with nickel and cobalt a distant second and third.

Occasionally, in the construction of buildings and homes you will note instances where iron nails have been used to secure aluminum siding to the building wall. When iron and aluminum are in good contact, which permits the flow of electrons from one to the other, the corrosion of the iron nails is enhanced. (Actually, aluminum is much easier to oxidize than iron, but the impervious oxide coating on the aluminum prevents this.) If copper nails are used to secure the aluminum, the aluminum will corrode slowly. Copper is more difficult to oxidize than aluminum even when the aluminum is coated with oxide. Sometimes in plumbing installations copper pipe is fastened to iron pipe; in such cases it is only a matter of time until the iron pipe corrodes away.

Prevention of corrosion

We have now considered three of the four major ways of preventing corrosion. Coat with paint, and repair any chips or holes at regular intervals. Rely on an adherent, impervious oxide coating; recently special steel alloys have been developed containing iron and carbon, with some phosphorus, nickel, chromium, and silicon which form such a coating of iron oxide. Stainless steel is an alloy of iron, nickel, and chromium that forms a protective coating of chromium(III) oxide. The third strategy is to coat the more easily oxidized metal with a less easily oxidized metal, such as tin over steel in tin cans.

This third method leads to a fourth. In the instance of tin coated steel, the iron is oxidized and the tin remains unaffected. If we were interested in preventing the corrosion of tin, it would make sense to coat a piece of tin with iron. In a corrosive situation, the tin would be preserved as long as any iron remained unoxidized. The iron would be "sacrificed" in order to save the tin. In practice, iron is protected from corrosion by coating *it* with a more easily oxidized metal. Typically, zinc is used. The product is called galvanized iron. As long as any unoxidized zinc remains, the iron in contact with that zinc will not be oxidized. Eventually, of course, the zinc will be gone. Only then will the iron be attacked. A thick zinc coating will protect iron for many years under usual conditions.

ACID-BASE REACTIONS

We will conclude this chapter with a consideration of another important, pervasive class of chemical reactions. You have already had a little experience with this type of reaction back in Chapter 3, if you poured a little vinegar on some baking soda. (Now would be a good time to try that if you missed doing it then.) Vinegar is a solution, a mixture of acetic acid, CH_3COOH, and water. In this mixture there are small amounts of other substances that impart specific flavors to vinegars, different in, say, cider vinegar compared to wine vinegar. White, or distilled, vinegar is almost pure acetic acid and water.

Acids and bases An **acid** is definable in different ways; for us the most useful idea is to think of an acid as a somewhat reluctant donor of protons, symbolized by H^+, that is, a hydrogen atom that has no electron. Complimentary, or **conjugate**, to an acid is another kind of substance, a **base**. It is helpful to think of a base as voracious; a molecule of base will steal, take, grab, any such strong word, a proton from an acid.

Some acids donate their protons more readily than others. Acids that donate their protons without much fuss, that is, easily, are called strong acids. Think of them, perhaps, as generous, and of a kind of "virtue" or strength in generosity.

There are many bases known as **strong bases**. Unfortunately, the word, strong, applied to bases does not carry the complimentary connotation. Most strong bases are indeed vigorous proton grabbers, but so are some weak bases. On the other hand, weak acids are not generous with protons. The words, strong and weak, as originally applied to acids and bases were related to the relative ease by which their solutions conducted an electric current, not to proton donation or taking. As a result, the commonly used terms strong acid, weak acid, strong base, weak base, tend to be confusing when we attempt to relate them to proton interactions. So, although you will find these terms used in other books, and properly so, we will use different terms in our discussion here. We will speak of generous acids which donate their proton or protons readily, to any old base. The other extreme would be a stingy acid. We will speak of greedy bases, which would take a proton from their own grandmother, and the other extreme, indifferent bases, which practically have to have a proton forced upon them.

Stingy acids Now for vinegar. Acetic acid is a stingy acid. Water is an indifferent base. Equation (7–23) describes what happens. (Only the H symbolized on the right end of the acetic acid formula is donatable.)

$$CH_3COOH(aq) + H_2O(l) \rightleftharpoons CH_3COO^-(aq) + H_3O^+(aq) \qquad (7\text{–}23)$$

Given the indifference of the water and the stinginess of the acetic acid, we don't expect much. Under conditions of dynamic equilibrium in vinegar, there are relatively few acetate ions, CH_3COO^-, and hydronium ions, H_3O^+, compared to acetic acid and water molecules.

We can look at this from another point of view. Acetate ions, CH_3COO^-, are very very greedy bases. Hydronium ions, H_3O^+, are generous acids. The net result is the same; not many acetate and hydronium ions compared to the molecules. This is fortunate in a way, because any solution with a high concentration of hydronium ions is poisonous. Vinegar is not poisonous, and we can enjoy the slightly sour taste of vinegar, due largely to those few hydronium ions, as a condiment.

Indifferent bases

Next, baking soda, sodium hydrogen carbonate, $NaHCO_3$. This is an ionic compound; the solid crystalline particles of baking soda consist of a three-dimensional lattice of sodium ions, Na^+, and hydrogen carbonate ions, HCO_3^-. Within the hydrogen carbonate ions themselves, as we know, the atom to atom bonds are covalent. When soda is dissolved in water, we get in solution

$$Na^+(aq) \quad \text{and} \quad HCO_3^-(aq)$$

Sodium ions are neither acidic or basic. You can tell that they are not acid because there is no proton for them to give up. (We ignore the protons inside the sodium nucleus because they are tightly bound in that nucleus.) Only by laboratory observation of its properties is it possible to discern whether the hydrogen carbonate ion is acidic or basic. It turns out to be both. In the presence of a base in the same solution, the hydrogen carbonate ion is a stingy acid.

$$X(aq) + HCO_3^-(aq) \rightleftharpoons XH^+(aq) + CO_3^{2-}(aq) \qquad (7\text{--}24)$$

In equation (7–24), X represents a base; only if X symbolizes an unusually greedy base would we expect at dynamic equilibrium to find a high concentration of XH^+ and CO_3^{2-} ions.

In the presence of an acid, even a stingy one, the hydrogen carbonate ion acts like a fairly greedy base.

$$YH(aq) + HCO_3^-(aq) \rightleftharpoons Y^-(aq) + H_2CO_3(aq) \qquad (7\text{--}25)$$

In equation (7–25) YH symbolizes any acid. The more generous the acid, the more we would expect to find at dynamic equilibrium a high concentration of products.

Reactions between stingy acids and indifferent bases

As we know, acetic acid is a stingy acid. Even so, it reacts with hydrogen carbonate ion.

$$CH_3COOH(aq) + HCO_3^-(aq) \rightleftharpoons CH_3COO^-(aq) + H_2CO_3(aq)$$

As it happens, hydrogen carbonate, also correctly called carbonic acid, is an unstable compound. It tends to decompose spontaneously into water and gaseous carbon dioxide.

$$H_2CO_3(aq) \rightleftharpoons H_2O(l) + CO_2(g)$$

If the gas, carbon dioxide, can escape, there is no possibility for dynamic equilibrium. So, when vinegar is poured over baking soda, we observe a fizzing, a bubbling, the evolution of gaseous carbon dioxide. In the liquid remaining behind there are dissolved sodium and acetate ions, and, perhaps, some as yet unused acetic acid and a few acetate and hydronium ions, these latter in dynamic equilibrium.

Soda water is a solution of carbon dioxide dissolved in water. Some, not all, of the dissolved carbon dioxide reacts with the water. Under conditions of dynamic equilibrium, such as would exist in a sealed bottle of soda water, not too much hydrogen carbonate is present.

$$CO_2(g) + H_2O(l) \rightleftharpoons H_2CO_3(aq)$$

As we know, hydrogen carbonate is an acid; it is a fairly stingy acid. Water is a base, an indifferent base. Therefore, under conditions of dynamic equilibrium we will find only a low concentration of product.

$$H_2CO_3(aq) + H_2O(l) \rightleftharpoons HCO_3^-(aq) + H_3O^+(aq)$$

The concentration of hydronium ion in soda water is low after the bottle is opened. Much of the hydrogen carbonate is decomposed into carbon dioxide and water, which thus lowers the hydronium ion concentration quite a bit from what it was in the sealed bottle. But even so, the sour taste of soda water is attributable to this low concentration of hydronium ion.

Generous acids   Few people have had any personal contact with generous acids, and this is just as well because they are hazardous to handle without prior instructional cautions. The generous acid you are most likely to have heard of is muratic acid. This is an old name for hydrochloric acid, or a water solution of hydrogen chloride, HCl.

Hydrogen chloride is a very generous acid, and water is a base, as we know. The dynamic equilibrium that ensues when hydrogen chloride is dissolved in water results in almost all product; the concentration of hydrogen chloride remaining as such in the mixture is so low that chemists generally show only one arrow in the equation.

$$HCl(aq) + H_2O(l) \rightarrow H_3O^+(aq) + Cl^-(aq)$$

Another way of saying the same thing is to allude to the fact that chloride ion, Cl^-, is a completely indifferent (almost) base. We could say that hydrogen chloride is a generous acid "because" chloride ion is such an indifferent base. This is analogous to saying that acetic acid is a stingy acid "because" the acetate ion is a greedy base. The hazard in working with hydrochloric acid, as we can now recognize, is due to the high concentration of hydronium ion in typical solutions of hydrogen chloride. The same conclusions apply to other generous acids, such as sulfuric acid, H_2SO_4 (rarely, though also correctly named hydrogen sulfate), nitric acid, HNO_3 (other rarely used name, hydrogen nitrate), and a few others.

Some household bleaches contain hydrochloric acid, in a relatively low concentration. Hydrogen chloride is itself not a bleaching agent; it is present in such bleaching solutions as a kind of chemically unavoidable side effect. These bleaching solutions contain hypochlorous acid, $HOCl$, which is both a reasonably generous acid and an oxidizing agent. (The oxidizing agent character is responsible for the bleaching action; soils and stains are oxidized to colorless products.)

Greedy bases

Ammonia, NH_3, is a base. Ammonia water is a common household cleansing agent made by dissolving ammonia in water. In this solution, water acts as an acid. (Water, it now turns out, is an indifferent base or a stingy acid, depending upon what is mixed with it.) Equation (7–26) shows the reaction

$$NH_3(aq) + H_2O(l) \rightleftharpoons NH_4^+(aq) + OH^-(aq) \qquad (7–26)$$

Since hydroxide ion is such a greedy base and ammonium ion, NH_4^+, is a reasonably generous acid, the concentrations of ammonium ion and hydroxide ion are low under dynamic equilibrium, compared to the concentrations of ammonia, NH_3, and water.

Reactions of generous acids and greedy bases

However, if we mixed a little of the bleaching solution containing hydrochloric and hypochlorous acids, that is, a solution which contains plenty of H_3O^+ ions, with a little ammonia water,* we get a lot of effect. That is, we are mixing a generous acid, hydronium ion, with a very greedy base, hydroxide ion, OH^-. The reaction is exoergic, without question!

It is true enough that there is only a low concentration of hydroxide ion in the ammonia water. But as the little that is there is consumed by reaction with the hydronium ion,

$$H_3O^+(aq) + OH^-(aq) \rightleftharpoons 2 H_2O(l) \qquad (7–27)$$

more hydroxide ion is formed as more ammonia reacts with water. This additional hydroxide ion of course then reacts with some of the not yet reacted hydronium ions.

Further, even though ammonia is a relatively indifferent base, it will certainly react with a generous acid, such as hydronium ion. So, in addition, the following very exoergic reaction shown in equation (7–28) occurs.

---

* **CAUTION:** Never mix household bleaches containing chlorine with ammonia water. In addition to the reactions discussed here, other reactions also occur. Some of the products of those other reactions, such as NCl_3, $NHCl_2$, and N_2H_4, are either very poisonous or explosive, or both.

$$H_3O^+(aq) + NH_3(aq) \rightleftharpoons NH_4^+(aq) + H_2O(l) \qquad (7\text{-}28)$$

In both reactions (7-27) and (7-28) hydronium ion is such a generous acid and, in the one case, hydroxide ion is so greedy a base, that at dynamic equilibrium the concentration of products is high compared to the concentration of the original substances. Of course, this result is less pronounced in the reaction of hydronium ion and ammonia. But in the case of the hydronium and hydroxide ions, only a trace of these ions is present at dynamic equilibrium. And thereby hangs another tale.

Acid-base properties of water

Water is both an acid and a base. It is a stingy acid and an indifferent base, but it is nevertheless both an acid and a base. In the presence of an acid, water will (reluctantly) take a proton. In the presence of a base, water will (with some hesitation) give up a proton. What will a water molecule do in the presence of another water molecule?

You can almost hear their conversation. One way or another one water molecule will act as a base and the other as an acid.

$$H_2O(l) + H_2O(l) \rightleftharpoons H_3O^+(aq) + OH^-(aq)$$

As we already know, at dynamic equilibrium, the concentration is mostly water, and a very, very low concentration of hydronium and hydroxide ions. But, in any amount of water, large or small, this dynamic equilibrium is continuously occurring. If you now think of water as a dynamic liquid, in constant acid-base reaction turmoil, you will have the right idea.

pH

This brings us to pH. (More symbols, naturally.) The symbol, p, is an operator, just like the symbol, +, in arithmetic. The plus symbol in "2+3" tells us what to *do* with the 2 and the 3. The symbol, p, is more complicated in its symbolic instruction. It tells us to take the logarithm of a number to the base ten and to then multiply that logarithm (which is a number, too) by -1. For example, the logarithm of 100 to the base ten is 2. So, p100 is -2. (Take the logarithm of 100, which is 2, and multiply it by -1 to get -2.) Similarly, if you know your logarithms, p1000 is -3. And p0.01 is 2; p0.001 is 3. (The logarithm to the base ten of 1000 is 3, of 0.01 is -2, and of 0.001 is -3.) The symbol pH directs us to find the concentration of hydronium ions, take the logarithm to the base ten of that concentration, and multiply the resulting number by -1.

In the laboratory, when we measure the concentration of hydronium ion in pure water, the number comes out to be 0.0000001 mole of hydronium ions in a liter (about a quart) of water. The logarithm to the base ten of this number is -7; so, multiplying by -1, the pH is calculated to be 7. The pH of pure water (at ordinary temperatures) is 7.

The pH of acids and bases

If we have a sample of vinegar, the acetic acid present reacts with the water to make some "extra" hydronium ions (and incidentally thereby decreases the hydroxide ion concentration, but we won't go into that). The pH will be some number other than 7. Laboratory measurements show that the hydronium ion concentration in a typical vinegar is about 0.0035 moles of hydronium ion in a liter of vinegar. The logarithm to the base ten of this number is -2.45 (rounded off a bit), so the pH of vinegar is typically 2.45. All acid solutions have pH's less than 7. Conversely, all solutions of bases in water, with a lower concentration of hydronium ions than pure water have pH's larger than 7.

pH and the living organism

Whole human blood is composed of red cells, leucocytes, platelets, and lipid (fatty) droplets, all supported by a watery solution called blood plasma. In a normal, resting human, the pH of this blood plasma solution typically ranges between 7.39 and 7.41. (This corresponds to a hydronium ion concentration of 4×10^{-8} moles/liter of blood plasma.) The pH of the blood plasma in venous blood is a little lower (more acidic, a slightly higher concentration of hydronium ions) than for arterial blood; typically venous blood has a pH of about 7.37. If the pH of the blood plasma exceeds these limits sufficiently, severe harm, even death, results. Vigorous exercise, for example, running up a stairway from the ground floor to the fourth floor of a typical building in, say, less than 50 sec reduces the pH of the blood plasma to about 7.2 a few minutes after reaching the fourth floor. About 2 hr are required to restore the pH to 7.4 or thereabouts. Hyperventilating or overbreathing, that is, rapidly breathing in and out, will produce a slightly dizzy feeling, a noticeable headache, and a blood plasma pH of about 7.5. Continue to hyperventilate and unconsciousness will result.

The control of pH. A pH range from 7.2 to 7.5 is not large, and the typical level that is usual is even more restrictive, 7.39 to 7.41. Yet we constantly assault our systems with various stresses and do not suffer harm. For example, the process of digesting various foods produces stress. Meat, cheese, bread, eggs, and potatoes all tend to lower the blood plasma pH; these foods are acidic. Basic foods include beans, cabbage, and citrus fruits, to name a few. Exercise and overbreathing also stress the pH control within our blood plasma. How is this control effected?

The control is obtained mostly by the presence of dissolved carbon dioxide, carbonic acid, and hydrogen carbonate ion in the blood plasma. Table 7–2 lists the major substances dissolved in blood plasma, with their typical concentrations in moles per liter. In addition to the dissolved carbon dioxide, carbonic acid, and hydrogen carbonate, other components such as the proteinaceous anions, the phosphate ions (both in the plasma), and the hemoglobin and oxyhemoglobin of the red cells play a part in controlling the pH of the blood plasma. However, in our discussion we will omit consideration of these others.

TABLE 7–2.   **Typical Substances in Blood Plasma**

Negatively Charged Ions (moles/liter)		Positively Charged ions (moles/liter)		Neutral Species (moles/liter)	
Chloride	0.103	Sodium	0.143	Carbon	
Hydrogen		Potassium	0.005	dioxide	0.0025
carbonate	0.027	Calcium	0.0025	Carbonic	
Proteinaceous		Magnesium	0.0015	acid	7×10^{-6}
anions	0.017	Hydronium	4×10^{-8}		
Other miscelleanous (phosphate, carbonate, hydroxide, and others)	0.009				

The dissolved carbon dioxide is in dynamic equilibrium with carbonic acid (water is also involved).

$$CO_2(aq) + H_2O(l) \rightleftharpoons H_2CO_3(aq)$$

Any base that gets into the blood plasma will react with either the hydronium ion which is present, thus changing the pH (and also changing the hydronium ion concentration, of course), or it will react with carbonic acid. Since there are about 200 times more moles of carbonic acid in the blood plasma than hydronium ions, as we can see from Table 7–2, it is more likely that the base will react with the carbonic acid.

$$X(aq) + H_2CO_3(aq) \rightleftharpoons XH^+(aq) + HCO_3^-(aq)$$

On the other hand, any acid that gets into the blood plasma will react either with some of the water present, thus increasing the concentration of hydronium ions by forming more of them (and thus changing the pH), or with the base, HCO_3^-. Roughly, there are about 2000 times more water molecules handy than hydrogen carbonate ions. But hydrogen carbonate ions are much more than 2000 times more greedy bases than water molecules, so the "foreign" acid tends to react with the hydrogen carbonate ions, thus preserving the pH at an approximately constant value.

$$YH(aq) + HCO_3^-(aq) \rightleftharpoons Y^-(aq) + H_2CO_3(aq)$$

Natural feedback mechanisms. When you exercise vigorously, carbon dioxide, which is a product of that muscular effort, is dissolved in the blood plasma at a rate that is momentarily greater than the rate of loss via the lungs. As a result, the concentration of carbon dioxide is greater than normal. In this reaction at dynamic equilibrium, then, more hydrogen carbonate than is normally present develops.

$$CO_2(aq) + H_2O \rightarrow H_2CO_3(aq)$$

We use only a rightward pointing arrow to emphasize the effect.

The "extra" hydrogen carbonate, in turn, tends to react more than normally with water, producing "extra" hydronium ions, thus lowering the pH.

$$H_2CO_3(aq) + H_2O(l) \rightarrow H_3O^+(aq) + HCO_3^-(aq)$$

Eventually, these effects are reversed and normal pH values restored.

On the other hand, if you overbreathe, you not only build up the oxyhemoglobin in the red cells but you also deplete some of the carbon dioxide that is supposed to remain dissolved in your blood plasma.

$$CO_2(aq) \rightarrow CO_2(g)$$

That decrease in the concentration of dissolved carbon dioxide promotes this next effect:

$$H_2CO_3(aq) \rightarrow H_2O(l) + CO_2(aq)$$

This tends to restore the "lost" dissolved carbon dioxide. The consequent decrease in hydrogen carbonate concentration promotes this third reaction:

$$H_3O^+(aq) + HCO_3^-(aq) \rightarrow H_2CO_3(aq) + H_2O(l)$$

And this symbolizes a decrease in hydronium ion concentration resulting in an increase in pH of the blood plasma brought about by overbreathing.

Failure of the mechanism in special circumstances. Overbreathing to increase the distance you are able to swim under water can be fatal. The chemical reasons are not yet fully understood, but we do know you deplete the concentration of carbon dioxide in the blood plasma by overbreathing.

$$CO_2(aq) \rightarrow CO_2(g)$$

Carbon dioxide, which was dissolved in the blood plasma is now exhaled excessively to the air, through the lungs. Ordinarily, breathing is a reflex action, and we feel that we simply must breathe if we deliberately hold our breath for longer than perhaps a minute or so. This reflex is triggered by an excess amount of carbon dioxide dissolved in our blood plasma. (How the trigger actually works is not fully understood.)

However, if you have overbreathed by taking too many long deep breaths in rapid succession, you have now reduced the carbon dioxide in your blood plasma way below normal. Meanwhile, here you are, swimming under water, exercising, building up that carbon dioxide in your blood plasma again. But if you overbreathed too much, the carbon dioxide you are now producing never gets up to the triggering level; hence, you do not feel so violent a need to breathe, but you have also, meanwhile, used up almost all of the extra oxygen you took in when you overbreathed.

With insufficient oxygen, especially in the blood in the brain area, death ensues. If you feel that you simply must try for a long distance record in underwater swimming, do not take more than three or four "good breaths" in rapid succession. And, while under the water, come up for air at the *first* slight feeling that you need more air; don't try for that last 10 ft more. You may have only 5 sec more of life left if you don't come up immediately.

SUGGESTIONS FOR DISCUSSION

1. Many but not all of the equations presented in this chapter describe oxidation-reduction reactions. For those that do, identify the species that is oxidized and the species that is reduced. (One of the reactions involving NO_2 may give you more trouble than anticipated.)

2. In your own words and as completely as possible, define the term dynamic equilibrium and cite at least one chemical and one nonchemical example of this concept.

3. List more than two reasons why it is more useful to you to know how to remove tarnish from silver than to know about the use of iodine and other substances in the decomposition of water to hydrogen and oxygen. List more than two reasons for the opposite comparison.

4. Without worrying excessively about producing the "right" answer, write chemical equations that describe what happens when tarnish is removed from silver in the presence of aluminum foil. Use an oxidation half-equation and a reduction half-equation.

5. Using one or more equations, describe what happens when an athlete deliberately overbreathes immediately prior to running a race, and then runs in that race. Do you recommend the practice of deliberate over-breathing just prior to vigorous exertion? Justify your answer.

6. In at least one paragraph, but not more than five paragraphs, list, describe, and summarize the uses of symbols in chemistry.

7. List, describe, and summarize the uses of symbols in an area different from chemistry with which you are familiar. Show similarities and differences in this use of symbols, comparing the other area of knowledge to chemistry.

8. Distinguish between an acid and a base, using several examples expressed in words or equations to illustrate your discussion.

9. (a) What are the formulas of the following compounds?

gallium chloride	mercury(II) perchlorate
rubidium oxide	titanium(III) acetate
cesium sulfide	titanium(IV) carbonate
boron bromide	chromium(III) hydrogen carbonate
boron nitrate	ammonium carbonate
boron sulfate	magnesium arsenate
barium sulfate	maganese(II) arsenite
zinc nitrite	phosphoric acid

(b) Name these compounds:

H_2SO_3	SnS_2	Ag_3PO_3
$CuC_2H_3O_2$	HNO_3	PbO_2
NH_4ClO_4	$Ca(NO_3)_2$	NiS
$Ti(Cr_2O_7)_2$	NH_4NO_2	SiO_2
$Cu(CN)_2$	HgS	$CoPO_4$
H_2SO_4	CO	$Cd(HSO_4)_2$
FeF_3	SbF_5	

10. Which would you expect to be more difficult, to write a complete and accurate one-sentence summary of this chapter or to write an essay of perhaps 3000–4000 words describing the information in this chapter? Justify your choice. Which, a single sentence or an essay, would be more useful to you, considering your personal reasons for taking this course? Explain your reasons for your choice. Now, either prepare a single sentence summary or a short descriptive essay.

Mezzologue

Although this is not quite the middle of this book, it is time for a few middle-words. The purpose of this book is to illuminate for you some topics that you might otherwise have chosen to overlook. Most textbooks, and this book is one, are written on the assumption that the student will begin on page one and continue serially to the last page, gaining understanding as they go along. It is further assumed that within some variations, each topic that is presented will be comprehended at about the same level of understanding.

Yet anyone who has read a book of any sort knows that those assumptions are generally false. Except for narrative expositions, a good mystery story, perhaps, or a fine novel, most people will dip and skip, choosing for their own reasons to read this intensely and to omit that entirely. This manner of using printed material is particularly exemplified by the way all of us use an encyclopedia; it is rare indeed to find someone who has read an encyclopedia from page one of volume one straight through to the last page of the last volume. Even if you were to do that, it is very likely that your knowledge, not to speak of your understanding, of the several topics would be markedly uneven.

This is a dip and skip textbook, and these middle-words are intended to help you get as much as you want, plus a little more, from your interactions with this book. To do this, a backward and a forward look is appropriate.

The first seven chapters have been about equally divided between a kind of standard chemistry textbook presentation and a discourse on the relation of chemistry to other areas of some importance. Chapters 3, 5, 6, and 7 are reasonably close to what is generally understood as "chemistry" by students (not their professors, necessarily) who have endured their first quarter or semester exposure to a traditional chemistry course. The major difference between that exposure and what has hopefully happened to students who have *studied* those four chapters in this book is that the traditional exposure generally includes a lot of algebra-arithmetic, called "stoichiometry." The other chapters preceding this mezzologue, Chapters 1, 2, and 4, attempted to show you why some people think chemistry is pretty darn interesting and how the concepts in chemistry do relate to some of your own interests.

Now, looking ahead, it is simply not possible to cover all of the interests students have and to show how each of these is chemically flavored. With the possible exception of Chapter 10, in which the approach blends chemistry with the interests of many students, and Chapter 15, which presents a kind of overview, the chapters that follow provide a sampling of topics with enough variety, it is hoped, to show you that almost any subject has chemical connotations.

Saying it differently, as you dip and skip in this book, you should expect to have more difficulty with Chapters 3, 5, 6, 7, and 10, and to enjoy the other chapters with less effort. There is no reason for me as the author to expect you even to glance at those five tougher chapters. They have been placed where they are, following and preceding the other chapters, for pedagogical reasons: You really ought to comprehend Chapters 3, 5, and 6 before you attempt Chapter 7, and it wouldn't hurt to have gone over Chapters 1, 2, and 4, as well. It would be even better, perhaps, to do them all in consecutive order.

However, notice that this kind of advice contains an implicit assumption: The way an author thinks a subject ought to be presented is therefore the best (or only) way to do it. This is patently false.

This kind of advice contains an even more, and terrifying, flaw: The purpose of getting an education is to become educated. One mark of educated people is that they each think for themselves, deciding for their own reasons to dip into this or to skip that. However, the experience of teachers demonstrates very clearly that some discipline is necessary in helping others to become educated, and the same experience also indicates that a lot of practice in picking and choosing is required before people can learn to dip and skip wisely on their own.

How then should you use this textbook? Dip and skip, under the guidance of your professor, for one thing. For another, since you also know that some discipline is in order, listen to the words of this book (that's not quite the same as reading those same words) and discover some of the reasons why the ordering of the topics is what it is. (I haven't discovered all of them yet myself.) Also, and this is very important, as you choose to skip here and dip in there, identify for yourself as specifically and as explicitly as you possibly can why you chose to study intensely or to overlook. Then, examine those reasons objectively, and decide to keep or to modify or reject those reasons.

That's what this book is really about—to help you discover and objectively examine your reasons. Chemistry is merely something to talk and think about as you undertake this never-ending task.

chapter 8

Energy:
Current Usage and
Future Limitations

Our first task in this chapter is to develop an awareness of energy and of its essential relation to society, today. Look around. Except for natural objects, such as trees, grass, and the clouds in the sky, every object you see has been affected by man-directed energy. The wooden furniture has been shaped, which requires energy. The metal in your surroundings came from the earth originally as an ore, which required energy to be mined, to be transported, to be extracted, and to be arranged into the shape it now has. The food you have recently eaten required energy to be produced, and transported, and prepared. The directed use of energy pervades every human activity.

A complete list would be excessive, but to make the point about the pervasiveness of our use of energy, each of these required energy, and each is useful, or necessary, or both, to all of us: Rubber, water available at the faucet, leather, cotton, wool, synthetic fibers, light bulbs, the emitted light itself, paint, cement, dyes, fertilizer, insecticides, pharmaceuticals, TV, radio, calculators, computers, tape players and recorders, soap and detergents, cosmetics, sewage treatment, explosives, propellants, wax, food additives, flavors, adhesives, fungicides, antifreeze, batteries, weed killers, analgesics, fire retardants, asphalt, corrosion preventatives, fabric finishes, anesthetics, lubricants, photographic devices and film, antiseptics, prosthetic devices, food preservatives, pest repellents, water softeners, bleaching agents, stain removers, dry cleaning fluids, fire extinguishers.

ENERGY
Definition of energy

What is energy? A good definition is discursive. Energy is not material, light is a form of energy; heat is a form of energy; the flow of electrons through wires is a form of energy (we call it electrical energy); energy can be stored in bonds between atoms, and released when a chemical reaction occurs (we could call this chemical energy). When gasoline burns in an atuomobile engine, the chemical energy in the bonds within the gasoline molecules and the oxygen in the air is released in the form of heat energy. Some of this heat energy is then transformed into energy of motion of the automobile and the passengers.

So, we have another form of energy, kinetic energy or energy of motion. Potential energy is still another form; no one will voluntarily stand under a person on a ladder who is holding an opened can of paint carelessly. Because of its height above the ground, the can of paint (and the painter too) have potential energy. In this case the potential energy would be converted into kinetic energy if the painter let go of the can. When the can, paint, and perhaps the painter hit the ground, their kinetic energy is converted into heat energy.

We also have light energy. Electrical energy from the power plant is converted partly into heat energy and partly into light energy within the light bulb. The light energy is eventually absorbed by the surfaces it hits and is converted into heat energy. The kinetic energy of a moving automobile and its occupants is converted into heat energy by applying the brakes. Just as the engine in an automobile can be thought of as an energy converting device (chemical energy into heat energy and part of that into kinetic energy), a brake is a device for converting kinetic energy into heat energy. The battery in a transistor radio converts chemical energy into electrical energy, and the radio converts the electrical energy into sound energy. The sound energy is eventually converted into heat energy when it is absorbed by the surfaces it hits.

We can define energy as that which is ultimately converted into heat, even though this is a bit circular because heat is one form of energy. We can define energy as the capacity for doing useful work, shaping a piece of lumber, moving an automobile, producing entertaining sounds from a radio, extracting a metal from its ore. We can describe, not define, energy as necessary for contemporary living. It is used in contemporary life in one form or another, as light, as kinetic energy, as chemical energy, as potential energy, electrical energy, and so on, all ultimately being converted into heat energy.

The first law of thermodynamics

The human experience with the various forms of energy is expressed in a general way by the first and second so-called laws of thermodynamics. The first law (which is not a "law" at all but a summary of human experience) can be stated this way: Under suitable conditions, any form of energy can be converted into any other form of energy, and then into still another form, and another, and another, and so on, and no energy is ever lost during conversion to another form.

The difficulty is that although the first law is undoubtedly correct, it is not very useful in the practical utilization of energy. According to the first law, we could convert the energy in sunlight directly into, say, electrical energy. Unfortunately no one knows how to do this efficiently in practice. The best we can do is to convert a few percent of it into electrical energy at very high cost. Theoretically, we could convert the chemical energy in gasoline and oxygen molecules directly into kinetic energy. The only way we know to do this involves an explosion like process that produces hazardous substances and is expensive besides. We can, and do, convert the potential energy in a reservoir of water into electrical energy, using the principle of the first law, and, except for frictional and other relatively minor losses, almost all of the potential energy of the water is converted into electrical energy. But the availability of suitable locations for water reservoirs is limited, and only a few percent of our total energy needs can possibly be supplied in this way.

The conversion of heat to useful energy: the second law of thermodynamics

In fact, given the entire history of man, only within the past 250 years has man developed a practical method for converting large amounts of energy from one form to another. This method involves the spontaneous flow of heat energy from a high temperature to a low temperature, and conversion of some of that heat to other forms of energy as it flows.

Let's translate that last paragraph. First, what is "heat at a high temperature?" Heat can be thought of as a quantity, an amount, of energy. Temperature can be thought of as an intensity, how much in a small, or large, space. Think about this imaginary experiment: Start with a bathtub full of water at 75°F and from it dip out one cup full of water. Now, take 50 matches and light them under the cup of water. The water in the cup will then be at some high temperature, say 120°F. Next, light 50 more matches under the bathtub of water, stirring the water while they burn. The bathtub water will be slightly warmer than it was, say 76°F. We have added the same amount of heat, by burning 50 matches in each case, to the cup of water and to the bathtub of water. The heat in the cup of water after the match treatment is at a higher temperature than the heat added to the bathtub of water. (You could say that the water in the cup is at a higher temperature too if you wish—but, although it is correct to say so, it doesn't get us where we want to go in this discussion.)

What will happen if we now lower that hot cup of water into the bathtub of water so that the rim of the cup is just above the water level in the bathtub and no water spills out of the cup? Easy, the water in the cup will get cooler. Heat, at the higher temperature in the water in the cup spontaneously leaves the water in the cup and enters the water, which is at a lower temperature, in the bathtub. The bathtub water will get slightly warmer, the water in the cup will cool down a lot. Eventually both will be at the same temperature.

Second law of thermodynamics. Statement one: Heat in an object at a higher temperature will spontaneously pass to any adjacent object, which is at a lower temperature, until the two are both at the same temperature. (The higher temperature object reaches a lesser, intermediate, temperature; the lower temperature object reaches a greater, same intermediate, temperature. Or, more succinctly: Heat at a higher temperature will pass spontaneously to a lower temperature.

Statement two: When heat passes spontaneously from a higher temperature to a lower temperature, some, but not all, of that heat can be converted into other forms of useful energy; the rest remains as heat at the lower temperature toward which it all "tried" to pass. Question: Why can't it all be converted into a useful form of energy, why does some of the heat *necessarily* achieve the lower temperature?

The answer is statement three of the second law: All the heat could, actually, be converted into a different, useful form of energy if the lower temperature to which it was going was absolute zero (−273.15°C, or −459.67°F.), but it is impossible for any real substance to have a temperature this low.

PRODUCTION AND USE OF ENERGY
Energy and the standard of living

A summary will be helpful here. We need energy in various useful forms and in the particular form of heat also in order to survive as human beings on this Earth. The entire history of man's development demonstrates an ever increasing use of energy and an ever improved condition of human life (though it is clear we have a long way to go, still). Beginning about 1700, and since then, we have found only one significant way to obtain large amounts of useful energy, other than heat energy. (We have known how to get heat energy for our purposes for thousands of years—merely burn wood or other fuels.) Since 1700 we have learned how to take the heat from burning fuels, at a high temperature, and by allowing it to flow to a lower temperature, extract a portion of it as other useful forms of energy, such as kinetic energy and electrical energy.

Clearly, if we are to make further progress, it will be necessary to obtain still more energy in useful forms. It is reasonable to predict that, if we had enough energy at our disposal, we could solve many of the other problems now facing mankind on this Earth. That is, our utilization of energy has provided us with all the things that surround us, our clothes, our food, fertilizer, asphalt, insecticides, transportation, water softeners, and all the rest mentioned or implied in the opening pages of this chapter. Given still more, much more, availability of useful forms of energy, we can expect to continue not only providing what is customary to those in the developed countries of the world but also to the others who now lack these amenities. The problem lies in the availability of energy.

A few facts and figures will illuminate the matter. The present world population is approximately 4×10^9 people. This total population now uses each year an amount of energy sufficient to boil the water in a lake one fourth the size of Lake Erie. This works out to an average usage equivalent to operating two 1000-W light bulbs continuously for each person alive. The United States, with an approximate population of 2×10^8 (less than 10% of the total) uses about one third of the annual energy available. This is equivalent, on a per person average, to the continuous operation of ten 1000-W light bulbs. Either we must resolve to make serious attempts to make available to others the benefits enjoyed in the developed countries, or, given the ever increasing world population trends, we must prepare to lose the standard of living to which we are now accustomed. If we choose to take the necessary steps to raise the standard of living in the developing countries, while maintaining the same level in the more fortunate countries, then about 100 years from now the total annual energy needs of the world will be equivalent to that required to boil the water in six Lake Eries.

Our energy reserves

At that rate of fuel consumption, some to produce useful energy and the rest not usable, alarming predictions ensue. The amount of heat to boil all the water in a lake the size of Lake Erie is symbolized by Q. A Q is 1.06×10^{21} joules (J), or 2.5×10^{17} kilocalories (kcal). (That's 2,500,000,000,000,000 marshmallows, at 100 kcal per marshmallow, as all good dieter's know.) In about 100 years, then, our world annual energy needs will be 6Q. Now we must compare this need with the energy sources available. Table 8–1 shows the world supply of known available fossil fuels and *estimates* of the amounts of these fuels that are within the Earth's crust, not yet discovered but discoverable if we look hard enough and recoverable.

At the rate of 6Q per year, our known supply would last less than 4 years, and the total known plus estimated recoverable will last only about 35 years. Of course, you can suggest that these figures are not realistic. How do we

TABLE 8-1. World Supply of Energy, in Q Units; Fossil Fuels Only

	Known	Estimated eventually recoverable, additionally
Coal, all forms	17.3	192
Crude oil	1.7	11
Natural gas	2.1	12 or less
Tar sand oil	0.2	2
Shale oil	0.9	1
Totals	22.5	216

decide, you may ask, that 6Q is to be the world's annual consumption not long from now? How can we be sure that the total recoverable and known supply is less than 300Q? We must respond that we cannot be sure; to some extent the numbers are estimates, but they are based upon past experience, which is the only guide we have, and they include the assumption that all citizens of the world have a right to a reasonably decent standard of living. At the minimum, the citizens of the developed countries have only the accident of their birthplace to "justify" their high per capita energy requirements and surely ought to strive to equalize such blessings.

But, unless we can find a source of energy in addition to fossil fuel, none of us will for long enjoy what we enjoy today. Notice that, even if we are foolish and immoral enough to refuse to aid the poorer nations, at our present rate of utilization of energy, 0.25Q each year, the 22.5Q known to exist will last less than 100 years. Considering that the annual rate of utilization of energy is constantly increasing, the 22.5Q will *all* be converted to heat and other forms of energy *within the lifetime of some of the readers of these words*! Clearly, unless some new solutions can be found, there will be some unusually undesirable results. What are the possible solutions?

Conservation

Various solutions to the energy problem have been suggested, but each one is associated with some qualifications. For one thing, we can begin to conserve energy by ceasing to be profligate in its use. We can walk or ride a bicycle instead of driving a car. Use a train or a bus instead of traveling by air. Use a train or bus instead of traveling by car. Turn out the lights when we are not in a room. Turn off the air conditioner in the summer, or at least set the thermostat at a higher temperature. Set the thermostat lower in the winter; wear a sweater or two, with long johns underneath. We can recycle our wastes instead of discarding them; much less energy is required to reuse the aluminum in a used aluminum can than to get that same amount of aluminum from its ore and convert it into a new aluminum can. Other discarded materials similarly require less energy to be recycled into new products compared to starting from scratch, and recycling provides the additional benefit of reducing our mountains of garbage. We can improve the effectiveness of insulation in our buildings; those who must use their own cars should accept the absence of air conditioners, no automatic shift, smaller horsepower engines instead of larger ones which need more gasoline per mile. These techniques will postpone the day when we would otherwise exhaust our known supplies of fossil fuel. As we shall see, if we can postpone that event long enough, there is some hope.

Postponement can be furthered by beginning immediately to exploit our offshore wells, even at the risk of ecological damage from a runaway well, according to some proponents. Others suggest that we try a little harder to find ways to prevent these runaways before we open up the offshore oil fields. Another partial solution suggests that we find out how to recover more of the oil from an oil field; at present we know only how to get a bit more than half of the oil that is there. We need improvements in the art of locating new, undiscovered, oil fields, also. Even though the total amount is small compared to our projected needs, we should begin, now, to learn how to extract fuel from tar sand oil and shale oil deposits. Currently, the art of extraction is primitive compared to what it might be, ideally.

It has been recommended that governments establish tax policies that would encourage conservation and funding policies that would enhance research into the better use of the fossil fuels we do have. Already, government and industrial supported studies of new methods for finding oil, of better utilization of coal, for example, have been established. Sooner or later, some predict, a surcharge tax will be imposed upon the sale or resale of all automobiles which travel less than (perhaps) 30 miles per gallon of gasoline; the poorer the gas mileage figure, the higher the tax.

Coal

Of all the fossil fuels, coal is the most abundant both with respect to the known and the estimated recoverable but not yet known amounts. If we include the estimated recoverable 192Q of coal, we will have enough of this fuel for a few centuries even at annual rates of utilization higher than our present 0.25Q. The trouble with coal, however, is that it is a "dirty" fuel. Not only is coal itself messy but the products of coal combustion pollute the atmosphere, largely because of the sulfur content of most coal. (If we restrict our usage to the coal low enough in sulfur not to pollute the atmosphere excessively, our known and estimated coal recoverable is dismally insufficient.) Some of the sulfur in coal is in the form of metal sulfides, such as FeS, iron sulfide. These can be removed by grinding the coal to a fine powder and removing the particles of metal sulfide directly. Unfortunately this kind of processing removes only about half the sulfur present in coal. The other half cannot be easily removed.

One solution is simply to burn the coal and remove the sulfur dioxide, SO_2, that forms during the combustion by means of various chemical reactions. If it is not removed, the SO_2, a gas, passes out of the stack to pollute the atmosphere. Several different chemical reactions have been suggested. Limestone, $CaCO_3$, can be mixed with the coal; the reaction is

$$SO_2(g) + CaCO_3(s) \rightarrow CaSO_3(s) + CO_2(g)$$

One product, calcium sulfite, $CaSO_3$, is a solid; the other, carbon dioxide, CO_2, is a nonpolluting gas.

Another suggestion is to oxidize the sulfur dioxide further. Under the proper conditions (involving a catalyst) sulfur dioxide will react with oxygen to form sulfur trioxide.

$$SO_2(g) + \tfrac{1}{2} O_2(g) \rightarrow SO_3(g)$$

All three substances are gases; the sulfur trioxide then reacts with water to form a solution of sulfuric acid; the solution is a liquid.

$$SO_3(g) + H_2O(l) \rightarrow H_2SO_4(l)$$

So far, these processes look fine on paper but have yet to be put into actual widespread practice.

Another way to remove most of the sulfur is by leaching it out in a hydrothermal process. At high temperatures and pressures, the sulfur containing components mostly dissolve in a water solution containing calcium hydroxide and sodium hydroxide.

Still another method takes a different tack. The coal is converted to a liquid or a gas by suitable chemical reactions. In liquid or gaseous form, fuels can be inexpensively transported by pipelines that are already in place, at least in North America and Europe. In the process of altering the fuel into a gas or liquid, the sulfur can be readily removed by already tried and true procedures. Years ago, before the advent of petroleum based liquid and gaseous fuels, every town and city of any size in the United States had a "coal gasification" plant to provide the housewife with gas for cooking. The techniques are known, although they can probably be improved. One improved process called the Hygas process is now producing a gaseous fuel called methane, CH_4, from coal and water as the primary materials. The process involves two reactions, the first with coal, symbolized as C (since coal is predominantly carbon) and steam, H_2O:

$$C(s) + H_2O(g) \rightarrow H_2(g) + CO(g)$$

This reaction is endoergic; the two products, hydrogen and carbon monoxide, are both gases. Both will burn in the air exoergically to form water and carbon dioxide. We could stop here and use this gas mixture as a fuel except that carbon monoxide is a poisonous gas and therefore not very suitable as a component in a fuel mixture.

Next, for each mole of hydrogen and of carbon monoxide, 2 moles more of hydrogen are required, 3 moles total. This extra hydrogen can be obtained from other sources, such as water, as we have noted in Chapter 7.

$$3\ H_2(g) + CO(g) \rightarrow H_2O(g) + CH_4(g)$$

This reaction is exoergic, but enough energy still remains in the carbon–hydrogen bonds of the methane to make it a useful gaseous fuel.

In the Hygas process the sulfur present in the coal is oxidized to become SO_2 (and a little as SO_3), and some carbon dioxide is formed as a side product as well. These three gases, SO_2, SO_3, and CO_2, are mixed with the methane, at this point. The gaseous mixture is bubbled through a solution of sodium hydroxide, NaOH, which takes care of the undesired gases. They are converted to dissolved ions by reacting with the OH^- ions in the sodium hydroxide solution:

$$SO_2(g) + OH^-(aq) \rightarrow HSO_3^-(aq)$$

$$SO_3(g) + OH^-(aq) \rightarrow HSO_4^-(aq)$$

$$CO_2(g) + OH^-(aq) \rightarrow HCO_3^-(aq)$$

The gas exiting from the sodium hydroxide solution is practically pure methane, quite suitable as a gaseous fuel, with approximately the same properties as natural gas.

Other processes for converting coal to a gaseous fuel are known and some of the more promising ones doubtless will be developed further. One of these simply involves reacting hydrogen directly with coal to form methane:

$$C(s) + 2\ H_2(g) \rightarrow CH_4(g)$$

Although simple to write as symbols on paper, the atoms and molecules themselves are more recalcitrant. Eventually, this process might become practical; presently, complex details involved make it a very expensive way to produce methane.

Similarly we can expect to convert coal to liquid fuels, such as octane, C_8H_{18}, and nonane, C_9H_{20}, these are two of the components of gasoline. However, efforts to make this conversion at a reasonable cost have not been successful. When the price of gasoline from petroleum becomes sufficiently high, of course, it will become economically possible to obtain the same kinds of molecules by converting coal. This may happen sooner, if a less expensive way to convert coal to octane, nonane, and other similar molecules is developed.

Summary

Let us summarize at this point. The amount of known and estimated recoverable petroleum and natural gas fuels is limited. At current rates of utilization, they will be exhausted in about a century; with the realistically anticipated ever increasing rate of utilization, they will surely be exhausted much sooner. Our recoverable estimated coal supply will last much longer; although coal is a dirty fuel, this problem can be solved eventually. Further, it is possible to convert coal to a gaseous or liquid fuel form, thus making it more convenient to use, although this conversion cuts the amount of heat energy we can obtain from the coal about in half. But even half of 192Q, should we convert all the estimated recoverable coal into gaseous and liquid fuels, will last quite a while. As we know, conservative habits of using our supply of fossil fuels, encouraged perhaps by tax and pricing incentives, will prolong the period before the supply is completely gone. All in all, it is reasonable to predict that we have until 2050 at the earliest or 2200 at the very latest to find a better, longer lasting answer for our energy needs. Prudence suggests that we should waste no time in seeking the answer; our known reserves are not likely to last even until 2050.

A different summary of our discourse thus far is given in Figure 8–1. On the left side the realtive widths of the identified areas are proportional to our primary sources of fuels, and the relative widths of the areas to the right indicate the proportional utilization, and the also necessary heat-flow-wastes involved for the United States in recent years.

ALTERNATIVE ENERGY SOURCES

Our current problem can be simply stated. We must have more energy soon and we must conserve our fossil fuel supply. We can, in principle, obtain some, or all, of our energy needs in plentiful supply from the Sun itself and a great deal, if not all, from other sources such as ocean thermal gradients, the winds, the tides, and geothermal sources. Unfortunately we do not know how to extract the amounts of energy needed, today. The principles are known, but their application, the practical way to obtain the energy we need from these sources, is not known. It will take time to find the practical answers. No one knows how long this might be. Meanwhile, our supply of fossil fuels shrinks daily.

Hydroelectric power

There are only two possibilities, for which we do have the known technology, which might supply us with our current, and steadily growing, energy needs. One of these is hydroelectric power.

Hydroelectric power is obtained by using the potential energy of water in an artificial or natural reservoir. The water is allowed to fall to a lower level and the kinetic energy of the moving water is convertable into electrical energy. There is no "required" loss of kinetic energy; in principle all of it could be converted to electrical energy. In practice, some is lost due to frictional and other effects. The problem with hydroelectric power is that over the whole

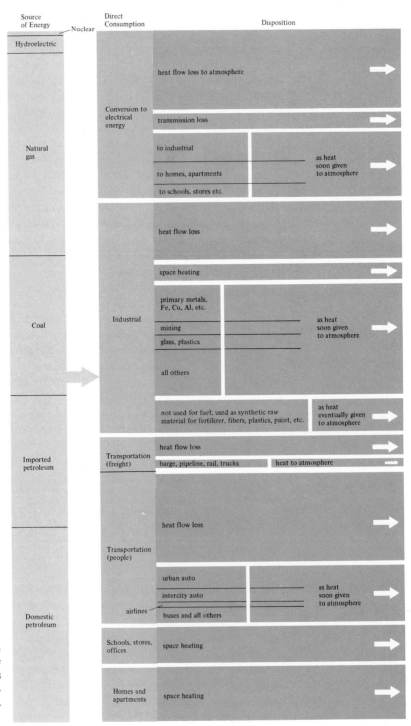

Figure 8–1. Energy, sources, usage, and ultimate disposition; United States average 1967–1977. Relative widths suggest proportional amounts.

surface of the Earth, even if it was available for the purpose, we could not find enough suitable locations for building artificial reservoirs to supply our energy needs in this manner. Natural lakes (reservoirs) are not available in sufficient number either. It is estimated that, at best, the energy which we could supply through hydroelectric power is less than three times more than the energy we now obtain from this source. This ultimate total of hydroelectric power is woefully insufficient even for today's needs.

Nuclear power

The light water reactor (LWR). The second alternative for which the technology is developed is nuclear power, using a light water reactor, usually identified by the letters, LWR. We know how to convert the energy released from nuclear fission into other forms of energy. The energy from nuclear fission begins as heat at a high temperature, and we know how to convert heat at a high temperature, with some necessary loss, to other forms of energy.

There are some important drawbacks to the use of LWR's. Let us dispose of the best known concern first, danger from radioactivity. We have already discussed this briefly in Chapter 6. Most of the radioactive products from an LWR have half-lives of 30 years or less. Translated, in about 600 or 700 years, these products will have lost their radioactive character, to all intents and purposes. It is therefore necessary to store these under rigid control for a long period. Fortunately, the volume is small; the amount we might attribute to one person per year occupies a volume about the size of an aspirin tablet, if that person were to use all the energy they needed as obtained from an LWR. That is, the entire amount of radioactive wastes which we would indeed wish to control carefully, accumulated over 700 years, would by themselves not even fill a pit 250 feet deep, 250 feet long and 250 feet wide.

Those wastes would be releasing energy, a lot of it, while they were stored in this pit, so we would actually need a lot of air flow, for example, to keep things cool. Also, we would need shielding (perhaps made from concrete) to reduce the effects of radiation on the surroundings. This makes the imaginary storage pit much larger, but not impossibly large, certainly much smaller than a cubic mile. This is all the storage we would need, because at the end of 700 years what was put in first could now be safely removed, making room for the waste from year 701. It is true that a small amount of the radioactive "ash" from an LWR has longer half-lives, up to thousands of years. But these are small in amount and can be recycled through the LWR and converted to shorter, 30 years or less, half-lived species.

The possibility of a large nuclear accident was discussed in Chapter 6 and need not be considered in detail here. But, briefly, in the WASH-1400, or Rasmussen, report commissioned by the Atomic Energy Commission shortly before it was dissolved by Congress, calculations based on actual operating experience with LWR's were used to predict the probability of fatality or serious injury to a person who did not work at an LWR facility to be less than the probability of being struck by a falling meteorite. The report also predicted less liklihood of serious injury or fatality for workers within an LWR facility compared to workers in a traditional electric power plant. These findings have been characterized as misleading by the Environmental Protection Agency.

All nuclear power plants now known or conceivable today have high waste heat losses in the conversion of their heat to other forms of energy. Unless this waste heat is somehow controlled, the resulting thermal pollution would cause local difficulties. To solve this will probably require the construction of costly cooling towers and other devices that eject the waste heat into the atmosphere harmlessly. (This effect could ultimately be quite serious, but not in the near future; we will consider this in more detail later in the chapter.) Further, evidence is accumulating that some of this wasted heat can be used beneficially, for example in warming a body of water slightly to enhance the production of fish and other forms of aquatic life, thus increasing slightly the total food production of the world.

The fuel, the uranium (or ultimately the thorium), used in an LWR is not processable into a bomb by a terrorist. To do so requires more sophisticated equipment than a single (or even a few) terrorists could have. The availability of the uranium or thorium fuel is another question. At present, the known reserves are only enough to provide a total of 0.7Q, less than 3 years world supply at present rates of energy utilization. However, this figure is based upon an artifically determined price for uranium (or thorium) ore which is unrealistically low with respect to our projected needs. As it happens, the cost of the uranium (or thorium) ore is all but insignificant in the determination of the cost of the electrical energy produced. The price of the ore could increase more than 50 times with little effect upon the cost of the electrical energy. At this much higher price, it is estimated conservatively that much more ore will be considered recoverable, up to $5 \times 10^4 Q$. This is enough to supply all our energy needs, even at greatly increased annual rates of utilization, for several centuries.

The CANDO and HTGR reactors. In addition to the LWR's now built and being built in the United States, there are two other technologically well established types of nuclear reactors. One type is known as a CANDU, developed in Canada, and the other is a high temperature gas cooled reactor, HTGR, popular in Europe. Although the details differ slightly, the preceding statements in reference to LWR's also apply to these other two types of reactors.

Breeder reactors. However, there is another reactor, a breeder reactor, that requires further discussion. In essence, breeder reactors can utilize other isotopes of uranium and thorium, in effect increasing the amount of estimated recoverable energy to as much as $10^7 Q$. This amounts to almost an unlimited supply of energy. Should breeders be developed (only a few are now in existence) as our primary source of energy, it would take so long to use $10^7 Q$

that any concern about what would then happen is pointless. However, there will be some serious trade-offs involved, so much so that many who are aware of these details feel that it would be a serious mistake to rely on breeder reactors as a source of energy for the world.

Natural uranium exists in the form of three isotopes, more than 99% is $^{238}_{92}U$, less than 1% is $^{235}_{92}U$, and a mere trace is $^{234}_{92}U$. Of these, only $^{235}_{92}U$ can be fissioned when it interacts with a slow moving neutron. In effect, less than 1% of the uranium available in the Earth can be used to produce heat energy at a high temperature in LWR's. If, somehow, we could utilize that $^{238}_{92}U$, at one stroke we could multiply our potential supply of energy by a factor of about 100. The function of a breeder reactor is to convert the $^{238}_{92}U$ into a different, fissionable, isotope.

The fissionable isotope is $^{239}_{94}Pu$.

To make $^{239}_{94}Pu$, all we need do is to put a supply of $^{238}_{92}U$ in a nuclear reactor where there is already a supply of neutrons. Some of the neutrons will interact with the $^{238}_{92}U$, producing, or breeding, the fissionable plutonium isotope:

$$^{1}_{0}n + ^{238}_{92}U \rightarrow ^{239}_{92}U$$

As we might guess the $^{239}_{92}U$ nucleus has "too many" neutrons and this high energy state decays to a lower state, a new nucleus, by β emission. The half-life is about 24 min.

$$^{239}_{92}U \rightarrow ^{239}_{93}Np + ^{0}_{-1}\beta$$

Even this neptunium nucleus is in a high enough energy state (too many neutrons) to decay further with a half-life of about 60 h:

$$^{239}_{93}Np \rightarrow ^{239}_{94}Pu + ^{0}_{-1}\beta$$

The plutonium isotope is reasonably stable. It decays by α emission with a half-life of about 24,000 years; but it is fissionable.

Another variety of breeder reactor begins with $^{232}_{90}Th$. These nucleii can absorb a neutron to form $^{233}_{90}Th$ at a high energy level. Then, this thorium isotope decays by a two-step β emission to form the fissionable isotope $^{233}_{92}U$. Since we have a large supply of estimated recoverable thorium in the Earth's crust, the thorium breeder reactor also seems promising. The problem lies in the long half-life of $^{239}_{94}Pu$ or of $^{233}_{92}U$, 160,000 years for the latter, and in the fact that both are α emitters. That is, in a breeder reactor, either $^{239}_{94}Pu$ or $^{233}_{92}U$ would be accumulating. Every so often this material would be extracted, shipped off to another nuclear reactor for fission, and more of the same bred further from a replenished stock of $^{238}_{92}U$ or $^{232}_{90}Th$, respectively.

Now there is nothing at all unsettling about large amounts of α emitters being merely shipped. Because α radiation cannot penetrate even a thin paper barrier, as long as it is outside a living body there is no danger. However, both of these isotopes are fissionable and concentrated. If there were a large nuclear accident involving a breeder reactor or if some of the isotope were diverted illegally during shipment or stolen by stealth or force from a reactor site, we risk the distribution of α emitting debris over a wide area in the form of fine dustlike particles that could be inhaled or ingested.

Of the two, uranium and plutonium, plutonium is far more poisonous. Inhalation or ingestion of two dust-sized particles of plutonium oxide, PuO_2, the most likely form under the conditions we are considering, could possibly cause cancer in one out of every thousand such instances. Three, or more, such particles increase the risk of cancer. In the limit, assuming a large nuclear accident or purposive terrorist type detonation, most of the persons within a few miles of the blast would almost certainly develop cancer. The risks are lower for a thorium to uranium breeder but, according to some who have considered the matter in detail, not enough lower to warrant their favoring even that kind of breeder reactor.

It is correct that the risk of an accidental disruption of a breeder reactor is much less than that of a LWR or CANDU. The design of a breeder reactor is inherently safer, and this is favorable. A breeder reactor, because of its design, will enable the heat at a high temperature to be more efficiently converted to useful energy, so it produces less waste heat. One type of breeder reactor called a molten salt breeder reactor would be best used for thorium-uranium breeding (which at least is less uncomfortable to think about than a uranium-plutonium breeder), but development work on this breeder is currently proceeding very slowly. This breeder would have several other important built-in safety provisions for the reactor workers, such as a minimized risk of radiation exposure when refueling the reactor. But taken all in all, we must balance the all but limitless energy potential a breeder reactor can indeed provide against the small but certain (if it happens) risk of health damage, accidental or terrorist inspired. Roughly, the problem is akin to that involved in contemplating multiplying zero by infinity; for that one there is no known answer.

Time for another summary. With conservation, we have enough fossil fuel to last for at least a few decades and we can gradually reduce our consumption of fossil fuels while building more LWR's and other equivalent reactors, which then begin to take up more and more of our energy load. The costs in dollars will be very high, of the order of hundreds of billions of dollars in worldwide terms. We can be sure that the cost of energy to the consumer will steadily increase, whether it is gasoline for his automobile or electricity by the kilowatt-hour; the era of relatively inexpensive energy is finished.

Meanwhile, mankind must make a decision about the distant future. If we do decide to allow breeders to be used to supply us with the energy we need in the very long pull, we need as much time as possible to develop technological, economic, and psychological safeguards against unexpected large nuclear accidents. Or, if we decide to mandate against the use of breeders, we need as much time as possible to develop other technologies for supplying us with the energy we will surely need.

Ocean thermal gradients

We can turn now to a brief description of the others, saving the best, direct conversion of solar energy, to the last. The next one to consider, for which there is no really good technology yet, is ocean thermal gradients. We are already familiar with the principle involved: heat at a higher temperature (from the Gulf Stream off the coast of Florida) will be brought into contact with a temperature lower by 20°C (the cooler ocean waters lying under the Gulf Stream). With only a 20° differential, the efficiency will be very poor and very little of the heat that would flow could be converted into useful energy. Even so, rough calculations indicate that a series of such devices located along each side of the Gulf Stream for a distance of about 1 mile would be able to supply all of the energy used by the United States annually at present.

Should this ever be done, there would be problems in transporting that quantity of electrical energy from one location to consumers all over the country. One solution is to convert the electrical energy into chemical energy in the form of hydrogen, which could be extracted on the spot from the ocean water. The hydrogen could then be transported, by pipelines and other means. A few interesting advantages are forseen: The cooler underlying ocean water, which is loaded with nutrients, would become warm enough by the forced heat flow to support aquatic life. Hence, near the ocean thermal gradient power installation there could be several thousands of acres of aquaculture, to provide large amounts of seafood for human consumption. On the adverse side, it is conceivable that the extraction of heat from the Gulf Stream, though only a small percentage of the total heat it carries, might be sufficient to trigger a marked change in the character of that ocean current and affect the climate of northwestern Europe and the British Isles detrimentally. This particular possibility will need special attention. Currently, a pilot project to test the practicability of extracting energy from ocean thermal gradients is being established. If successful, it will be several years before any practical results will be obtained.

Growing our energy

It might be useful to pause for reflection at this point. The source of the heat energy in the ocean thermal gradients is, of course, the Sun which heats the ocean water. The source of the energy in fossil fuels is (or better, was) the Sun, eons ago. In those times the incident energy of the Sun was used by growing plants to make high energy chemical bonds in the cellulosic constituents of the growing plants. Then, through poorly understood geological processes, some of these plants became coal. Our oil deposits were formed in at least a grossly similar way, although the formation of these is even less well understood. That is, except for nuclear energy and, as we shall see, geothermal energy, all of the energy man has used, or plans to use, came from or will come from the Sun. Even our nuclear energy sources derive from the stars, as you will recall from what was learned in Chapter 1.

It has been suggested that we use what might be called "immediate coal;" that is, presently growing plants, to supply our energy needs. After all, for many years, man burned wood as his major fuel. Why not follow this example again? In principle, this could be done, although naturally we would use a plant that converts the incident solar energy more efficiently (that is, it grows faster) than trees do. Sugar cane has been suggested; it grows faster than wood and, in addition, the sugar in the cane and the bagasse (the fibrous part) can be converted readily to ethyl alcohol by fermentation. Ethyl alcohol is a very satisfactory liquid fuel; it can be used in an automobile instead of gasoline by making only a few minor adjustments to the carburetor, for example—although you will get fewer miles per gallon. Because it would take hundreds of millions of acres of land to grow the sugar cane required to fill all the world's energy needs, this suggestion will necessarily provide only a partial solution to the total problem.

Wind power

Another possibility is wind-powered electrical generators. For one thing these do not suffer any heat loss; except for frictional and other similar losses, all of the energy captured can be converted into electrical energy. The problem lies in capturing the wind's energy, neglecting the other problems associated with the on-again–off-again quality of the winds. Depending upon which optimist or pessimist you listen to, the estimates will vary; however the best realistic proposals describe the possibility of supplying perhaps 1–2% of our total energy needs at great cost through this means. Even this would require some delay as technological solutions to particular problems were developed one by one. Here again, notice our reliance ultimately upon energy from the Sun.

Ocean tides

An unusual nonradiant solar energy possibility is to use the potential and kinetic energy in the oceanic tides. The essential technology is known; a tide-power electric generating plant has been built and is now operating on the coast of France. In principle, more energy than we are ever likely to need is available in the tides. The problem is that except for the location in France there are no other places where the configuration of the coastal shore line permits the economic construction of a tide-power plant. Further, given enough funds to build an artificially configured "shore line" we do not yet know how to construct it.

Geothermal power

For several years in Italy and in New Zealand, and recently in northern California, geothermal electric power plants have been converting some of the heat energy in naturally occurring steam and hot water into electrical energy. As we know, the interior of the Earth is very hot, due to the release of energy as radioactive atoms in the interior of the earth decay to lower energy levels. At various locations, such as Iceland, northern Italy, northern California, Yellowstone Park, and others, where water is also present, some of this heat escapes from the Earth's interior as steam. This steam can be used as a source of high temperature heat.

As long as the conditions are such as to produce complete evaporation of the water and the water contains little or no dissolved substances, as in California for example, we are able to extract the energy in a practical way. However, there are many other locations where the water is only partially converted to steam, producing steam with drops of water in it, or where the water contains dissolved and corrosive constituents. These locations can in principle be used, but the technology for their practical application has not yet been fully developed. To do so will take time.

Other locations have "hot rocks" near enough to the surface but no natural water. Here, of course, water can be added from an external supply if one is available, and be converted to steam at a high temperature, which can be used to produce other useful forms of energy after the customary losses. The technology to do this is almost all known, and this source of energy could be made practical within a few years.

Estimates of the recoverable energy from geothermal sources vary, depending on which kind of expert you listen to, from a minimum equal to the annual utilization of the United States to a maximum of a few thousand Q. Clearly, this source of energy should be investigated thoroughly, both to develop the necessary technology and to establish more reliable estimates of the total recoverable energy which might be available.

This brings us to two hypothetical alternative sources of energy for the future. One is the fission of boron nucleii by protons. In principle, the scheme would work because the products, three helium nucleii, weigh less than the initial nucleii, one boron and one hydrogen nucleus, each. A very high temperature is required for this nuclear fission, but beyond that little is known. The probable advantages lie in the fact that no radioactive "ash" would be produced, and there is an ample supply of boron and hydrogen to last for several tens or hundreds of years if this source were used exclusively to supply the world's energy needs.

Fusion

The second possibility is nuclear fusion. As we learned in Chapter 6, and in Chapter 1, nuclear fusion processes with the products having less mass than the original reactants, can release energy. Because the process works in the Sun and other stars, we know it can be done. So far, man has been able to produce large amounts of energy in this manner only by means of an atomic fusion bomb, which is too dramatic to be practical as an on going energy source. The fusion process for producing useful energy, if developed, will take several years. Even then, for all we know now, it may turn out to be impractical here on Earth. Clearly, efforts to develop this source of energy should continue, but we cannot rely on it to solve our problems today nor in the near future.

Solar energy

This brings us to the direct conversion of solar energy. For example, sunlight could fall upon water in pipes or troughs and the resulting hot water could be used to heat a building or home. Reasonably practical technologies have been developed to perform this function; more efforts would bring about improved results. Clearly, this is one almost immediately available solution to meeting some of our energy needs.

It would be nice if we could expose water to sunlight and, under the conditions we might devise, the water would decompose into hydrogen and oxygen. The hydrogen could then be distributed where needed, as a fuel, by pipelines. This can be done today, but the utilization of the Sun's incident energy is so inefficient as to make this method impractical. What is needed is more research to find a way, to increase substantially the efficiency of such a process.

Another technology we have is photovoltaic cells. These are flat plates that, when exposed to sunlight, directly convert the energy from the Sun into electrical energy. As yet their efficiency is poor, but these devices are used to power some satellites even now. The cost of such devices is high, more than 200 times more costly per unit of electrical energy than a typical nuclear power plant. However, there is reason to anticipate that further research and development will sharply lower the cost and, hopefully, increase the efficiency.

The advantage of a method for direct conversion of solar energy is that it is, in itself, pollution free. The only ecological considerations involved would be related to the manufacture of the devices themselves and, to a small extent, the fact that if such devices were placed in open fields (instead of on the roofs and walls of existing buildings) they would tend to shade the ground beneath, thus preventing some growth of plant life in that shaded area. The present disadvantages are low efficiency of energy conversion and high dollar cost compared to other sources of useful energy, despite their higher ecological costs.

ENERGY AND THE FUTURE

In concluding this chapter we ought to look at two unpleasant predictions. The first is not too bad; the second ought to terrify every reader.

What will life be like by about the turn of the century? Probably, our standard of living will be slightly lower, and the price we will pay for energy (not counting inflation) will be several times higher than it is now. As best as can reasonably be predicted, we will just squeak by, barely missing a severe world-wide energy crisis. Our petroleum reserves will be all but gone, and what remains will be rationed for specialty uses. By then the end of our coal supplies will be in sight, although there will be plenty left for the time being. Meanwhile, the delays of the 1950s, 1960s and 1970s in the construction of nuclear power plants will show up as a shortage of energy. Probably, voluntary or enforced conservation practices will prevent undue hardships except for several isolated instances.

You will be fortunate if you have an automobile with gas in the tank. It will most certainly be a light passenger carrying device, with no air conditioning or power brakes, perhaps not even a fan driven heater. You will probably not live in a separate dwelling because they are less efficient in their utilization of space heating. In any event, your dwelling will be well insulated, including storm windows and doors. Many people will choose to do without air conditioning because the cost of electricity to operate that convenience will be high. Some will even choose to do without much TV for the same reason. You will probably choose to live near a bus stop or train station or learn to enjoy riding a bicycle from your home to that location. Very few people will travel by air; the cost will be very high. Most will choose trains or buses for long distance travel. Taxis may become almost as extinct as a horse and carriage is now (and these may come back, to some extent). When you purchase an item, one of the questions uppermost in your mind will not be "What does it cost?" Instead you will want to know "How much electricity does it use? When I return it for recycling after it is worn out, how much tax credit will I get?" You, or your children, will mow the grass by human power, should you be fortunate enough to live in a separate dwelling with grass around it to be mowed. If you are living in the northern hemisphere, most of the windows in your dwelling will face the south. Since it will be costly to travel even short distances and TV will be expensive to use, you will be reading more books, playing more chess or checkers, and finding other ways to entertain yourself.

The discussion in this chapter quite obviously represents one person's views. Because no single person can hope to encompass for himself, much less explain to others, all the ramifications of the energy problems we face, this discussion has necessarily been subjective. I have tried to be as objective as possible, but as it was being written I noticed a few places where my own views were apparent, at least to me. The only way for anyone to make a decision that is right for themselves is to read and discuss what several informed people have to say on all sides of the question. To this end, and more than in any of the preceding chapters, you are urged to read the references cited (some of them, at least) at the end of this chapter and to read others which come to your attention. As you do this, you will notice that experts honestly disagree. When you find very well informed and honest people disagreeing, what kind of decision can be made? For that, I offer this advice, which I have tried to follow in composing this chapter:

1. Assume that expert A is correct, and that we follow his advice. What would happen in case expert A was wrong?
2. Assume that expert B is correct, and that we follow his advice. What would happen in case expert B was wrong?
3. Now, make your own decision.

It would be nice to close this chapter on a happy note, but to do so would be dishonest. Now the world uses about 0.25Q of energy, which ultimately is returned to outer space after passing through our atmosphere. The present population of the world is about 4×10^9 people. Most of these people do not get their "fair share" of energy, compared to residents in the United States or other developed countries.

At some point in the not too distant future, the world population will be 1×10^{10} people—that's more than 20 people for each one now living. There is no question, given enough energy to use, every one of those people can have the standard of living that we enjoy now. The world will be more crowded, but not much more—there *is* plenty of empty space around. All it takes is energy to make it very comfortable. All it takes is energy to produce the food that will be needed. All it takes is a *lot* of energy, and therein lies the terrifying problem.

Let's look at some numbers and see if we can find a solution and, as well, point out forcefully why a solution must be found, else the population of the world will be zero people. Figure 8–2 summarizes the situation as it is today. Each year, we receive 5200Q from the Sun, incident upon our upper atomosphere; 1768Q are reflected immediately. Of the 3432Q remaining, 988Q are absorbed by the upper atmosphere, mostly by breaking molecular bonds; molecules become atoms. When, as eventually happens, those atoms form molecular bonds again, that 988Q is released, and passes to outer space.

This leaves 2444Q coming into our lower atmosphere. Most of this, 1144Q, is absorbed by surface water, oceans, lakes, rivers, ponds, swamp water, and so on, by evaporation. Sooner or later, this water vapor condenses again, forming liquid water (it rains or snows), and the 1144Q are released, eventually passing to outer space. Further, 1040Q, almost as much as is absorbed for a time by the evaporation of water, is absorbed by the molecules in our lower atmosphere. These molecules, especially water (vapor) and carbon dioxide, but also including the nitrogen and oxygen molecules, absorb the 1040Q by vibrating and rotating more vigorously than before. The energy from the Sun is converted into molecular rotational and vibrational energy. Eventually, these molecules lose that energy, they don't vibrate and rotate as much as they did, and that 1040Q passes to outer space.

The 260Q that is still unaccounted for is used in two major ways, 26Q is absorbed by growing plants, mostly as chemical energy in the bonds between

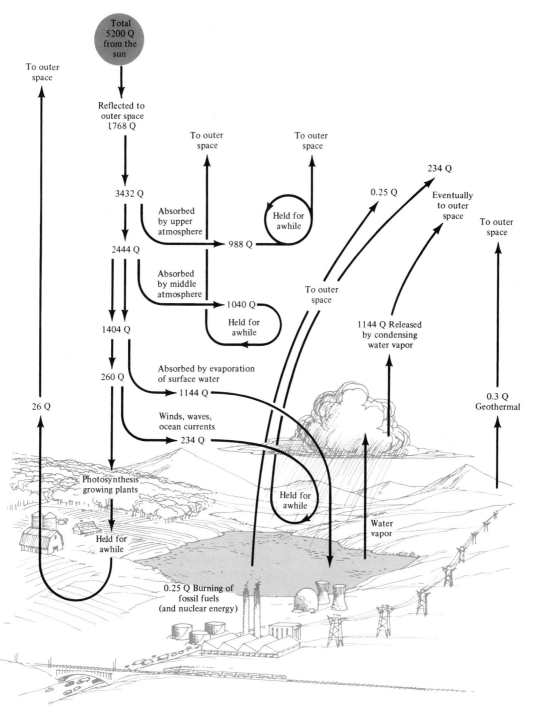

Figure 8–2. Energy balance of the Earth, with figures for annual amounts, 1975 expressed in Q units.

the atoms in molecules of the compounds (starch, cellulose, and sugar) that make up the major part of any growing plant. The plant eventually dies or is harvested, and, sooner or later, by digestive processes in animals that eat some or all of the plant, by burning, and by decaying, the 26Q is released, and goes to outer space. As we know, in past eons some of that energy was trapped in coal and oil and is only now being released by the burning of fossil fuels. The remaining 234Q is converted into kinetic energy, into convection, the movement of winds, ocean currents, and waves. Sooner or later, it too is dissipated and returns to outer space.

All in all, the Earth is in balance with respect to the energy received from the Sun. Of the total energy incident upon the Earth, about one third is reflected immediately and the other two thirds is absorbed, for a varying time, and then released. We return as much as we get in the long term view.

Actually we return a bit more than we get. From geothermal sources, the Earth itself radiates energy to outer space, perhaps about 0.3Q per year. Some of this 0.3Q, before it leaves, is absorbed by the air, as kinetic energy in the winds, and then eventually finds its way to outer space. It is correct to say that, to a very slight extent, we have more wind on the Earth, due to this release of geothermal energy and its temporary absorption as kinetic energy than we would otherwise have. The effect is slight; we are comparing 234Q without geothermal energy to 234.3Q with the added geothermal energy involved in convection effects.

Also, as Figure 8–2 shows diagrammatically, we also add 0.25Q per year into the system from human activity. This is largely the energy released by our burning of fossil fuels, as we know. There is a small fraction of this, less than 0.001Q, from nuclear power. Again, as with the geothermal energy, the effect is small but present. The winds are a bit more energetic than they would be otherwise. We are now comparing 234.3Q with 234.55Q.

There is one additional important factor to be considered. For thousands of thousands of years the Earth climate system was in a steady state; what was absorbed was eventually released plus a tiny bit more released. Now, an additional tiny bit more is being released. It is probably correct to guess that, if our tiny bit more, 0.25Q, were released more or less uniformly over the surface of the Earth, there would be no noticeable effect. However, man concentrates in geographically restricted locations. The additional 0.25Q is released in those places, not uniformly over the whole Earth.

There is some suggestive evidence that this has already disturbed the balance, although it is not at all certain. That is, we do know that the temperature of the atmosphere near a large city is slightly higher than in the open country. We do know that the weather pattern near industrial locations is changing slightly. There is no conclusive evidence whatsoever that this is caused by some of that 0.25Q being released each year in a few places.

As we have already learned, however, at some time in the future the energy released by man's activities will be 6Q per year or even more. When that happens, we would compare 234.3Q with 240.3Q, quite a different comparison. It is reasonable to conjecture that this additional 6Q per year would indeed alter the climatic pattern of the Earth, especially since it will be released in special locations, not evenly distributed. If 6Q will not alter the climatic pattern, perhaps 8Q would; perhaps 3Q or 4Q would. We do not know.

Whatever the critical amount of additional energy, the first effects will probably be a change in the familiar seasonal rainfall. The change will be large enough to be noticeable to everyone. If nothing is done, the next effect may be a change in the climate of large regions; what was a desert might become an oasis and what was fertile may become a desert. In that event, the result would be the movement of thousands of people away from the infertile regions toward those places that were still fertile or had recently become so. Such movements of large numbers of people alters their cultural habits, usually unfavorably. If nothing is done even then, the final result will be climatic changes affecting the whole Earth. The change could be beneficial; it could be the opposite. There is no way to predict on the basis of our present knowledge. Therefore, using the principle enunciated earlier in this chapter, it would be wise to see if we have an alternative. We do.

Instead of *adding* 6Q (or whatever the critical amount is) to the 234Q and the 0.3Q from the Sun and geothermal sources, we can *borrow* 6Q by direct conversion of solar energy, and *return* the borrowed amount, taking care that it is returned in a well distributed manner over the surface of the Earth. Actually, the 6Q that we might add from nuclear fission (or fusion) a century or so from now includes the waste heat involved as lost during the flow of heat from a high temperature to a lower temperature. Approximately, 2.5Q of that 6Q is involved as lost heat energy. By utilizing direct conversion of solar energy to useful energy, we avoid the necessity of any high temperature to low temperature heat flow; we would then need approximately 3.5Q of solar energy to be converted. It would be easier to return 3.5Q in a well distributed manner than to similarly return 6Q.

In principle, there is yet another alternative. The problem will arise because of the addition of the 6Q (or other critical amount) to the lower atmosphere as heat and its conversion into kinetic energy of the winds, for the most part. If some way could be found (none is now known) to release that 6Q directly to outer space or to store it (much as some of the Sun's energy was stored in the geologic process of coal and oil formation) there would be no problem.

As far as we know now, the effect of adding 0.25Q is negligible. As each year passes, however, this amount is increasing. It is reasonable on the basis of past and present meterological facts to anticipate that several years will pass before the first, possibly adverse, effects will be openly noticeable. At that time, it will be very desirable to begin to *reduce* the amount added and to *increase* the amount borrowed (or transmitted directly to outer space or stored) by conversion of solar energy. We do not know how long it will take to develop the technology for converting large amounts of solar energy directly into useful energy. If it turns out to require only a few years, we could begin to develop the technology for that conversion when we first notice the minor climatic changes. If it turns out to require several years, it may be too late. Clearly, now is the time to beign active, costly, and hopefully successful efforts to develop technological means of direct conversion of solar energy to useful energy.

If this course is followed, historians of the future will look back on our times today describing them as a brief period in human history when man did not rely directly on the Sun as the source of energy for life and its amenities. If this course is not followed, there may be no historians of the future to make that observation.

FOR FURTHER INFORMATION

To keep this chapter within reasonable bounds, there was no mention of secondary but important alternatives to present sources of energy, such as the utilization of some waste heat for space heating or other purposes, or the oxidation of sewage and other wastes as sources of energy, or the conversion of organic wastes and other refuse into fuel by fermentation processes. Also, ecological side effects were only briefly mentioned; some that were not alluded to include strip mining of coal, nitrogen oxides as a (minor) pollutant from the burning of fossil fuels, and acreage needs of nuclear power plants. The possibility of direct conversion of chemical energy into electrical energy (with no substantial waste of heat) by means of fuel cells, not yet a developed technology, was ommitted.

These matters, and others not identified here, as well as more detailed treatments of the topics in this chapter are discussed in the references that follow. Additionally, the journals containing the cited references and other publications as well should be consulted for recently developing advances as we continue in our attempts to solve the problems on which the fate of man in this world may well depend.

(The order of listing of the references, below, is random, no attribution of relative significance or any other quality is intended or implied.)

C. J. Hitch: Unfreezing the future. *Science*, **195**:825 (1977). In an editorial statement, Dr. Hitch argues for flexibility in proposed solutions to the energy crisis, pointing out that our knowledge is not sufficient to forsee an effectively planned program at this time.

J. R. Platt: What we must do. *Science*, **166**:1115 (1969). Written before most people had even heard of reasonable estimates of our future energy needs, Dr. Platt outlines a plan of organization of people intended to guide our use of resources wisely over the next several years.

A. L. Hammond: Solar power, promising new developments. *Science*,

184:1359 (1974). A brief semitechnical description of the manufacture of silicon in single-crystal ribbons for use in photovoltaic conversion of solar energy to electrical energy.

An Assessment of Accident Risks in U.S. Commercial Nuclear Power Plants, Atomic Energy Commission, document WASH 1400, 1974. A controversial evaluation of the risks to the public and to workers involved with the operation of LWR's, based on data from actual operating experience. Several authoritative groups criticized the original report. See for example B. J. Luberoff: *Chemtech*, **5**:385 (1975) and R. Gillette, *Science*, **186**:1008 (1975), so a new and revised report was issued as *Reactor Safety Study*, WASH 1400 (NUREG 75/014) National Technical Information Service, Springfield, VA 1975.

E. I. Shaw: Radiation and society. *J. Coll. Sci. Teach.*, **4**:88 (Nov. 1974). Another account of the radiation risks associated with nuclear power plants and some comments on how we decide what to do.

M. E. Bailey: The chemistry of coal and its constituents. *J. Chem. Educ.*, **51**:446 (1946). A chemical description of coal.

Anon.: *A National Plan for Energy Research, Development, and Administration: Creating Energy Choices for the Future*, Energy Research and Development Administration, Washington, D.C., 1976. What the choices for alternate sources of energy might be, and how to get from here to there, maybe.

C. S. Cook: Energy: planning for the future. *Amer. Sci.*, **61**:61 (1973). On the interrelation between energy and pollution.

J. Darmstadter et al.: *Energy in the world economy*. Johns Hopkins Press, Baltimore, 1971. A statistical documentation of the utilization of energy throughout the world in the period 1925 to 1965.

P. Kruger and C. Otte (eds.): *Geothermal Energy*. Stanford University Press, Stanford, Calif., 1973. An optimistic treatment of the topic of the title.

J. Wei: Energy, the ultimate raw material. *Chemtech*, **2**:142 (1972). A broad but not superficial discussion of man's present and projected future utilization of energy, with a less than wholly optimistic prediction of what we shall eventually be able to do with our waste heat.

T. L. Brown: *Energy and the Environment.* Charles Merrill, Columbus, Ohio, 1971. Essentially similar to Wei's treatment, but expanded and detailed. A clearly written exposition.

J. H. Kreiger: Energy: the squeeze begins. *Chem. Eng. News*, Nov. 13, 1972, p. 20. A brief survey of our energy problem.

E. Hirst and J. C. Moyers: Efficiency of energy use in the United States. *Science*, **179**:1299 (1973). A survey, with data, emphasizing the need for energy conservation.

G. A. Lincoln: Energy conservation. *Science*, **180**:155 (1973). We can no longer afford our lavish use of energy, according to this author, who supports his case well.

C. Holden: Energy: shortages loom but conservation lags. *Science*, **180**:1155 (1973). A report with ominous overtones.

A. L. Hammond: Solar energy: proposal for a major research program. *Science*, **179**:1116 (1973). Also see: Anon., *Chem. Eng. News*, March 5,

1973, p. 17. Both brief notes, now out of date, suggest an optimism that is yet to be more than a hope.

E. A. Walters: and E. W. Wewerka: An overview of the energy crisis. *J. Chem. Educ.*, **52**:282 (1975). How it happened in the first place and its effect upon the future.

W. M. Brown: A huge new reserve of natural gas comes within reach. *Fortune*, October 1976. According to Dr. Brown, a staff member of the sometimes controversial Hudson Institute, a U.S. Geological Survey indicates that enough methane, the major constituent of natural gas, lies dissolved in brine under the coastal regions of Mexico, Texas, and Louisiana to supply all United States energy needs, alone, for more than 300 years at current rates of consumption. If the nearby regions underlying the waters in the Gulf of Mexico are included, estimates indicate more than four times this much methane, total.

Not all of the published materials on current energy problems appear as formal articles. Some of the more lively publications are letters to the editor. As an outstanding example of one among many, marked by its objective tone and questioning attitude, see J. Selbin: *Science*, **187**:789 (1975).

G. Hueckel: A historical approach to future economic growth. *Science*, **187**:925 (1975). A little more than 100 years ago crude oil was considered to be a nuisance. Not too much longer ago than that, the utility of coal was unknown, and its first uses, a little later on, were for the smelting of iron ore, not as a fuel. Perhaps the processes by which technology generated the utility of oil and coal suggest the ways in which resources not now recognized as useful could become so. A decidedly optimistic article.

M. L. Kastens: Productivity in world food supply system. *Chemtech*, **5**:675 (1975). According to Kastens, in terms of technology, processing, and preservation of food, we could provide ample food for up to at least four times the present world population. The inability to do so lies in the intracacies of energy production and distribution. The problem can be solved.

Low-cost, Abundant Energy: Paradise Lost? An anonymous report in the annual report of Resources for the Future, Washington, D.C., 1973. An overall summary with interesting graphs, not unlike this chapter.

R. H. Essenhigh: The case for fuel conservation. *Chemtech*, **5**:112 (1975). Our ever increasing consumption of energy must stop somewhere, it cannot continue forever.

W. B. Lewis: Nuclear fission energy. *Chemtech*, **4**:531 (1974). It is quite evident that nuclear reactors are the safe and available answer to our energy needs.

R. Gillette: *Science*, **185**:1027 and 1140 (1974). Two informative articles on the hazards (or not too hazardous?) in the use of plutonium as a nuclear reactor fuel.

D. L. Klass: Synthetic crude oil from shale and coal. *Chemtech*, **5**:499 (1975). A description of the liquefaction processes now known for making crude oil from shale and coal and the problems that these entail.

W. Hafele: Systems approach to energy. *Amer. Sci.*, **62**:438 (1974). One of the best, most forthright, detailed, well supported with data discussions of the whole energy problem. Hafele uses the term "embedding of energy" to describe the problem associated with excessive release of energy, by man, into the atmosphere.

D. P. Gregory: The hydrogen economy. *Sci. Amer.*, **228**:13 (Jan. 1973); T. H. Maugh: Hydrogen: Synthetic Fuel of the Future. *Science*, **178**:849 (1972); W. E. Winsche et al.: Hydrogen: its future role, etc. *Science*, **180**:1325 (1973); R. J. Schoeppel: Prospects for hydrogen-fueled vehicles. *Chemtech*, **2**:476 (1972); Anon.: Hydrogen: likely fuel of the future. *Chem. Eng. News*, June 26, 1972, p. 14; Anon.: Hydrogen economy concept gains credence. *Chem. Eng. News*, April 1, 1974, p. 15. A series of selected discussions on the use of this secondary source of energy; also see the citations in Chapter 14.

Energy and power. *Sci. Amer.*, **225**, entire issue (Sept. 1971). A thorough, exceptionally well-illustrated, discussion of the problems involved in the utilization of energy. Highly recommended to those who wish a clear though somewhat superficial presentation.

Energy. *Science*, **184**, entire issue (April 19, 1974). A thorough discussion of the problems involved in the utilization of energy. Highly recommended to those who wish a clear and sophisticated, objective, presentation. Since it is approximately 3 years more recent, the information is therefore likely to be more reliable than that in *Scientific American*, **225**.

G. R. Yohe: Coal. *Chemistry*, **40**:18 (Jan. 1967). A well written description of and about coal.

A. L. Hammond: Geothermal energy: an emerging major resource. *Science*, **177**:978 (1972). Mr. Hammond, a staff writer for Science, always does a thorough job.

W. D. Metz: Ocean temperature gradients: solar power from the sea. *Science*, **180**:1266 (1973). This article, by another competent writer on the staff of *Science*, presents this topic in a hopeful tone.

W. E. Heronemus: Wind power: a significant solar energy source. *Chemtech*, **6**:498 (1976). B. Sorensen: Dependability of wind energy generators with short-term energy storage. *Science*, **194**:935 (1976). Two informative and optimistic articles on the utility of the winds as one alternate source of energy.

M. McCormack: Energy crisis—an in depth view. *Chem. Eng. News*, April 2, 1973, p. 1. An editorial; informative, thoughtful, and important, by a competent congressman who is also a scientist.

A. L. Hammond: Energy and the future. *Science*, **179**:164 (1973). Similar to Congressman McCormack's editorial, but with enough differences to make comparisons fruitful.

R. Roberts: Energy sources and conversion techniques. *Amer. Sci.*, **61**:66 (1973). On our ability to meet the energy needs of the future; a thorough discussion.

J. W. Landis: Fusion power. *J. Chem. Educ.*, **50**:658 (1973). A look into the distant future, by the president of one of the industrial organizations that is committed to a thorough study of the development of fusion power.

G. A. Mills and H. Perry: Fossil fuel, power and pollution. *Chemtech*, **3**:53 (1973). A look into the immediate future, by the Chief, Division of Coal, Energy Research and the Director of Coal Research, U.S. Bureau of Mines.

A. M. Weinberg: Some views of the energy crisis. *Amer. Sci.*, **61**:59 (1973). Dr. Weinberg is well known as a thoughful, insightful, and objectively oriented proponent of the use of nuclear power. His views are respected by all, even by those who disagree with his interpretation of the facts and necessary assumptions.

G. C. Szego and C. C. Kemp: Energy forests and fuel plantations. *Chemtech*, **3**:275 (1973). An imaginative suggestion for the culture of forests to be burned as a renewable source of fuel, thus purportedly eliminating the need for further development of nuclear power. See also, B. J. Luberoff, W. P. Bebbington, H. Tarko, W. C. Walker, A. J. Moll, B. M. Holloway, G. C. Szego and C. C. Kemp (again), and R. P. Hammond in *Chemtech*, **3**:257, 391, 392, 445, and 635 (1973) for comments, pro and con, regarding this interesting suggestion.

R. Gillette: Nuclear safety: AEC report makes the best of it. *Science*, **179**:360 (1973). A somewhat critical evaluation of the "not as safe as one would like" and "were some facts perhaps withheld?" discourses which have occurred with some justice in recent years.

A. L. Hammond: Fission: the pro's and con's of nuclear power. *Science*, **178**:147 (1972). The widespread use of breeder reactors may, or may not, present unsolvable problems.

L. A. Sagan: Human costs of nuclear power. *Science*, **177**:487 (1972). An overprotective governmental policy restricts the exposure of the public to radiation to a greater extent than is justified by the costs of reducing other exposures to radiation. Provocative.

A. V. Kneese: The Faustian bargain. *Resources for the Future*, no. 44 (September, 1973). An interesting mixture of fact and outright fictional assertions, cleverly prefaced by a statement in which the author admits he is not well informed, but placed so far in advance of the negativisms presented, that, to the uninitiated, the use of nuclear power would mean certain catastrophe. Clever (and misleading); much of the author's concern is based on reality, but not all of his "facts" are factual.

L. J. Carter: Radioactive wastes: some urgent unfinished business. *Science*, **195**:661 (1977); and G. I. Rochlin: Nuclear waste disposal: two social criteria. *Science*, **195**:23 (1977). Two thoughtful contributions on the problem of nuclear waste and possible solutions. Not entirely optimistic.

R. P. Hammond: Nuclear power risks. *Amer. Sci.*, **62**:155 (1974). A discussion by a real expert with years of experience with radioactive hazards, one who is well informed, on whether there is, or is not, a Faustian bargain involved in our proposed extensive use of nuclear power. Also see H. A. Bethe: The necessity of fission power. *Sci. Amer.*, **234**:21 (Jan. 1976) who says there is no other choice that we can make now. Rebuttals to Bethe appear in the letters columns of *Sci. Amer.*, **234**:8, 10, 15, and a rerebuttal: 234:15, 16, 20, and 21 (April 1976).

D. R. Inglis: *Nuclear Energy: Its Physics and its Social Challenge*. Addison-

Wesley, Reading, 1973. On nuclear energy and its social implications. More or less readable, but has some tough maths and physics in it. Skip the first three chapters and begin reading Chapter 4 of this book.

A. L. Hammond and W. D. Metz: *Energy and the Future.* AAAS, Washington, D.C., 1973. In addition to this excellent summary, the American Association for the Advancement of Science has published a series of objective books, reprints, and audiotapes on the energy problem. Each past and future publication can be recommended strongly.

A. L. Hammond: *Science,* **189**:128 (1975) and W. Worthy: *Chem. Eng. News,* July 7, 1975, p. 24. Two reports, quite similar, on a promising development in the desulfurization of coal, the Battelle hydrothermal process. Also see D. P. Burke: *Chem. Week,* Sept. 11, 1974, p. 38.

B. Hannon: Energy conservation and the consumer. *Science,* **189**:95 (1975). A labor intensive economy uses less fuel than a capital intensive economy. One way to ease the cultural shock of a return to a labor intensive economy is to make the change gradually through a taxing structure on energy use, adjusted to wage levels.

A. H. Brown: Bioconversion of solar energy. *Chemtech,* **5**:434 (1975). Urges studies to obtain enough data so that we can decide whether or not bioconversion of solar energy is one answer. Dr. Brown, a plant physiologist, may or may not be a cry in the wilderness; unheard until it is too late.

E. S. Cheney: Limits on energy supply. *Chemtech,* **5**:371 (1975). No matter how much coal, oil, and gas is now undiscovered, and assuming that we discover and recover all of it, it is still not enough.

K. Brooks: Energy: technology is the chemical process industry's key resource. *Chem. Week,* May 29, 1974, p. 10. Our knowledge of technological processes enable us both to conserve current energy consumption and to obtain energy from new sources.

W. D. Metz: *Science,* **188**:136 (1975) and L. J. Carter: *Science,* **188**:996 (1975). Two reports on the proposal that hydrogen bombs be used as sources of energy by underground detonations.

R. E. Lapp: The cortical connection. *Sci. Teacher,* **40**:29 (March 1973). An address to science teachers by a well known and concerned one-time nuclear physicist who has more recently followed his conscience in successful attempts to communicate his well founded opinions to the public. The scope of this address is approximately that of this chapter (though more briefly presented). Recommended.

The energy crisis. *Bull. At. Sci.,* **27** (no. 7, 8, 9, whole issues) (1971). Several authors contribute a balanced view of the topic in the title.

R. E. Lapp: *A Citizen's Guide to Nuclear Power.* New Republic, Washington, D.C., 1971. A small book, clearly written, to inform the general public about the issues involved.

G. T. Seaborg and J. L. Bloom: Fast breeder reactors. *Sci. Amer.,* **223**:13 (Nov. 1970). Although a bit out of date, this well illustrated article describes the details of operation of a fast breeder reactor.

L. Rocks and R. P. Runyon: *The Energy Crisis.* Crown, New York, 1972. An alarmist position, supported by a lack of technical knowledge in the field of interest: the distinction between geological "reserves" and resources that will become recoverable as economic forces render them available through alterations in the pricing and use structures.

F. Daniels: *Direct Use of the Sun's Energy.* Yale University Press, New Haven, 1964. By this time, an old but classic description of the need and the means to effect direct and indirect conversion of solar energy to other forms of energy. Nontechnical, well illustrated, with tables galore; recommended. If man ever is able to utilize solar energy, Dr. Daniels stands alone as the distinguished chemist who should receive credit for initiating the steps toward that goal.

S. Novick: *The Careless Atom.* Houghton-Mifflin, Boston, 1959. Although on a different topic, this citation is placed in juxtaposition to that of Daniels' to call attention to the difference between careless mixtures of fact and fiction, also well mixed with bias, in this generally negative book, compared to the careful presentation by an equally sincere author on the utilization of solar energy. Both books have the same theme: nuclear energy is of questionable value, but from two quite different points of view.

A. M. Squires: Clean power from dirty fuels. *Sci. Amer.,* **227**:26 (Oct. 1972). As usual in the *Scientific American,* this article is well illustrated and informative.

O. H. Hammond and R. E. Baron: Synthetic fuels: prices, prospects, and prior art. *Amer. Sci.,* **64**:407 (1976). It is all very well to talk glibly about the eventual production of the fuels we need from exotic sources. But if made from these sources in large quantities there will be some adverse effects upon our society unless we prepare for these in advance.

J. Barnes: *Geothermal Power. Sci. Amer.,* **226**:70 (1972). By the director of resources and transport division of the UN, urging increased research and exploration for the further development of geothermal power.

P. L. Auer (organizer): *Report of the Cornell Workshops on the Major Issues of a National Energy Research and Development Program.* College of Engineering, Cornell University, Ithaca, 1973. The development of energy sources for the future will require close cooperation between government and industry and the acceptance of changes in life style by society. This document has already influenced decisions at the highest levels, and rightly so; it is authorative. Highly recommended.

A. J. Fritsch and B. I. Castleman: *Lifestyle Index.* Center for the Public Interest, Washington, D.C., 1974. Unlike so many of the publications from "environmentally concerned" groups (which shrilly allude to catastrophic events that will happen tomorrow, or even today basing their conclusions on their own misinterpretations of fact and fiction) this publication is a constructive attempt to help us all conserve the energy resources we have now so that, somehow, the future will be assured. Read it! And follow its recommendations.

J. C. Fisher: *Energy Crises in Perspective.* Wiley, New York, 1974. An optimistic book, well written; perhaps too optimistic, perhaps not. Recommended for its thorough treatment.

J. W. Moore and E. A. Moore: Resources in environmental chemistry. *J. Chem. Educ.*, **52**:288 (1975). (Also see Chapter 14, in this book.) An annotated bibliography on energy and on topics related to energy. Loaded.

Exploring Energy Choices and *A Time to Choose: America's Energy Future.* Energy Policy Project of the Ford Foundation, Ballinger, Cambridge, Mass., 1974. Two related publications, with "scenarios" describing what might happen if, unless. Recommended.

Bulletin of the Atomic Scientists, **31**:whole issue (September 1975). Almost the entire issue consists of a series of articles on nuclear reactor safety. The selection is balanced pro and con about as well as could be expected. Very highly recommended.

M. Willrich: *Energy and World Politics.* Free Press for the American Society of International Law, New York, 1975. Global political dynamics and its interaction with the energy crisis, including effects upon national security and the environment.

J. M. Dukert: *So What's New.* ERDA, Office of Public Affairs, Washington, D.C., 1975. A history of energy sources and uses published in time for the United States bicentennial year, consisting of a large, very interesting, wall chart with accompanying explanatory booklet. Recommended.

G. A. Vendryes: Superphenix: a full-scale breeder reactor. *Sci. Amer.*, **236**:26 (1977). In France it has been decided. A large breeder reactor will be constructed to furnish electrical power. How it will work technologically, the anatomy of that reactor, and some prospects for the future are discussed in this article.

The indirect utilization of solar energy may be a promising alternative to fossil fuels. Three articles provide a background for further reading: M. Calvin: Photosynthesis as a resource for energy and materials. *Amer. Sci.*, **64**:270 (1976); W. G. Pollard: The long-range prospects for solar-derived fuels. *Amer. Sci.*, **64**:508 (1976); and M. J. Povich: Fuel farming. *Chemtech*, **6**:434 (1976). In general the three articles are cautiously optimistic.

D. J. Rose, P. W. Walsh, and L. L. Leskovjan: Nuclear power—compared to what? *Amer. Sci.*, **64**:291 (1976). According to these authors, there is only one answer to the energy crisis: nuclear power, and we had best get on with its development instead of wasting time on other alternates to fossil fuels.

R. N. Adams: *Energy and Structure: A Theory of Social Power.* University of Texas Press, Austin 1975. The control of energy is a prerequisite to the control of political and social power, according to Adams. If true, the implications ought to concern everyone who values his/her freedom.

N. P. Cochran: Oil and gas from coal. *Sci. Amer.*, **234**:24 (May 1976). A description of several processes for the production of oil and gas from coal, with a discussion of the economic factors involved.

K. Schmidt-Nielsen: Locomotion: energy cost of swimming, flying, and running. *Science*, **177**:222 (1972). Not really related to the other citations in this list, but awfully interesting nevertheless. For example, even though the frictional and viscosity effects are much less in running compared to swimming, much more energy must be expended to run than to swim. Tables and graphs present these and other data with accompanying discussions. Also see R. C. Plumb: *J. Chem. Educ.*, **49**:112 (1972), for a related thermodynamic exposition on life-saving.

Here are four articles on the direct conversion of solar energy to electrical energy: B. Chalmers: The photovoltaic generation of electricity. *Sci. Amer.*, **235**:34 October (1976), on the fabrication of inexpensive solar cells. J. G. Asbury and R. O. Mueller: Solar energy and electric utilities: should they be interfaced? *Science*, **195**:445 (1977), maybe we should not combine the two technologies; it may well turn out to be too expensive. W. G. Pollard: The long-range prospects for solar energy. *Amer. Sci.*, **64**:424 (1976), solar energy may never be more than slightly utilizable as an alternate source of energy. Dr. Pollard is a recognized expert in the field, his pessimism is disturbing. So, the fourth reference is more cheerful, by E. Faltermayer in *Fortune* for February, 1976, titled Solar energy is here, but it's not yet utopia.

Chapter 14, in the event that you have not yet looked at it, treats topics that are closely related to the content of this chapter. Some of the cited articles in Chapter 14 are identical to a few cited here, for instance. You may wish to look at Chapter 14 now to find other references that supplement those listed here.

chapter 9

Food, People, and Chemistry

For animals, including humans, food is required to sustain life. Among other things, food supplies energy. This energy can be thought of as held or contained within the chemical bonds, between the atoms, in food molecules.

THE ENERGY OF CHEMICAL BONDS

For example, sugar is a food, a carbohydrate, with molecules composed of carbon, hydrogen, and oxygen atoms. In the plant (such as sugar cane or sugar beet) carbon dioxide and water molecules are reformed into sugar molecules and oxygen molecules. The oxygen is released into the air, the sugar remains in the plant. However, the bonds in the product molecules, the sugar and the oxygen, contain more energy than the bonds in the original carbon dioxide and water. This extra energy is obtained by the plant from sunlight.

When sugar is metabolized in the body by reacting with oxygen in the air to form carbon dioxide and water again, this extra energy is released and used for life processes within our bodies. You might wish to burn a sugar cube in the air and notice the energy released as heat and light. (This is the same procedure described in Chapter 4, where the emphasis was upon catalysis.)

Obviously, when sugar reacts with oxygen within our bodies the release of energy is necessarily less dramatic; the heat is released little by little in a series of controlled stepwise reactions rather than rapidly, as when sugar is burned in the open air. The several stepwise reactions are controlled by a series of catalysts, called **enzymes**, and much of the energy that is released is immediately stored again in other chemical bonds.

Storage and release of chemical bond energy by the body

The details of the process are not completely known, and what is known is sufficiently complicated so that a complete description is confusing to everyone except specialists. In broad terms, the released chemical bond energy is trapped in a molecule called adenosine triphosphate, which is diagrammatically represented in Figure 9–1. (H symbolizes hydrogen atoms; N, nitrogen atoms; O,

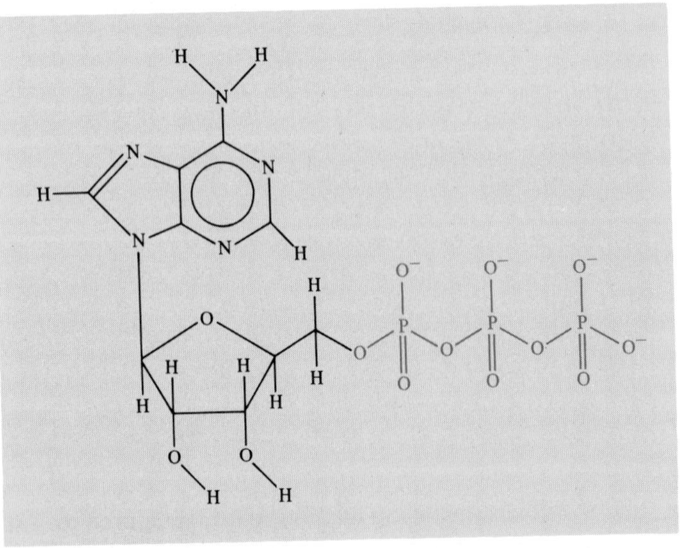

Figure 9–1. Adenosine triphosphate, ATP, or A℗℗℗.

oxygen atoms; and P, phosphorus atoms. Carbon atoms are represented by line intersections; the lines themselves represent covalent bonds. The minus signs near the oxygen atoms symbolize negative charges located largely at those atoms.) Notice in particular the three phosphorus atoms, linked by oxygen bridges one to another. The energy released by the "burning" of food can be thought of as being stored within the bonds of the phosphorus–oxygen grouping at one end of the molecule. To emphasize this, we will use the symbol A℗℗℗ for the adenosine triphosphate molecule (another common symbol for this molecule which you will see in other books is ATP).

In our bodies, the A℗℗℗ molecule is transported from its region of formation to other places where the energy is needed. At those locations, it is changed into adenosine diphosphate, which is shown schematically in Figure 9–2. We will use the symbol A℗℗ for adenosine diphosphate (other references often use the symbol ADP). As you may have guessed, when A℗℗℗ gives up its energy to become A℗℗, the phosphorus–oxygen grouping on the end is severed. The overall generalized chemical equation is

$$A℗℗℗ \rightarrow A℗℗ + \text{energy}$$

This reaction occurs in reverse at those regions in the body where sugar and oxygen react. We can show this in two generalized reactions:

$$\text{sugar} + O_2 \rightarrow CO_2 + H_2O + \text{energy} \qquad \text{(in a series of steps)}$$

$$\text{energy} + ℗ + A℗℗ \rightarrow A℗℗℗$$

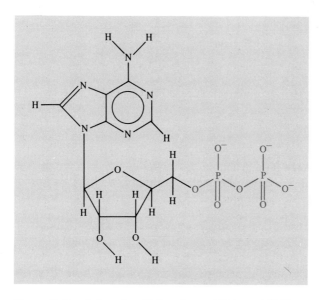

Figure 9–2. Adenosine diphosphate, ADP, or A℗℗.

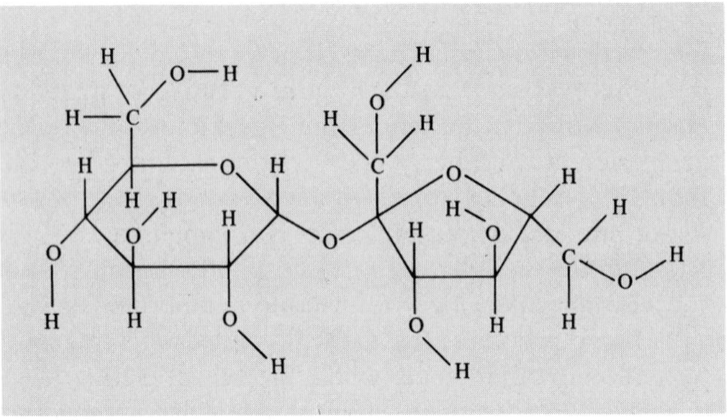

Figure 9–3. Table sugar (sucrose).

METABOLISM OF CARBO- HYDRATES

With this introduction as our very general guide, we can now look at some of the more important details. In the first place ordinary table sugar is itself a moderately complicated molecule, as we noted in Chapter 5. The ring on the left of Figure 9–3 is essentially a simpler sugar, called α-D-glucose, and the ring on the right is another sugar called β-fructose. In the digestive tract there is an enzyme that enhances the reaction of ordinary sugar (called sucrose) with water to form two simpler sugars, glucose and fructose:

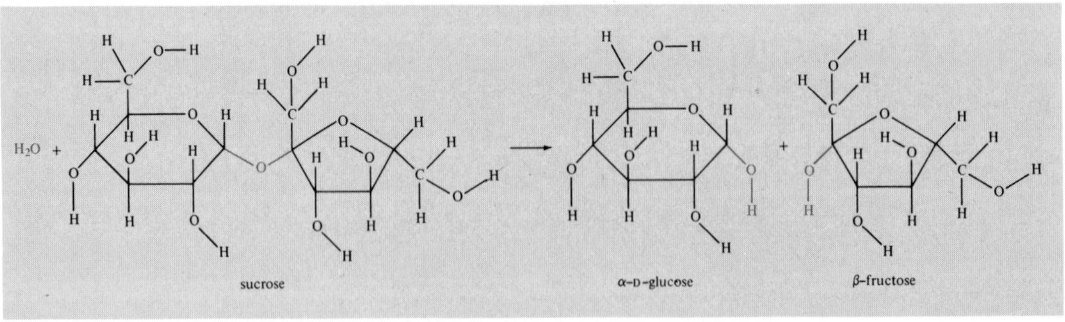

Glucose and fructose are actually the species that undergo reaction with oxygen in our bodies, releasing energy to the Aℙℙ (and the other ℙ) resulting in the formation of the higher energy compound, Aℙℙℙ.

Starches

Sucrose or other sugars such as fructose and glucose (there are many others also) are known as carbohydrates. For humans however, the most important carbohydrate is starch. There are many different kinds of starch, wheat starch, potato starch, corn starch, and so on. All of these, without exception, are composed of glucose units linked together one after another, something like the four units are linked in Figure 9–4. The difference between one kind of

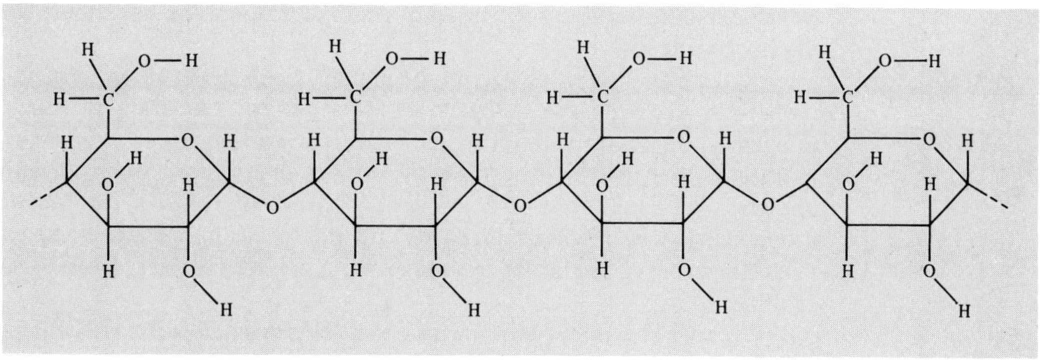

Figure 9–4.   Section of a starch molecule.

starch and another lies in the length of the glucose chain, in the branching of some chains, and in the three-dimensional arrangement of the chain. Most starches are long coiled molecules; some are highly branched; some are bushy shaped, and so on. (Chapter 10 treats some of these details.)

When starch is eaten, enzymes enhance the reaction of starch with water to reform individual glucose molecules. The glucose molecules are then utilized as a source of energy by reacting with oxygen in a series of steps within the body. Each step of the series of reactions involves one or more different enzymes that catalyze that particular step.

Metabolism of glucose   Essentially, the glucose molecule is broken into two smaller molecules and two APPP molecules are produced. This is easier to see if the glucose molecule is portrayed in its chain, not ring, structure. (Most of the time a glucose molecule is in a ring form, but it can, and does, momentarily exist in a chain structure.):

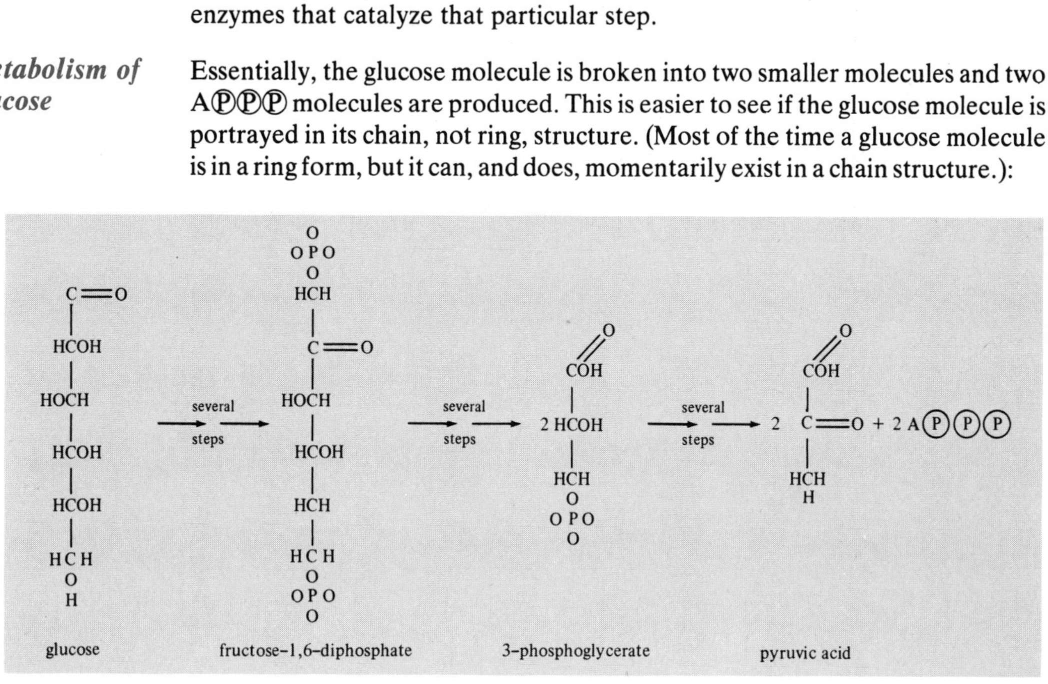

(Since our purpose it to present an overall description, the equations, above, are not balanced.)

Actually, this is merely the beginning. Each molecule of pyruvic acid (two from each glucose molecule) now undergoes oxidation to form two molecules of water and three molecules of carbon dioxide and several molecules of A℗℗℗. The process involves many more steps collectively known as the **Krebs cycle** or **TCA cycle**, or **tricarboxylic acid cycle**. One of the other substances involved in catalyzing this cycle is **coenzyme A**, called **CoA-SH** also. Schematically, it is shown in Figure 9–5 (all the carbon atoms are symbolized by C's).

CoA-SH contains within its structure a residue of pantothenic acid. Without this vitamin our bodies cannot make CoA-SH, and without CoA-SH there would be no way to obtain energy from carbohydrate food. In fact, as we shall see, there would be no way to obtain energy from fat nor to make use of the protein food we eat. The CoA-SH reacts with pyruvic acid to form **acetyl coenzyme A**, also called **acetyl-SCoA** or **acetyl CoA** (Figure 9–6).

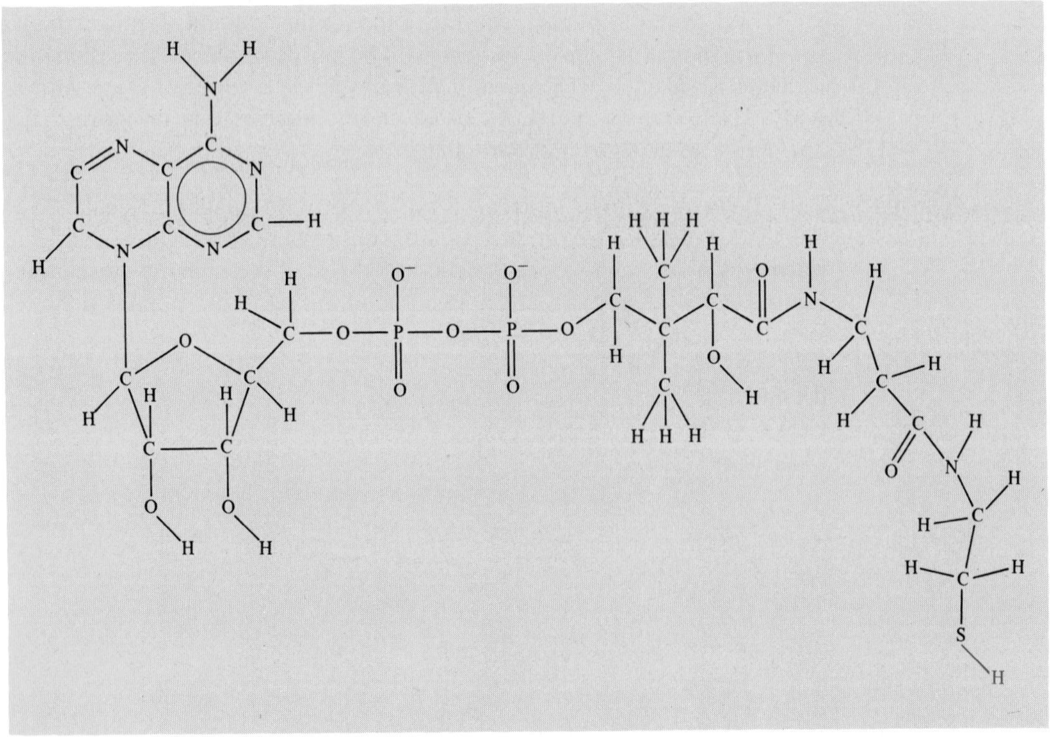

Figure 9–5.   Coenzyme A, CoA–SH.

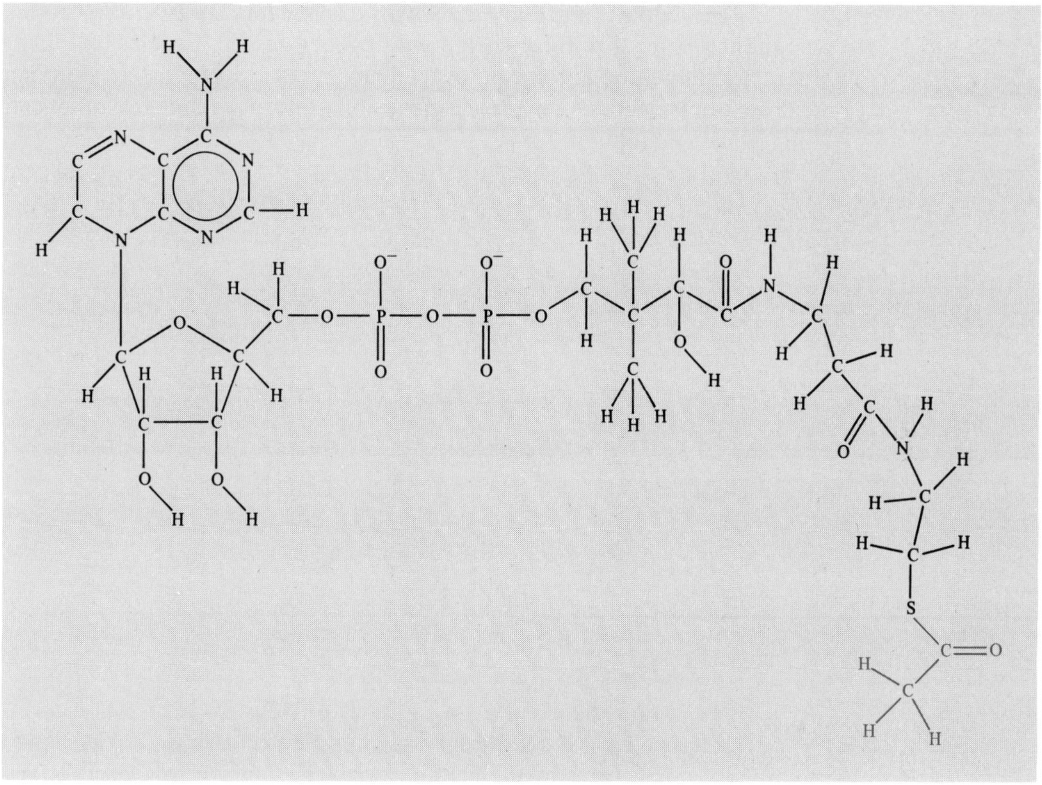

Figure 9–6.   Acetyl coenzyme A, acetyl CoA.

Notice that the only difference between CoA-SH and acetyl CoA is at the end of the long tail of this molecule. The reaction between pyruvic acid and CoA-SH is shown in the following summarizing equation.

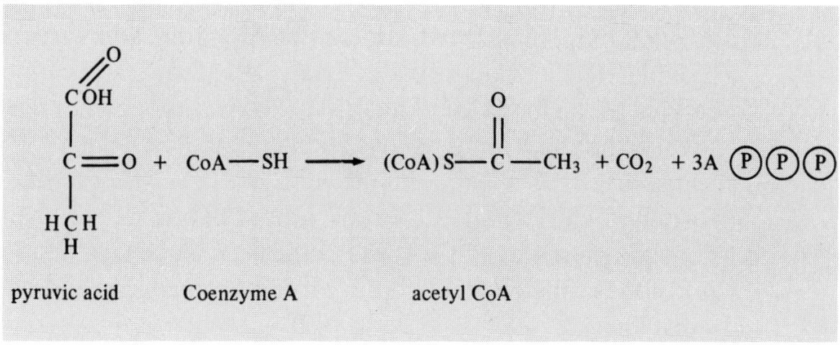

The acetyl CoA then reacts with oxaloacetate (this is a new one, not mentioned before) to form citrate and regenerates the CoA-SH. Notice that there are four carbon atoms in the oxaloacetate and six in the citrate. The two carbon atoms needed come from the acetyl-CoA, as is shown symbolically in the following equation:

oxaloacetate   acetyl CoA   citrate   Coenzyme A

The citrate then undergoes a series of stepwise reactions (see Figure 9–7), losing two carbon atoms (as two molecules of carbon dioxide) and four hydrogen atoms (as two molecules of water). During this process several molecules of A(P)(P) are converted into A(P)(P)(P), which now contains the energy released in the total process. Eventually, some of the products from the citrate are transformed once more into oxaloacetate, and the whole thing happens again, over and over and over. Each new cycle begins of course with a new pyruvic acid molecule reacting with CoA-SH. Remember we get two pyruvic acid molecules every time a glucose molecule is "broken" into those two pieces.

Thus, from each glucose molecule we get two molecules of A(P)(P)(P) and two molecules of pyruvic acid. Then, each pyruvic acid molecule yields 15 more A(P)(P)(P) molecules, two molecules of water, and three molecules of carbon dioxide. The carbon dioxide, eventually, reaches the blood stream where it is carried to the lungs for exhalation. The water becomes part of the water system within the body and is eventually excreted. The A(P)(P)(P) molecules are transported to a site of muscular activity, where their energy is released forming A(P)(P) and (P), or to other places where their energy is used to help form molecules of the various substances needed for life, such as hormones, nucleic acids, fats, proteins, glycogen, and several other different substances.

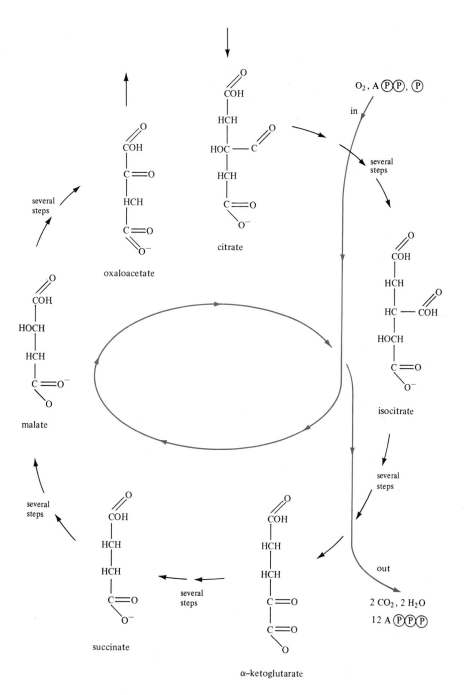

Figure 9–7. The tricarboxylic acid cycle, summarized.

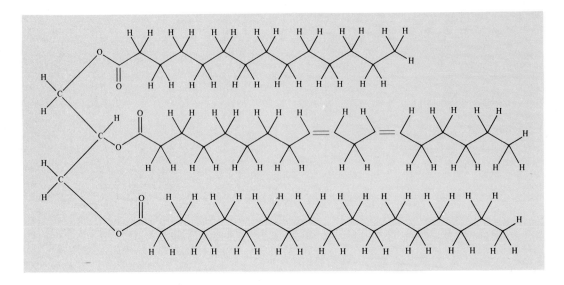

Figure 9–8. Typical unsaturated fat molecule.

METABOLISM OF FATS
Chemical composition of fats

Fats and oils are composed of another type of molecule. Just as we can describe carbohydrates or starches as units of glucose (or other sugars) linked together, we can describe all fats and oils as three long chains of carbon atoms, each attached by oxygen atom links to a three-carbon atom combination. Figure 9–8 shows a typical fat molecule, similar to the schematic formula shown in Chapter 5 but with a different number of carbon atoms and hydrogen atoms in the three **fatty acid residues**. Other fat molecules are different in the length of the three carbon atom chains or in the presence and location of the double bonds within any single chain, or both. In general, oil molecules have more double bonds within the chains. The chains that have double bonds are said to be **unsaturated**, and fats or oils with such double bonds in the chains are known as unsaturated. **Polyunsaturated** fats and oils have more than one double bond per chain.

The breakdown of fats

In the body, under the influence of appropriate enzymes, a fat molecule reacts with three water molecules to form a glycerol (sometimes called glycerine) molecule and three **fatty acid** molecules. The glycerol is eventually converted into 3-phosphoglycerate, then transformed into pyruvic acid, and finally incorporated into the same reaction system that metabolizes glycose. Each fatty acid molecule is taken apart, two carbon atoms at a time, forming one

molecule of acetyl-CoA for each two carbon atoms by reacting with CoA-SH
And, as you may remember the acetyl-CoA is also involved in the metabolism
of glucose.

acetyl CoA

***Buildup and
storage of fats***
Suppose that the body already has enough energy available in the form of
A℗℗℗ molecules. In that case, acetyl-CoA may be used to make fatty acids,
building these up two carbon atoms at a time. The 3-phosphoglycerate is
converted back to glycerol and together with three fatty acid molecules
produces one molecule of fat within the body which may be stored for future
use. Similarly, any excess starch that is ingested may be converted into fat
within the body. That is, from glucose the body can form 3-phosphoglycerate
and acetyl-CoA, and, from these, whether they were generated from fat or
starch, fat molecules can be made. In general, when the body receives only as
much starch and fat as is needed for daily activities, no extra fat is stored; if too

little is ingested, the fat stored already within the body is used up; and, unfortunately, if too much is ingested, more fat is produced and stored.

METABOLISM OF PROTEINS
Composition of proteins

As we learned in Chapter 5, proteins are large molecules. The unit for carbohydrates or starch is a single substance (for the most part) glucose. The units for fats are glycerol and three fatty acids; any three fatty acids, although the four most common in human fatty tissue are palmitic, stearic, oleic, and linoleic acids. The units in proteins are α-amino acids, and there are 20 of these commonly found in human proteins. Some of these are shown diagramatically in Figure 9–9.

All of these structures have a common characteristic. Notice the carbon next to the $-C\overset{\displaystyle O}{\underset{\displaystyle OH}{}}$ grouping. That carbon is attached to an amino group, $-N\overset{\displaystyle H}{\underset{\displaystyle H}{}}$; it is called an α carbon. Since these substances are all acidic, to some

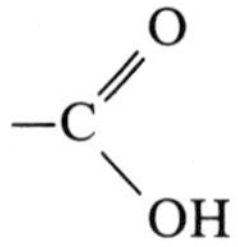

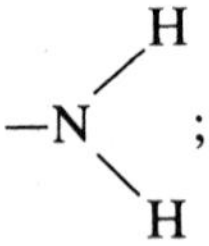

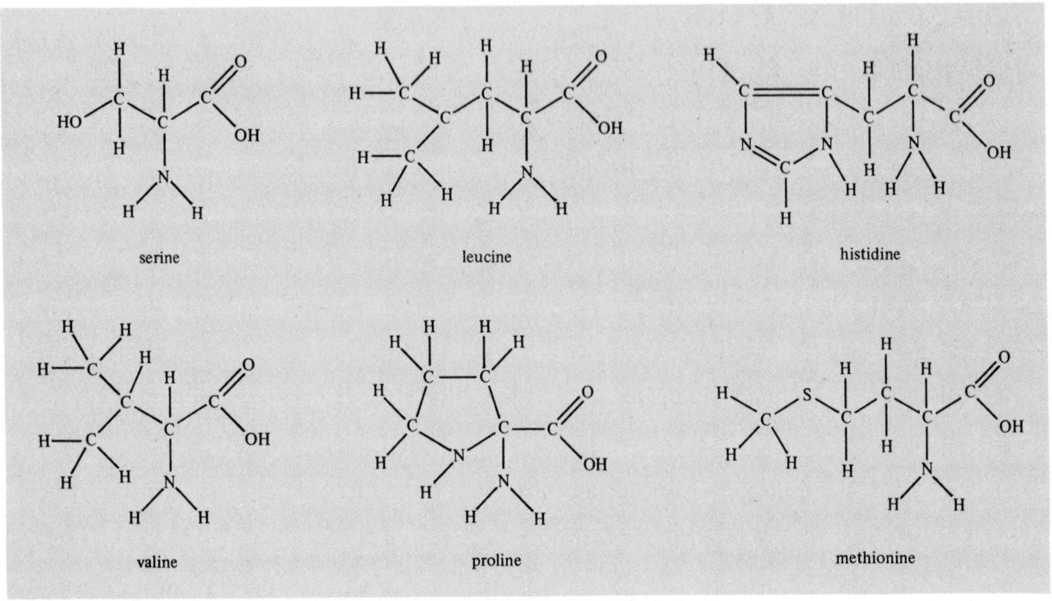

Figure 9–9.   Six α-amino acids.

extent, and since all have an amino group attached to the α carbon atom, they are called α-amino acids. Two α-amino acids can react with each other.

alanine   serine   a dipeptide

A third can react with the product from the first reaction.

glycine   a tripeptide

A fourth can react with that product (see Figure 9–10), and so on. A protein molecule is formed by a series of reactions involving hundreds, or more, of α-amino acids. There are 20 α-amino acids, and any combination of these that involves a few hundred or more individual α-amino acid molecules is a protein. So, an uncountable number of different protein molecules is possible. Any of these protein molecules can be broken down into the constituent α-amino acids by reaction with water under the influence of specific enzymes.

Digestion of proteins

One of the enzymes that will break down protein is found in raw pineapple juice. If you enjoy raw pineapple, you may have noticed after eating it that your tongue felt peculiar. It should; some of the protein that is now in your stomach as separate α-amino acid molecules was originally part of your tongue. When protein is digested, it is broken up into separate α-amino acid molecules. These

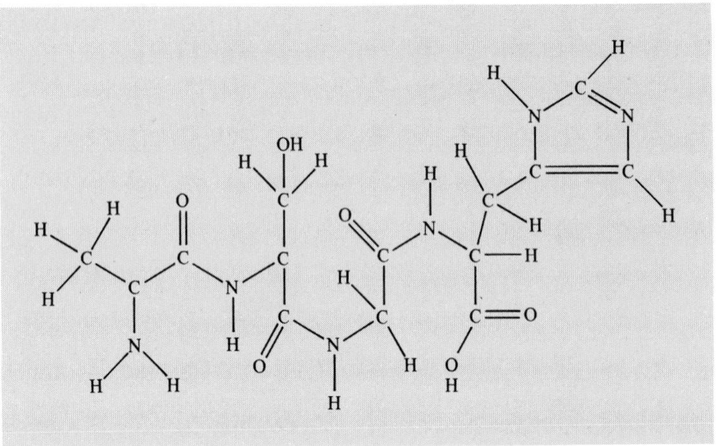

Figure 9–10. A polypeptide showing (left to right) alanine, serine, glycine, and histidine residues.

are then mostly transported to other sites in the body where protein molecules are needed. Provided that the diet has furnished the right proportion of the different α-amino acids, they can be reassembled in the proper number and order to form the kind of protein molecules need at that spot.

However, not all α-amino acid molecules are used to make new protein. Some are used for a different purpose. The nitrogen atoms are removed in a series of reactions, ultimately forming urea which is excreted in the urine. The residue, a chain of carbon atoms with hydrogen and oxygen atoms attached, is just right for participating in the Krebs cycle we described under the Metabolism of Glucose to form carbon dioxide and water and energetic A℗℗℗ molecules. The residue might also be converted, two carbon atom units at a time, into fatty acids, and eventually into fat stored in the body. (Unfortunately, any excess of carbohydrate, fat, or protein, seems destined to become fat.)

FOOD
Energy from food

A summary will help at this point. The energy we need to survive is obtained indirectly from the Sun in the form of chemical bond energy in the molecules produced by plant life. Either we ingest these molecules directly by eating vegetables or indirectly by eating the flesh of animals which have ingested the plant products. Ultimately, this energy is utilized by our bodies in the form of transportable A℗℗℗ or it is stored as fat for future use.

The general process involves a breaking up of food molecules into smaller pieces. The pieces from starches and fats are directly usable as energy sources. Starch in particular is used for energy, being metabolized via the Krebs cycle to produce A℗℗℗, carbon dioxide, and water. The A℗℗℗ molecules provide energy for muscular activity as well as energy for building complicated and necessary molecules within our bodies, and for other energy absorbing purposes, such as forming concentrated solutions of dissolved substances within the cells and tissues of our bodies. Fats can be used in the same way as starches, but they need not enter the Krebs cycle if there is a sufficient or excessive intake of carbohydrate. In this case the fats are broken down only as far as glycerol and fatty acids, and these pieces are reconstructed into our own special kind of fat molecules and are stored within our bodies.

The pieces from proteins are α-amino acids. Usually they are not broken down further but are used to rebuild and maintain our own protein structure. They are broken down into smaller pieces and incorporated into the Krebs cycle to be used for energy only when our intake of starches and fats is insufficient. However, proteins can also be converted into fats to be stored when the intake of all food is in excess of what is needed. In this case or when protein is used for energy, the nitrogen is removed and excreted in the form of urea molecules.

The normal adult should have approximately 50–75 g of fat per day. If less than this is habitually eaten, the body will require a disproportionate amount of carbohydrate and protein to satisfy the energy requirements. Further, the fat must contain linoleic acid, since the human body cannot make this indispensable acid. Also essential and found in many fatty foods are vitamins A, D, E, and K. For those who wish to lose weight, some fatty foods are recommended because they help to produce a feeling of satiety (translated: less desire to nibble on snacks between meals). However, fats are high in energy; 1 g of fat furnishes about 9 kilocalories (kcal).

In a chemistry text we do not use the old-fashioned Calorie (with a capital "C") that was once popular with nutritionists. A calorie (small "c") is the amount of energy needed to raise the temperature of 1 g of water from 14.5°C to 15.5°C (about 57°F to 59°F). One gram of water is about one fifth of a teaspoon, so you see that 1 calorie (cal) is a very small amount of energy. The gram of fat we mentioned before, when burned in air, produces 9000 cal. Obviously we need a larger unit of energy. A kilocalorie (kcal) is used as the energy unit. This is 1000 times larger than a calorie. Roughly, 1 quart (qt) of water weighs 1000 g. So, one kilocalorie is the approximate amount of heat energy absorbed by a quart of water when you raise its temperature from 57°F to 59°F (strictly, 1000 g of water from 14.5°C to 15·5°C). In the chemical bonds of 1 g of fat, burned to form carbon dioxide and water, there is enough energy released to raise the temperature of 1 qt of water 18°F or 9 qt of water 2°F. Common sources of food fats include vegetable oils, shortening, butter, margine, nuts, meat, fish, potato chips, some candies, cake, cookies and even some breads.

Carbohydrates or starches provide about 4 kcal/g, less than half as much energy as 1 g of fat. As most people know, sources of carbohydrates include sugar, candy, cake, cookies, bread, cereal, potatoes and other tuberous plant foods, beans, peas, and fruit. Many of these foods also provide other nutrients in addition.

Proteins also provide about 4 kcal/g. Sources include cereal, meat, fish, milk, cheese, eggs, beans. Except for water, most of the weight of our bodies is proteinaceous. This includes muscles, skin, the network lattice in which calcium and phosphorus are held in our bones and teeth, parts of individual cell membranes throughout the body, blood vessels, blood plasma, hair, and finger and toe nails. A normal adult requires about 50 g of protein each day. This daily protein intake is used to replace some of the protein molecules already in place within our bodies, and to provide the raw materials for the internal synthesis of hormones, enzymes, and antibodies.

Eight of the 20 common α-amino acids in proteinaceous food are said to be "essential" on human nutrition. That is, they must be supplied from plants that we eat or animal products that we eat; the human body cannot make these eight from other α-amino acids. The names of these eight essential α-amino acids are isoleucine, leucine, lysine, methionine, threonine, tryptophan, phenylalanine, and valine. If these are present in the food we eat in sufficient quantity, the body uses them directly and also constructs from them the other 12, alanine, arginine, aspartic acid, citrulline, cystine, glutamic acid, glycine, histidine, hydroxypyroline, proline, serine, and tyrosine.

If even one of the essential α-amino acids is deficient, or absent from the diet, serious consequences quickly ensue, and eventually death. Few single sources of protein contain desirable proportions of all the essential α-amino acids. Therefore, several different kinds of proteinaceous food should be consumed as good dietary practice. Of the eight, the human body needs smaller amounts of tryptophan and larger amounts of leucine (about three or four times as much leucine as typtophan). The other six are intermediate, each needed in amounts from about twice to three times the need for tryptophan. In a general sense, the best balance in amino acids is obtained when the diet contains foods from plants and from animals Of course, the so-called nonessential amino acids in protein are valuable in the diet because they can be utilized.

MINERALS AND VITAMINS

In addition to our nutritional need for starch (carbohydrate), fats, and proteins, we also need minerals and vitamins from our food. Vitamins are complicated molecules that, for the most part, are involved in enzyme control of the reactions in our bodies. The minerals fill various needs, such as phosphorus in A℗℗℗ molecules, calcium and phosphorus in bones and teeth, iron in red blood cells, iodine in thyroxine, and so on.

All in all, the foods we eat supply about three dozen elements. Four of these, oxygen, carbon, hydrogen, and nitrogen, make up about 95% of the weight of the human body and are the basis of starch, fats, and proteins.

Another approximate 4% of our weight is owing to calcium, phosphorus, potassium, sulfur (present in virtually all proteins), sodium, chlorine, magnesium, and iron. All the rest of the elements found in the human body constitutes approximately 1% of our total weight. This 1% includes copper, manganese, iodine, zinc, cobalt, aluminum, arsenic, barium, boron, bromine, selenium, chromium, fluorine, tin, silicon, molybdenum, and still others. Most of these are known to be essential to life; perhaps all the rest are also, or perhaps they merely happen to be present because they were in the food we normally eat or entered the body by accident.

Minerals

Minerals are essential in the human diet. The word, "mineral" has a special meaning in a nutritional context. It refers to any element other than oxygen, carbon, hydrogen, or nitrogen. In all cases, these other elements are present as bonded atoms (such as phosphorus in adenosine triphosphate) or as charged ions (such as sodium ions in the blood plasma); they are never present as free, uncombined atoms, except by accident or deliberate intention, in which case they are not nutritionally useful.

Calcium, the most common mineral in human tissue, is needed for our bones and teeth, as most people know. In these tissues the calcium is present, along with phosphorus and oxygen, as the compound $Ca_5(PO_4)_3OH$. This substance is a solid, chemically very similar to the rock-mineral apatite, which occurs abundantly in nature. Calcium is also present in the blood and is involved in muscle contraction processes. The first symptoms of calcium deficiency include hypersensitivity, twitching, and cramps. Sources of calcium include milk, cheese, and green leafy vegetables.

Phosphorus is present throughout the body in many forms such as adenosine diphosphate and adenosine triphosphate and in our bones and teeth. Because it is also pervasive in the foods we eat, phosphorus deficiency is quite rare. In fact, the only common way to induce a phosphorus deficiency is to ingest substances that tend to remove the phosphorus we have already obtained. This particularly includes excessive and habitual use of "antacid" concoctions taken for "upset stomach" disorders. Symptoms of phosphorus deficiency are a general weakness, anorexia, malaise, and "pain in the bones." Phosphorus is available in meat, fish, milk, cheese, eggs, cereals, soft drinks, and, to a slight extent, in fruits and vegetables. There is some tendency for people to ingest more phosphorus than they need.

Before we go further, a word of caution will be in order for some readers. As you note the symptoms associated with a deficiency of this or that nutritional substance, do not immediately conclude that you suffer from the associated deficiency. The same symptoms can be caused by other factors, not the least of which is suggestibility. A diagnosis of a nutritional deficiency cannot properly be made without more knowledge and experience than this book provides.

Iron is present conspicuously in our blood, as we shall see in a following chapter. It is also notable in all muscular tissue, in the liver and spleen, and in bone marrow. The human body has a limited capacity to compensate for an excess or a deficiency of iron and excretes this element constantly but slowly. Therefore, it is necessary to include iron containing food in our diet continuously. However, excessive ingestion of iron is also to be avoided, since an excess of iron is toxic. Major nutritional sources of iron include liver and meat, eggs, fruits, unrefined cereals, and green leafy vegetables.

Iodine is present in the thyroid gland and in one of the hormones produced by this gland, thyroxine. Thyroxine and other hormones promote a variety of reactions within the body. Of all the hormones, thyroxine is one of the most important because it "controls" metabolic processes. The adult human requires approximately 3×10^{-4} g of iodine daily. This is no problem for those who live within a few hundred miles of the ocean. The ocean contains dissolved iodide ion and, when droplets of ocean spray in the air evaporate, the iodide ion and other constituents are carried inland by winds. Ultimately they fall upon plants and the soil. So, ordinary consumption of food supplies the necessary iodine. For those living further inland, this source is insufficient or unavailable.

Beginning several years ago and continuing today, distributors of table salt add about 1 oz of potassium iodide, KI, to each 600 lb of salt, well mixed; the mixture is called iodized salt and can be purchased in any grocery store. If you live near the ocean and use iodized salt, you ingest more iodine than you need, but except for rare instances no harm whatsoever is done. If you live well inland, iodized salt can be said to be an essential component of your diet. Symptoms of iodine deficiency include lassitude and an enlarged thyroid gland, known as goiter. As you have guessed, iodine is available in sea food and to some extent (depending upon where the food was produced) in milk, meat, and fruit.

Sodium is the principal positively charged ion present in blood plasma and other extracellular fluids. Each normal adult ingests several grams of sodium, as sodium ion, daily, largely obtained from the salt we use on our food. The salt balance in our body is markedly affected by sweating. Excess sodium is rapidly excreted, unlike iron, for example. A high rate of iron intake may have an adverse effect because the substance is secreted only slowly, but our bodies control the sodium level more effectively. However, the habitual use of salt in excess may lead to high blood pressure.

We need several grams of **chlorine**, as chloride ion, daily and obtain this from salt also. Chloride ions are present in blood plasma, in other extracellular fluids, and in intracellular fluids as well. Chloride ion is also present in saliva, in gastric fluid, and in all other body fluids.

A few grams of **potassium** per day is necessary. Like sodium if an excess is ingested, it is excreted. Almost all relatively unrefined foods we eat contain considerable potassium. For us, it is the predominant positively charged ion present in intracellular fluids and in particular is involved in the transmission of nerve impulses and in some enzyme systems.

Magnesium is present in our bones and to a lesser extent in all other tissues. It is also involved in some enzyme systems. Next to potassium, magnesium is the predominant positively charged ion in intracellular fluids. A magnesium deficiency produces neuromuscular dysfunction, which is as unpleasant as it sounds. This deficiency is quite rare, since the foods we normally eat, meat, cereals, milk, and cheese, contain enough magnesium for our needs.

Every cell in our bodies contains some **sulfur**, mostly as a component of sulfur containing, α-amino acids in the cell proteins. It is also present in all the extracellular proteins, including those in our saliva, blood, and bile fluid. Sulfur is in insulin and in all the other protein containing hormones. We obtain the sulfur we need from the protein in foods we eat, including especially meat, fish, cereals, milk, cheese, beans, nuts, and eggs.

Our intake of **manganese** probably should be about 0.003 to 0.007 g/day; major sources include nuts and cereals. (Whenever "cereal" is mentioned in this chapter, whole grain cereal is meant, not refined cereal products such as white flour or polished rice.) Manganese is found in our bones and other tissues and organs, and in many enzyme systems.

Copper deficiency is rare in man; symptoms include anemia, skeletal defects, and cardiovascular lesions. Copper is found in the largest amounts in the brain, liver, heart, and kidney. It is critically important in the formation of hemoglobin. Dietary sources of copper include nuts, legumes, shell fish, liver, and raisins.

Zinc is found in many of the enzymes that govern metabolic processes, such as those discussed under the metabolism of glucose in this chapter. Zinc has other functions, but the details are not well understood. Symptoms of zinc deficiency include loss of appetite, dwarfism, loss of taste acuity, and slow healing of wounds. In some areas of the United States, it is suspected, only marginal amounts of zinc are available in typical diets. Sources of this mineral include meat, fish, eggs, milk, and cereals.

Cobalt is a component of vitamin B_{12}, to which we will refer later in this chapter.

Molybdenum is a component of the enzyme, xanthine oxidase, which is part of the reaction system involved in the formation of uric acid. Normal dietary sources provide sufficient molybdenum.

Fluorine, as fluoride, a negatively charged ion, is believed to be an essential mineral. At any rate, we typically ingest about 0.0003 to 0.003 g/day, and it is incorporated in our bones and teeth. This element is present in fish, tea, and in drinking water obtained from deep wells in limestone strata. It has been well established that fluorine, as fluoride ion, in the diet tends to prevent dental caries (cavities). In many regions where fluoride ion is present in very low concentrations in the drinking water, it is deliberately added to the water supply. The optimum amount is 0.001 g/liter or 1 part per million (ppm).

This politico-chemical-nutritional decision has been the source of considerable controversy over the years. Those favoring the addition of fluoride ion point to the facts with respect to decreased incidence of dental caries. Those opposing point to the toxic effect of excess fluoride ion asserting that too much fluoride in early childhood causes unsightly mottled teeth, which is true. The opponents of fluoridation also point out that a large amount could be inadvertently (or even deliberately) dissolved in the water supply, seriously poisoning an entire community. This is a conceivable possibility, although perhaps very unlikely. Others are concerned with the invasion of privacy. They assert that the addition of fluoride ion to their water is tantamount to administering medication to them without their consent. Those favoring fluoridation point out, in rebuttal, that municipal water supplies have, for years, been treated with chlorine or ozone to kill harmful bacteria otherwise present. The controversy has not yet been settled to the satisfaction of all concerned in many communities.

Another way to minimize dental caries is to apply the fluoride ion topically, directly on the surface of the teeth. This can be done routinely by a dentist or dental technician, or more commonly but less effectively by daily use of a fluoride containing dentifrice. Many brands of this kind of toothpaste are available, some carrying a certificate of approval granted by responsible dental authorities. If during use some of this toothpaste is swallowed, no harm from the fluoride in the concoction is ordinarily expected because the concentration is low.

Chromium, the same metal we see on shiny plated automobile bumpers and trim, is a "cofactor" in the insulin hormone system. Chromium deficiency produces poor metabolism of glucose, and in some areas of the United States it is suspected the amount of chromium available in a normal diet is marginal. Chromium is typically present in meat, cereals, and brewer's yeast.

Selenium, the last to be described in this listing, is essential, but in very small amounts. The function of selenium is not well understood; sufficient amounts apparently are supplied by a normal diet. In a very few and isolated regions selenium is present as a negatively charged ion in drinking water in privately drilled wells, such as on a farm. Humans who ingest excessive amounts of selenium suffer from minor discomforts, such as severe headaches and general malaise, and major difficulties including severe skin disorders and insanity. As a small recompense, when selenium is present in drinking water in concentrations sufficient to produce these symptoms, it is very likely that the surrounding soil is rich in deposits of precious metals, especially, gold, silver, and platinum.

Time for a summary. The eleven major elements in our nutrition are oxygen, carbon, hydrogen, nitrogen, calcium, phosphorus, potassium, sulfur, sodium, chlorine, and magnesium, in decreasing order of percent by weight in the human body. Generally speaking, these are present in sufficient amounts in a normal diet. The next nine, in the same order, are iron, copper, zinc, manganese, iodine, cobalt, chromium, molybdenum, and fluorine. As a general statement, these nine tend to occur in low amounts in refined foods, such as white bread and fabricated foods like breakfast cereals unless these foods are "fortified" by the deliberate addition of one or more of the constituents removed by the refining process.

The preceding statement should not be interpretable as an indictment of the food industry. Unrefined foods, such as whole wheat flour, for example, are often not acceptable to the housewife. Whole wheat flour has a strong taste and is not suitable for all types of baking. Refined flour, on the other hand, has a taste we have grown accustomed to, even though only some of the nutritional substances that were removed by refining have been replaced. The food industry in a capitalistic economy tends to supply what is desired and purchased, although it is generally true that whole, unrefined, "natural" foods are nutritionally superior because they are more likely (not guaranteed) to contain a larger proportion of the nutrients we need. It is also generally true that "enriched" or "fortified" refined foods are not restored to their original nutritional condition, at least as this pertains to the minerals and vitamins that may have been present initially. However it is not necessarily true that refined and fabricated foods are bad for you. It is merely necessary to be sure to supplement a diet that consists of refined and fabricated foods with other sources of the nutrients we need.

A few words about "organically grown" foods are in order here. It is true that some soils are depleted if we compare the minerals they now contain to those which they may have had some years ago. It is also true that "artificial" fertilizers may not contain enough of these depleted minerals to replace completely what may have been lost over the years. The conclusion that foods grown on soils treated only with artificial fertilizer are therefore deficient in these needed minerals is naive. If a soil were so much depleted of those minerals, the plants would not grow either. Although there are some exceptions, we do not obtain a plant with "more" minerals in it simply because we take care to fertilize the soil with organic fertilizer, even if that material happens to contain "more" of those minerals. Too often, persons who are quite properly concerned about good nutrition accept without question the so-called arguments favoring the exclusive or major use of organically grown foods. There is no factual evidence to support the exaggerated claims of many of the proponents of organic foods. Such foods are certainly not nutritionally harmful, but, typically, they are overpriced in terms of their real nutritional value.

Other arguments favoring the use of organic foods are related to the use of pesticides in the production of "artificially" nurtured plants. This can be a serious objection; recent federal regulations require that if present such materials be identified on package labels, thus forewarning the consumer.

Other minerals must be listed in this summary simply to be complete. The human body contains trace amounts of aluminum, arsenic, barium, boron, bromine, cadmium, silicon, vanadium, tin, and nickel. Some of these, perhaps all, may be essential, although their specific functions are not yet known. Again, they are available in sufficient amount, in so far as we can guess, in the normal human diet.

Vitamins

We turn next to the vitamins. Four vitamins are soluble in fats and oils: A, D, E, and K. Therefore, we might expect to obtain them in fatty foods and, in a general sense, this is correct. Several of the water soluble vitamins were at some time known as members of the "B complex" vitamins. Many different designations and names have been used, and changed, for these factors over the years. For example, the terms vitamin F, vitamin G, or vitamin H appear in older literature; these designations are no longer used.

A vitamin is a specific, known organic substance essential for life in small amounts. Vitamins may be obtained from the food we eat or may be synthesized from specific precursors in our body, or both. They are necessary for metabolic processes in our body. The known vitamins vary considerably in their structures. It is possible that several other substances, not yet identified as such, will be designated as vitamins in the future. Here, we will describe each of the known vitamins, beginning with the fat soluble vitamins.

Vitamin A is also called **retinol**. Figure 9–11 shows its schematic structure. Originally, the structure was unknown. Vitamin A was recognized only by the fact that characteristic deficiency symptoms appeared when certain

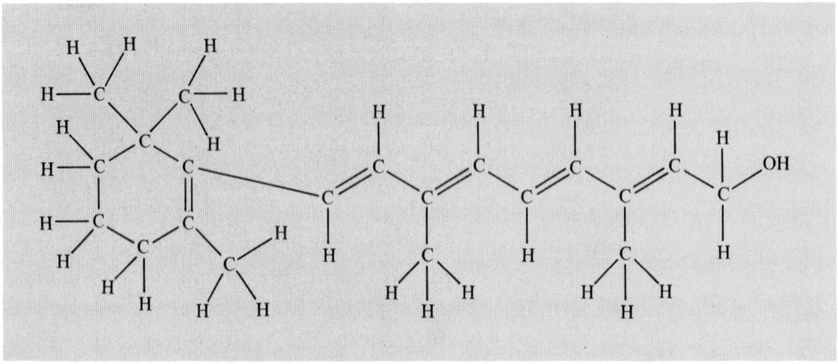

Figure 9–11.   Retinol (vitamin A).

types of foods were lacking in the diet. When those foods were restored, the symptoms vanished. Hence, it was concluded, those foods contained an unknown substance or substances necessary for good health. The other vitamins were identified in the same general way initially and were given alphabetical letter names since this was the most practical way, then, to designate them. Today, the structures of all the known vitamins have been determined, and chemical names have been assigned to those molecular species. It is still generally acceptable to refer to a vitamin by its alphabetical letter designation, but the chemical name is preferable for most of the vitamins.

Vitamin A is involved in night vision and in the development of teeth in young children. Vitamin A deficiency notably affects the outer layers of skin and the mucous membranes associated with body openings, such as the eye, mouth, respiratory tract, digestive tract, and genitourinary tract. Presumably, this vitamin aids these mucous membranes in their action as barriers to infection.

Vitamin A occurs only in foods of animal origin, particularly in fish liver oils, butter, and egg yolk. However, a precursor substance, **β-carotene**, occurs in some plants used as food. β-Carotene can be converted by the body into vitamin A; note the structural relationship between β-carotene, shown in Figure 9–12, and Figure 9–11, which shows the structure for retinol. For all practical purposes, foods which contain β-carotene can be considered as though they contained vitamin A. These include almost all yellowish colored vegetables, such as carrots, sweet potatoes, yams, corn, and even spinach (more greenish than yellowish). Because vitamin A is indeed necessary for night vision, some people have assumed that their night vision would be enhanced if, for example, they eat a lot of carrots. The extra carrots are no doubt usually beneficial, but eating several more does not further improve one's ability to see well in the dark. Some foods are fortified by addition of vitamin A; the extra vitamin A is obtained by extraction from fish liver oils or made synthetically.

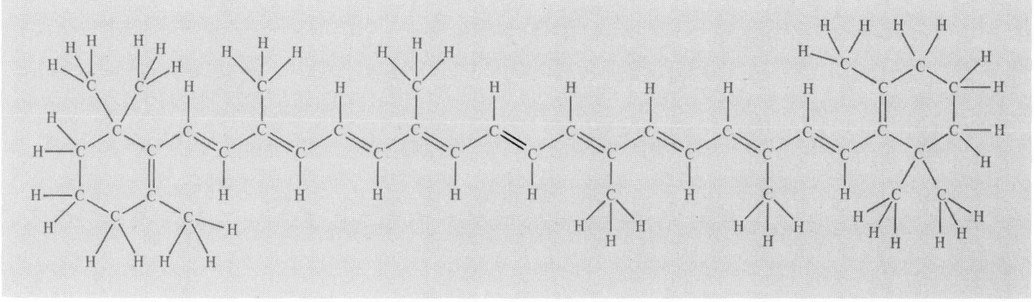

Figure 9–12.   β-Carotene.

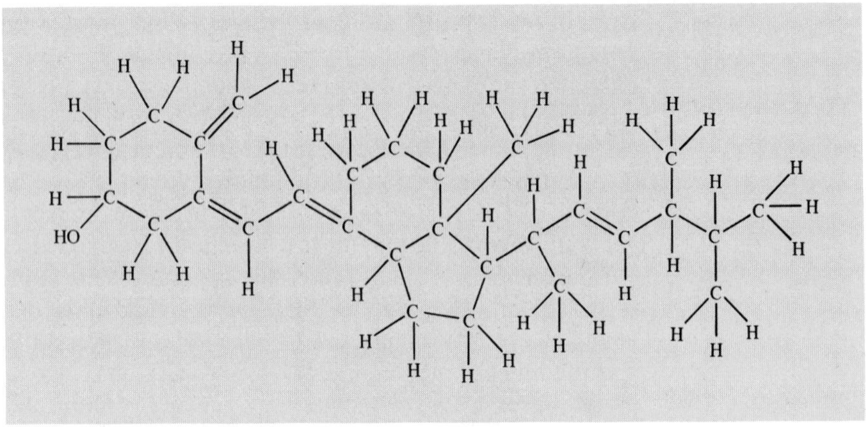

Figure 9–13.   Ergocalciferol (vitamin D₂).

When foods containing vitamin A or β-carotene are cooked, there is no resultant loss of either one, although on long exposure to the air, some is lost. An excess of vitamin A causes headaches and high blood pressure, but this is quite rare in a normal diet.

Under the influence of sunlight, ergosterol is converted to **vitamin D₂** or **ergocalciferol** (Figure 9–13). Similarly, **7-dehydrocholesterol**, normally on the surface of the skin, is converted to **vitamin D₃**, or **cholecalciferol** (Figure 9–14). Very few foods naturally contain sufficient vitamin D₂ or D₃. Either form serves the same function in the body, enabling it to absorb and metabolize calcium and phosphorus properly. Generally, the precursor, 7-dehydrocholesterol, is present in sufficient amounts, but in the winter in cloudy weather or at other times if an individual spends most of his time indoors, the body will not be able to make enough vitamin D₃. This applies especially to children since their bones and teeth are growing and their need for calcium and

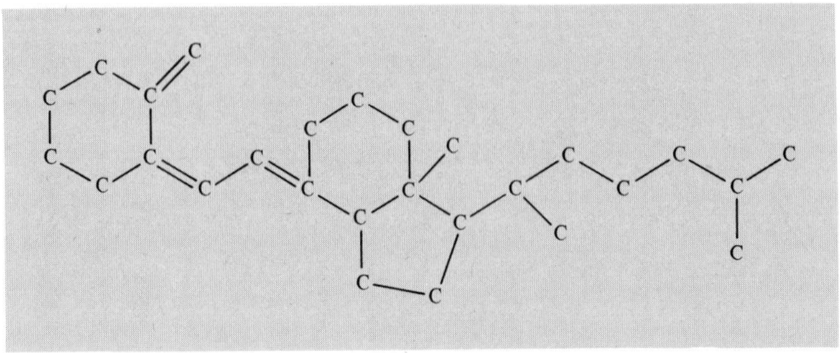

Figure 9–14.   Cholecalciferol (vitamin D₃) showing bonded carbon atoms, only.

phosphorus is greater. Most adults obtain enough exposure to sunlight the year round through normal exercise and other outdoor activity. (Night workers and sheltered elderly persons may not get enough exposure to the sunlight.) Since milk is a common food for children, it is expedient to provide vitamin D_2 or D_3 as a supplement, added to milk.

The best natural sources of the D vitamins are fatty salt water fish, especially the liver. A deficiency of either form of this vitamin produces rickets in young children; their bones become curved and stunted. Rickets is rare in the tropics where sunlight is abundant or where salt water fish form a normal part of the usual diet. Like vitamin A, vitamin D_2 or D_3 is stable under cooking processes. An excess of this vitamin produces loss of appetite, calcification of normally soft tissues, and is suspected as a possible cause of mental retardation.

A name for one form of **vitamin E** is **α-tocopherol**. It is not to be confused with β-tocopherol; although the names are similar, β-tocopherol has only about one-third the biological activity of the α-form. Many not completely reputable "food supply" stores tout all the tocopherols as beneficial, and indeed they are—to the merchant. Recent federal regulations require that nonnutritional components of a "health food" be clearly identified as nonnutritional in advertising and on the label of the container. Read the fine print before purchase unless you like to spend money foolishly. Figure 9–15 shows the most active, α-tocopherol.

The lack of vitamin E in other animals, such as rats, is associated with sterility in the male and poor fetal development in the uterus of the female. There is no established evidence that similarly applies to human males. Vitamin E is essential for fetal and infant development in humans, and a prolonged deficiency of this vitamin in human diet is harmful to adults. An excessive intake may be harmful, but there is no solid evidence. Vitamin E occurs naturally in lettuce and other leafy vegetables, in wheat germ, and in

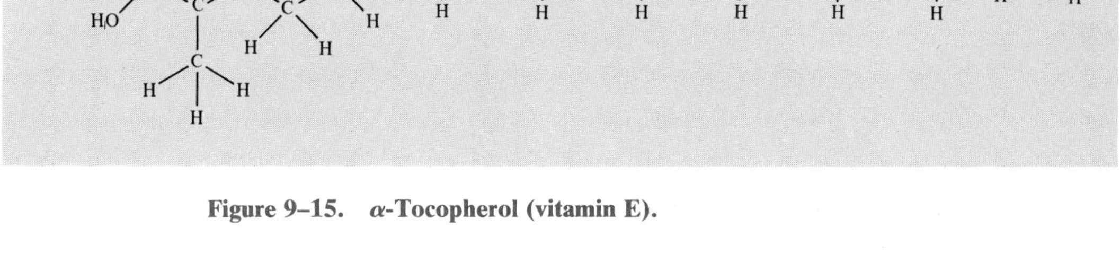

Figure 9–15. α-Tocopherol (vitamin E).

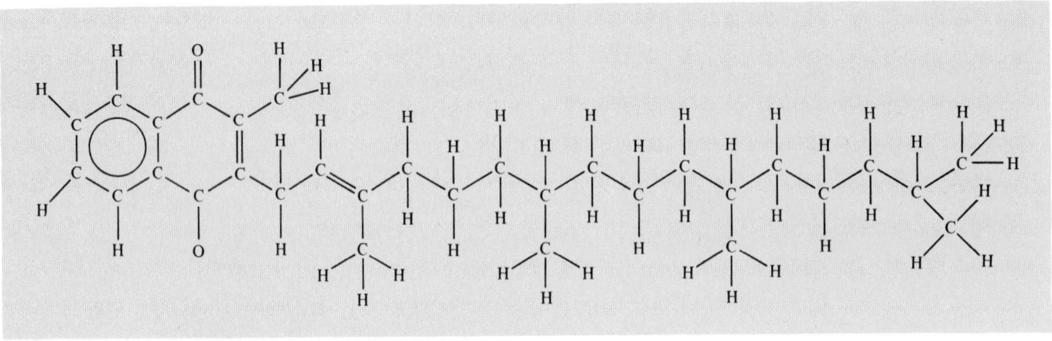

Figure 9–16. Vitamin K₁.

nuts and vegetable oils. It is not destroyed by normal cooking processes, but there is some loss upon exposure of foods containing it to the air.

Lack of **vitamin K** in the body leads to slow clotting of blood and internal hemorrages. This vitamin is found in all green leafy vegetables and even in cauliflower. One form, vitamin K_2, is conveniently synthesized for us by intestinal flora. Prolonged treatment with antibiotics has the adverse effect of drastically reducing or killing those flora, and that source of vitamin K is temporarily lost. However, vitamin K deficiency is rare in adults. Figure 9–16 shows vitamin K_1.

Vitamin B is now known to be a group of substances. A lack of one or more results in an unpleasant list of consequences, including paralysis, anemia, nausea, depression, pellegra, and seborrheic dermatitis. The group of B vitamins are each water soluble, which means in a practical sense that some of them, at least, will dissolve in the water used for cooking foods that contain these substances. Additionally, they are slightly unstable when heated.

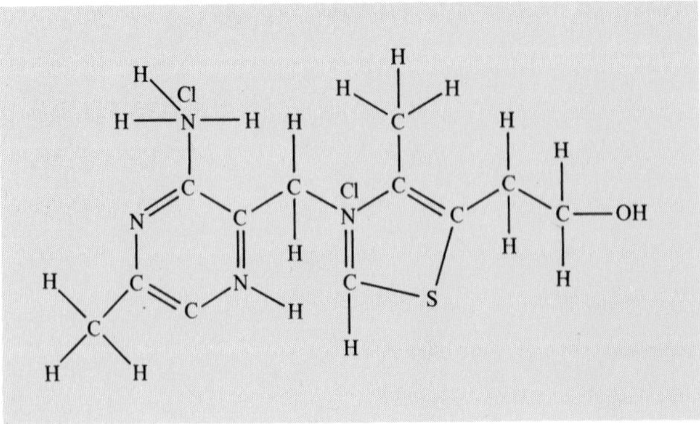

Figure 9–17. Thiamin (vitamin B₁).

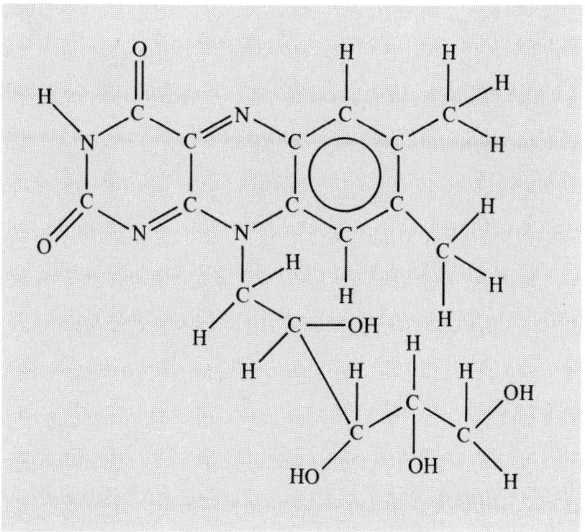

Figure 9–18.   Riboflavin (vitamin B₂).

Prolonged cooking at high temperatures and discarding of the cooking water is to be avoided if maximum amounts of these vitamins are desired.

   Thiamin, or **vitamin B₁** (Figure 9–17), is found in whole wheat, lean pork, beans, peas, and a wide variety of other unrefined foods. **Riboflavin**, **vitamin B₂** (Figure 9–18), occurs naturally in milk, liver, eggs, and leafy vegetables. Riboflavin was once known as vitamin G. **Niacin** is a generic term for both **nicotinic acid** and **nicotinamide** (Figure 9–19, not to be confused with or equated to nicotine). This member of the vitamin B complex is present in lean meat, peanuts, fish, beans, and peas. It can be formed in the body from tryptophan, an amino acid. **Vitamin B₆** (**pyridoxine** or **pyridoxal**, Figure 9–20, and **pyridoxamine**) is obtained by ingesting vegetables, lean meat, milk, eggs,

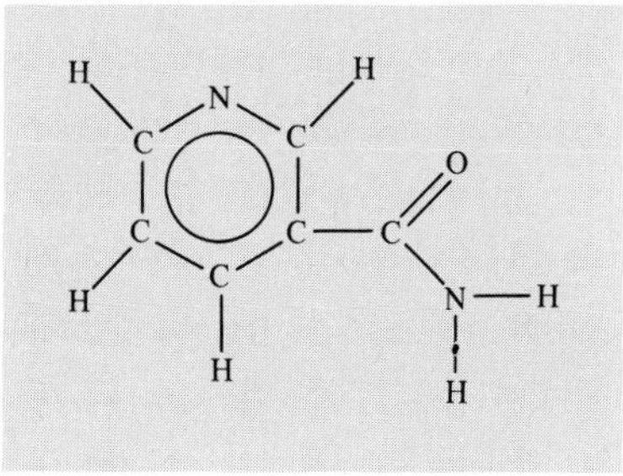

Figure 9–19.   Nicotinamide.

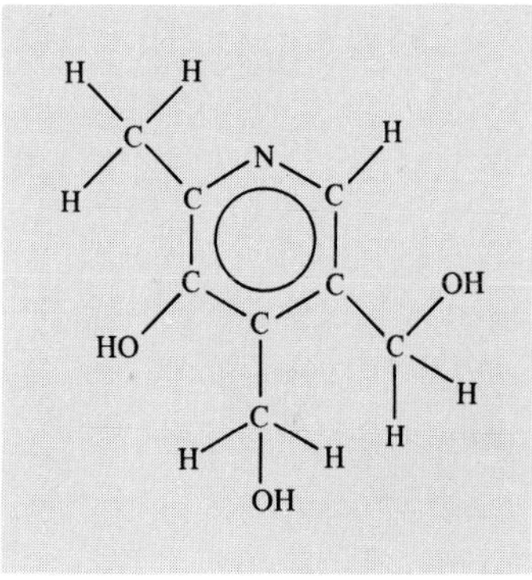

Figure 9–20.   Pyridoxine.

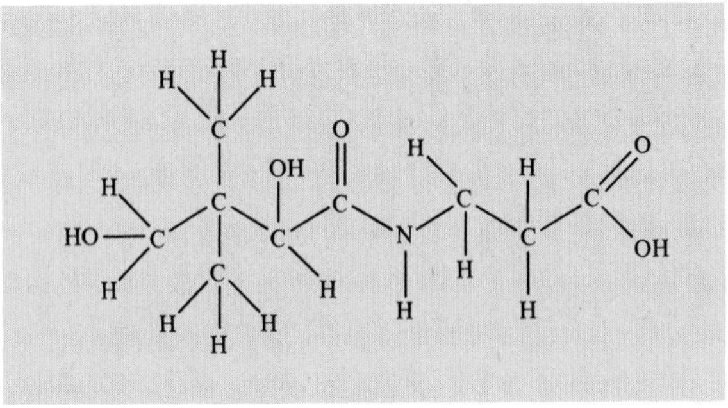

Figure 9–21.   Pantothenic acid.

and liver. **Pantothenic acid** (Figure 9–21) is found in many different foods including milk, liver, eggs, and vegetables. Compare its structure with that of acetyl CoA and CoA-SH shown in Figure 9–5. **Folacin** is a generic term for **folic acid** (Figure 9–22) and related compounds. This B vitamin is found in liver, milk, leafy vegetables, and in a wide variety of other foods. **Vitamin B_{12}** is a family of compounds, all of which contain cobalt. **Cyanocobalamin** will serve here as the representative substance. It is obtained from liver, lean meat, milk, cheese, fish, and eggs. In the schematic diagram of a vitamin B_{12} given in Figure 9–23, hydrogen atoms are not shown. Also, note two of the nitrogen atom symbols with arrows; this is intended to indicate that these two nitrogen atoms are bonded to the cobalt atom, shown as centered among four other nitrogen atoms; that is, in the real structure, the cobalt atom is centered among six nitrogen atoms. The cobalt is at the center of an octahedron and the six nitrogen atoms are at the apices of the octahedron.

Choline (Figure 9–24) is not exactly a B vitamin because our bodies can make it from scratch. It is present in many foods including milk, egg yolk, nuts,

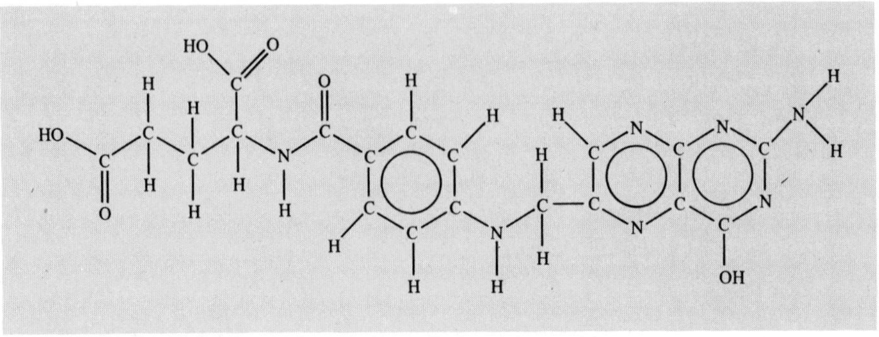

Figure 9–22.   Folic acid.

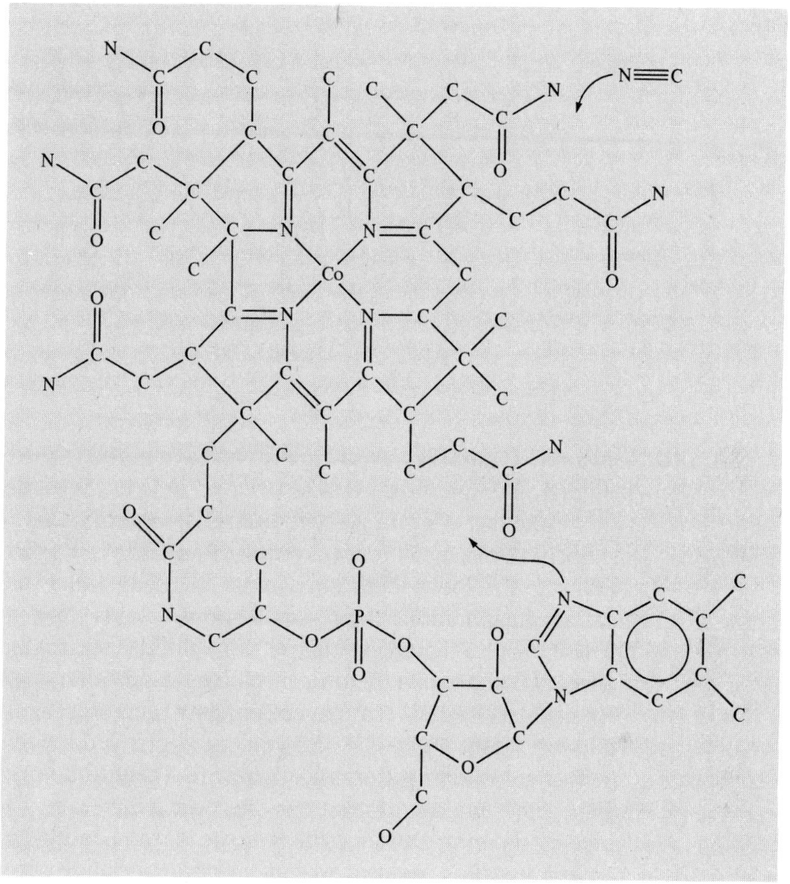

Figure 9–23. Vitamin B_{12} (schematic, hydrogens not shown).

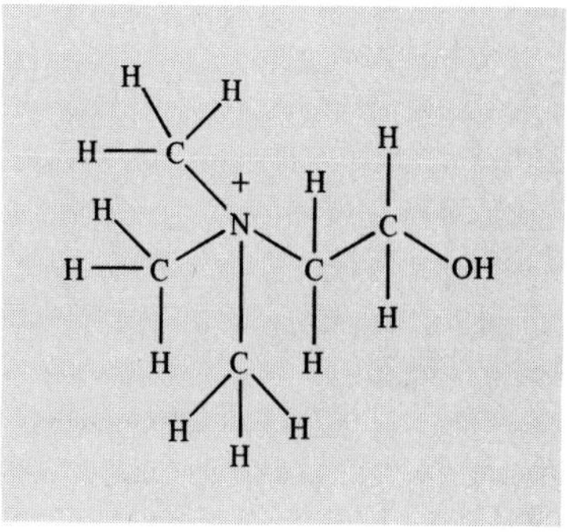

Figure 9–24. Choline.

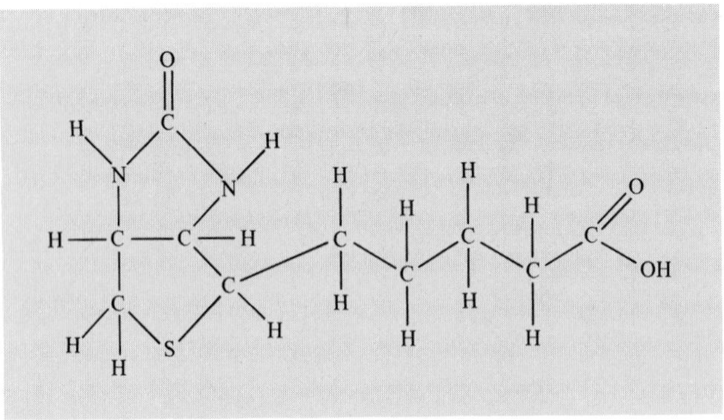

Figure 9–25.   Biotin.

and many vegetables. And, finally, we have **biotin** (Figure 9–25) made for us by friendly intestinal flora. It is also found in liver, eggs, beans and peas.

The last vitamin we will discuss is **vitamin C, ascorbic acid**. An extreme lack produces scurvy, a weakening of collagenous tissues. The symptoms are swollen gums, loose teeth, sore joints, bleeding under the skin, and slow wound healing. In metabolic processes, vitamin C aids the formation of collagen, a form of protein that occurs in our skin, blood vessels, hair, and bone structures. Sources of vitamin C include citrus fruits and other fresh fruits, some vegetables such as squash, cabbage, spinach, tomatoes, and potatoes. Of all the vitamins, vitamin C is the most readily destroyed by heat and exposure to the air. It is soluble in water, and therefore likely to be present in cooking water when, say, tomatoes or potatoes are cooked in a watery medium. In citrus fruit, however, vitamin C is not as likely to be lost by heating or exposure to the air.

Vitamin C is also present in other foodlike substances, such as "rose hips" and "acerola berries," and no doubt other exotic sounding sources promoted by "health food" stores. Unless you like to ingest such things, there is no chemical or nutritional reason to prefer these as a supplemental source of vitamin C; no matter where it comes from, the molecule looks like Figure 9–26, and performs the same functions in the body.

A large excess of vitamin C produces diarrhea. Except for isolated elderly adults who tend to use convenience foods low in vitamin C and infants

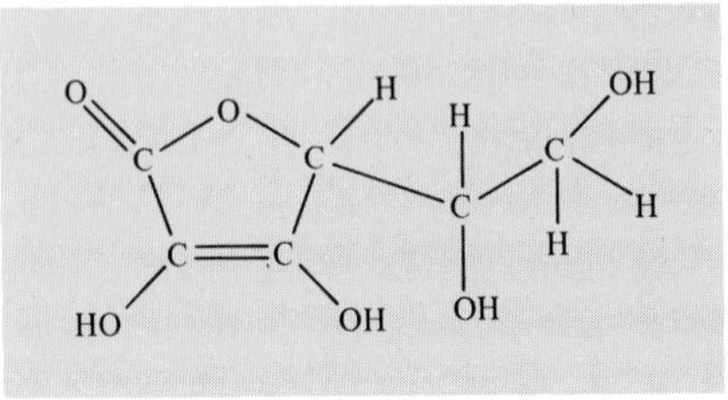

Figure 9–26.   Ascorbic acid (vitamin C).

fed reconstructed milk substitutes without citrus fruit juice supplement, vitamin C deficiency is rare in the United States. The normal adult requires about 0.045 g/day. A safe maximum dose is approximately 0.100 g, daily.

There is some evidence, not accepted by all authorities, but which is interpreted by other competent people to suggest that a high, regular, daily intake of ascorbic acid will tend to prevent the incidence and severity of colds. It is known that ingestion of 5 g of ascorbic acid daily by the normal adult human will almost always cause diarrhea. If you wish to undertake this regimen in the hope of preventing colds, consultation with a physician is strongly recommended.

In general, an excess of any vitamin or other nutrient is to be avoided. Multivitamin pills or special self-determined daily use of single vitamin supplements can cause severe and permanent disabling effects. Annually, 4000 cases of excess vitamin poisoning are reported in the United States, and without doubt many more unreported cases occur. Good nutritional practice involving foods that contain the vitamins and other nutritional essentials without supplements except under a physician's direction will more than suffice for almost everyone. A varied diet is the key, not supplements from a bottle, or tablets of rose-hips and β-tocopherol.

CALORIC NEEDS

In addition to other nutritional essentials, a proper diet must also furnish the energy we need, the calories. A survey of the daily and weekly habits of a group of typical college students was undertaken to provide realistic estimates of caloric needs. The results are summarized in Table 9–1.

The information in the table should be intelligently applied. Since it presents averaged data, the numbers can only be used as a guide for any individual reader. The hours you spend in each category may be quite different; you may exercise, or work, more intensely than the average, you may not ride a bicycle, your weight is probably different. In general, the calories consumed are proportional to the time spent in a given activity and also to the weight of the individual.

The energy you use must be supplied from somewhere, either in the food you eat, now, or in the fat stored in your body. If you consume more calories of food energy than you use, your weight will increase; if you consume less, your weight will decrease. There is no magic diet that furnishes many calories of food energy but which, somehow, does not produce a gain in weight, or a loss of weight, when the calories used in daily activities are not in balance with those furnished. Especially, young adults should be advised not to undertake a special diet for weight loss or gain without first consulting a physician. Many special diets are deficient in essential nutrients as well as deficient, or excessive, in calories.

When wisely planned, weight loss or gain is readily and gradually achieved. For example, a daily intake of 100 kcal less than is used produces a

TABLE 9–1.   Energy Requirements for Typical Healthy College Student (in kilocalories)

			Time and Caloric Requirements			
			135 lb Male		**120 lb Female**	
Daily	**Activities on Class Days**	**Weekends in Between**	**hr/wk**	**kcal**	**hr/wk**	**kcal**
Arising dressing			5.0	620	6.0	525
Eating breakfast			3.5	360	3.0	225
	Bicycling to class		1.0	250	1.1	200
	Sitting in class		18.0	1,850	18.0	1,325
	Walking between classes and labs		5.5	1,570	5.8	1,165
Eating lunch			4.9	505	5.2	395
	Standing in lab (1 or 2 days per week)		6.0	700	6.0	525
		Active exercise or work	6.0	2,950	4.0	1,215
		Milder exercise	3.0	1,025	2.0	410
Studying, relaxing, TV, movies general conversation			38.5	3,240	36.3	2,610
	Bicycling back		1.0	250	1.1	200
Eating evening meal			5.6	580	7.0	480
Undressing retiring			4.0	500	4.5	370
		Varied personal errands, shopping, snack eating, dating, etc.	10.0	2,750	12.0	2,150
Sleep			56.0	4,200	56.0	3,025
	Weekly totals		168.0	21,350	168.0	14,820
	Average daily calorie need (rounded off)			3,000		2,100

Note: The SDE (specific dynamic effect) of less than 200 kcal/day per person is included in the figures shown, but without identification as such explicitly.

TABLE 9–2.   **Table of Calories, Selected Foods**

Almonds, 1 cup	850	pork, ham	800
Apple, medium	70	turkey	500
Asparagus, 1 cup	35	veal	500
Banana, medium	90	Milk, 1 cup	
Beans, snap, 1 cup	25	whole	165
Bread, 1 slice		skim	90
white	60	Mushrooms, 1 cup	30
whole wheat	55	Noodles, 1 cup	200
Broccoli, 1 cup	45	Oatmeal, 1 cup	150
Butter, 1 pat	50	Olive, 1 green	5
Cabbage, 1 cup	40	Onion, 1 medium	50
Cake, 1 medium serving		Orange, 1 medium	70
angel food	110	Pancake, 1 medium	100
chocolate	130	Peach, 1 medium	45
very rich	300	Peanut butter, 1 oz	200
Carrot, 1 medium	20	Pear, 1 medium	95
Cheese, 1oz		Pickel, 1 medium	10
cheddar	115	Pie, 1 medium serving	
cottage	30	apple, cherry, etc.	350
swiss	100	custard	265
Cocoa, 1 cup	330	lemon meringue	310
Corn, 1 medium ear	85	mince	350
Cream, light, 1 cup	480	pumpkin	275
Doughnut, 1 medium	140	Popcorn, 1 cup	55
Egg, large		Potatoes	
boiled	80	baked, 1 medium	100
fried, scrambled	110	boiled, 1 medium	100
Fish, shell fish, 8 oz		fried, 1 medium serving	150
bluefish, baked	360	mashed, 1 cup	145
clams, raw	170	Rice, 1 cup	200
haddock, fried	360	Roll, 1	
oysters, raw	200	plain	90
salmon	320	sweet	180
shrimp	290	Sherbert, 1 cup	240
tuna, canned in oil, drained	450	Soup, 1 cup	
Gelatin, 1 cup	155	bouillon	10
Grapefruit, $\frac{1}{2}$ medium	75	meat	80
Honey, 1 tablespoon	60	tomato, cream of	90
Ice cream, 1 oz	40	vegetable	80
Ice milk, 1 oz	35	Spaghetti, 1 cup	220
Jam, jelly, 1 oz	400	Spinach, 1 cup	45
Lettuce, 2 large leaves	5	Squash, 1 cup	95
Macaroni, 1 cup	210	Strawberries, 1 cup	55
Meat, poultry, 8 oz normal,		Sugar, 1 rounded teaspoon	20
not fat, not lean, boneless		Tomato, 1 medium	30
beef	800	Turnips, 1 cup	40
chicken	300	Walnuts, 1 cup	655
lamb	700		

weight loss of 10 lb in 1 year. The converse is also true: 100 kcal extra of food energy daily makes you 10 lb heavier in 1 year.

Table 9–2 lists amounts of various foods and their calorie equivalents (in kilocalories). More extensive tables can be found in many other sources. If you compare one table with another, you may notice some differences in the calorie equivalent figures. Except for typographical errors, these reflect differences in the results of laboratory studies; in such cases, take either number as correct, or if you wish, average the discordant data.

A different kind of caloric table is more useful to some people. The amount of food in *each* line of Table 9–3 will furnish one fourth of the daily calories needed for normal activity. Or, the amounts of food in any four lines would furnish all of the calories needed normally for 1 day.

TABLE 9–3. Amounts of Food, Each Furnishing One Fourth of Normal Daily Caloric Needs

135-lb Male		120-lb Female
1 qt and 1 cup	Whole milk	$3\frac{1}{4}$ cups
10 oz	Hamburger	7 oz
14 slices	Whole wheat bread	10 slices
12 slices	White bread	8 slices
7 oz	Cheddar cheese	5 oz
$1\frac{1}{2}$ lb	Cottage cheese, creamed	1 lb
$1\frac{1}{2}$ cups	Mashed potatoes	1 cup
10 oz	Cherry pie	7 oz
$1\frac{1}{2}$ cup	Ice cream	1 cup
3 lb	Apples	2 lb

Note that none of these foods, or any other food taken singly, can provide a nutritionally adequate diet. Except for infants fed their own mother's milk, no single food is nutritionally complete; variety is essential.

Students who eat regularly at an institutional table, which is supervised by a dietitican, can generally be assured of an adequate diet if they eat all that is presented. For reasons that are more easily understood by their friends than by the dietitician, such a procedure seems to be difficult for students to follow. The scoring list of foods that follows may therefore be useful. Give yourself a score credit of one point each day for eating any one item or items listed in the same grouping. Score an additional half-point for second helpings. To be assured of an adequate diet, your minimum score should be 9 points on any single day, and at least 11 or 12 points three times a week. A score of 6 points of less, at any time, identifies a nutritionally deficient diet. Water is not listed, although it is also necessary.

A good breakfast: with whole grain cereal and milk or cream, or meat, or one egg.
One serving: green vegetable or leafy vegetable
One serving: any *other* vegetable
One: citrus fruit or tomato
One: *other* fruit
One serving: potato, or rice, or pasta
One: glass of milk or large piece of cheese
One serving: meat (including poultry) or fish
One serving: any high protein food such as meat, fish, beans, peas, or large amount of peanut butter
One: egg today or one egg yesterday
Two: slices of whole grain bread with butter or margerine
One serving: a favorite food, in moderation, such as cake, cookies, pie, other snacks, ice cream, etc. (coffee, tea, beer, wine, etc., do not count)

Food additives   Much of the prepared food available today contains food additives. A food additive is an added food, food product, food extract, or a non food substance, synthetic or natural, that is present in a product sold for human consumption as a food. Since a food additive can be potentially harmful, federal regulations have been established to protect the consumer. In the strict practical sense this protection cannot be perfect.

Until early in this century, there were no regulations. Over the years that followed, the regulations established have, indeed, substantially increased the protection of the consumer. Also over the years, a large body of experience has been developed, and we now know with some assurance that some substances added to foods are probably not harmful. (That is, no one has been harmed, yet.) These food additives are generally recognized as safe; an official list of "GRAS" (generally recognized as safe) additives has been promulgated. From time to time, the list is revised as substances are deleted or added.

For example, cyclamate, an artificial sweetening agent, was on the GRAS list until recently. It was removed when it was reported that cyclamate produces cancer when given in very large amounts to a few experimental animals. Whether it has the same effect in a human body is unknown, although the little evidence available indicates that it is not harmful to humans. It is conceivable that one or more other substances now on the GRAS list are in fact harmful, but the evidence is quite clear that most are not.

Today, no substance that is not on the GRAS list can be added to any food for human consumption unless and until the manufacturer proves that it is safe and demonstrates that its addition to the food will have an intended effect—such as making potato chips crispier, or less likely to become rancid. It is not difficult to show that an additive will have an intended effect. It is

TABLE 9–4. Food and Nutrition Board, National Academy of Sciences–National Research Council Recommended Daily Dietary Allowances[a] (Revised 1974) (Designed for the maintenance of good nutrition of practically all healthy people in the U.S.A.)

	Age (years)	Weight (kg)	(lb)	Height (cm)	(in.)	Energy (kcal)[b]	Protein (g)	Vitamin A Activity (RE)[c]	(IU)	Vitamin D (IU)	Vitamin E Activity[e] (IU)
Infants	0.0–0.5	6	14	60	24	kg×117	kg×2.2	420[d]	1,400	400	4
	0.5–1.0	9	20	71	28	kg×108	kg×2.0	400	2,000	400	5
Children	1–3	13	28	86	34	1,300	23	400	2,000	400	7
	4–6	20	44	110	44	1,800	30	500	2,500	400	9
	7–10	30	66	135	54	2,400	36	700	3,300	400	10
Males	11–14	44	97	158	63	2,800	44	1,000	5,000	400	12
	15–18	61	134	172	69	3,000	54	1,000	5,000	400	15
	19–22	67	147	172	69	3,000	54	1,000	5,000	400	15
	23–50	70	154	172	69	2,700	56	1,000	5,000		15
	51+	70	154	172	69	2,400	56	1,000	5,000		15
Females	11–14	44	97	155	62	2,400	44	800	4,000	400	12
	15–18	54	119	162	65	2,100	48	800	4,000	400	12
	19–22	58	128	162	65	2,100	46	800	4,000	400	12
	23–50	58	128	162	65	2,000	46	800	4,000		12
	51+	58	128	162	65	1,800	46	800	4,000		12
Pregnant						+300	+30	1,000	5,000	400	15
Lactating						+500	+20	1,200	6,000	400	15

[a] The allowances are intended to provide for individual variations among most normal persons as they live in the United States under usual environmental stresses. Diets should be based on a variety of common foods in order to provide other nutrients for which human requirements have been less well defined. See text for more detailed discussion of allowances and of nutrients not tabulated. See Table I (p. 6) for weights and heights by individual year of age.

[b] Kilojoules (kJ) = 4.2×kcal.

[c] Retinol equivalents.

[d] Assumed to be all as retinol in milk during the first six months of life. All subsequent intakes are assumed to be half as retinol and half as β-carotene when calculated from international units. As retinol equivalents, three fourths are as retinol and one fourth as β-carotene.

impossible to prove beyond all doubt that an additive is not harmful. To prove perfect harmlessness would require that the additive be administered to human subjects who would then, hopefully, survive. Practical harmlessness is proved by administering the proposed additive in excessive doses to experimental animals; usually several different species are used. If no harmful effects are noted in any of the test animals, the additive is presumed to be safe for human consumption.

Water-Soluble Vitamins							Minerals					
Ascorbic Acid (mg)	Folacin[f] (μg)	Niacin[g] (mg)	Riboflavin (mg)	Thiamin (mg)	Vitamin B_6 (mg)	Vitamin B_{12} (μg)	Calcium (mg)	Phosphorus (mg)	Iodine (μg)	Iron (mg)	Magnesium (mg)	Zinc (mg)
35	50	5	0.4	0.3	0.3	0.3	360	240	35	10	60	3
35	50	8	0.6	0.5	0.4	0.3	540	400	45	15	70	5
10	100	9	0.8	0.7	0.6	1.0	800	800	60	15	150	10
10	200	12	1.1	0.9	0.9	1.5	800	800	80	10	200	10
10	300	16	1.2	1.2	1.2	2.0	800	800	110	10	250	10
45	400	18	1.5	1.4	1.6	3.0	1,200	1,200	130	18	350	15
45	400	20	1.8	1.5	2.0	3.0	1,200	1,200	150	18	400	15
45	400	20	1.8	1.5	2.0	3.0	800	800	140	10	350	15
45	400	18	1.6	1.4	2.0	3.0	800	800	130	10	350	15
45	400	16	1.5	1.2	2.0	3.0	800	800	110	10	350	15
45	400	16	1.3	1.2	1.6	3.0	1,200	1,200	115	18	300	15
45	400	14	1.4	1.1	2.0	3.0	1,200	1,200	115	18	300	15
45	400	14	1.4	1.1	2.0	3.0	800	800	100	18	300	15
45	400	13	1.2	1.0	2.0	3.0	800	800	100	18	300	15
45	400	12	1.1	1.0	2.0	3.0	800	800	80	10	300	15
60	800	+2	+0.3	+0.3	2.5	4.0	1,200	1,200	125	18+[h]	450	20
80	600	+4	+0.5	+0.3	2.5	4.0	1,200	1,200	150	18	450	25

[e] Total vitamin E activity, estimated to be 80% as α-tocopherol and 20% other tocopherols. See text for variation in allowances.

[f] The folacin allowances refer to dietary sources as determined by *Lactobacillus casei* assay. Pure forms of folacin may be effective in doses less than one fourth of the recommended dietary allowance.

[g] Although allowances are expressed as niacin, it is recognized that on the average 1 mg of niacin is derived from each 60 mg of dietary tryptophan.

[h] This increased requirement cannot be met by ordinary diets; therefore, the use of supplemental iron is recommended.

SOURCE: *Recommended Dietary Allowances*, 8th rev. ed., National Academy of Sciences, Washington, D.C., 1974.

Without doubt this procedure is questionable. This, however, begs the real question. Some additives, for example, are used to enhance flavor. Most of these are food products or have a long history of trouble-free usage and are almost certainly harmless, as used. Still other additives enhance shelf-life, tending to prevent spoilage of one kind or another, such as rancidity, growth of molds, deterioration in appearance, and so on. Without these additives the food products would be unacceptable to the consumer and food that could have been made wholesome and nutritious would be wasted. In these times we cannot afford to waste food for this reason. So, one reason for using food additives that we cannot always be certain are perfectly harmless is to decrease food waste. The argument then is whether a definitely slight risk of harm is worth the conservation of more food.

Additives are used either directly to prevent spoilage or deterioration in appearance (almost the same thing from the consumer's standpoint) or otherwise to enhance the desirability of the food product. Further detailed information about food additives can be found in the references cited at the end of this chapter. However, one kind of food additive deserves further treatment here, nutritional supplements.

Nutritional labeling

Food manufacturers are now required by law to provide nutritional information about their product on the label whenever either a nutrient is added or whenever the manufacturer makes a claim about the dietetic quality of the food product. Other food products may also carry such information at the option of the manufacturer. The information is provided as a "Nutrition Information" panel on the package wrapper. Within that panel area, the listing includes serving size, number of servings in the package, and the calories, protein, carbohydrate, and fat per serving.

The percentage of the U.S. Recommended Daily Allowance (U.S. RDA) is also given for protein and seven of the sixteen vitamins and minerals for which a U.S. RDA has been established. Included are five vitamins (A, C, B_1, B_2, and niacin) and calcium and iron (the two minerals). Those for which a U.S. RDA has been established but which may or may not be listed in part or whole on the Nutrition Information panel are vitamins D, E, folacin, B_6, and B_{12}, and phosphorus, iodine, magnesium, and zinc as the minerals. Nutritional information on the content of copper, biotin, and pantothenic acid, for which U.S. RDA figures are tentative, and on the content of sodium, cholesterol, polyunsaturated fat, and saturated fat, for which no U.S. RDA figures are yet established, may also appear at the option of the processor.

The base reference for the U.S. RDA figures used in nutritional labeling is a single set of numbers obtained by a weighted average adjustment of the several sets of numbers listed in Table 9–4. The information in the table represents current conservative opinion, based on facts obtained from a long series of nutritional and dietary experiments carried out on men and other animals for the past several decades. As more information is accumulated, the list will be up-dated and enlarged. At present, perhaps another 60 or 70, or even more, substances are suspected to be nutritionally essential or are known to be, but in either case, the data are insufficient to establish a reasonable daily allowance.

Finally, the numbers in the table describe the recommended ingested and absorbed daily allowance, while the figures in a listing of nutrition information on a package label describe the content within the food; some of that, unfortunately, may not become available to you when the food is consumed.

FOR FURTHER INFORMATION

This chapter had two parts, a brief treatment of some special aspects of metabolism (catabolism and anabolism) and a broad look at nutrition. Either of these topics could occupy a lifetime of delightful study; we will take another swing at biochemistry (of which metabolism is only one important part) in another chapter. Here, you will find a list of references to get you started in the literature dealing with nutrition; at the end of the later chapter on biochemistry, many of the cited references will take you further into metabolism, should you wish to make the trip.

M. S. Chaney and M. L. Ross: *Nutrition*, 8th ed. Houghton Mifflin, Boston, 1971. A well accepted text.

C. H. Robinson: *Normal and Therapeutic Nutrition*, 14th ed. Macmillan, New York, 1972. A text written especially for nursing and dietetic students, but more than good enough to be useful and interesting to almost everyone.

G. H. Beaton and E. W. McHenry: *Nutrition, A Comprehensive Treatise*, Vols. I, II, and III, Academic Press, New York, 1964, 1964, and 1966. Technical, authoritative; if it isn't in here somewhere, it will be hard to find somewhere else.

Scientific American, **223** (Sept. 1970). "The Biosphere," a series of articles, many dealing with topics (energy cycle, water cycle, nitrogen cycle, mineral cycles, human food production, etc.) directly related to this chapter.

H. A. Guthrie: *Introductory Nutrition*, 2nd ed. C. V. Mosby, St. Louis, 1971. Another fine introductory text.

E. D. Wilson, et al.: *Principles of Nutrition*, 2nd ed. Wiley, New York, 1965. Still another fine text.

H. Fleck: *Introduction to Nutrition*, 2nd ed. Macmillan, New York, 1971. Yet another fine text.

M. T. Arlin: *The Science of Nutrition*. Macmillan, New York, 1972. One more fine text.

E. A. Martin: *Nutrition in Action*, 3rd ed. Holt, Rinehart and Winston, New York, 1971. This also is a fine text.

L. J. Bogert et al.: *Nutrition and Physical Fitness* 9th ed. W. B. Saunders, Philadelphia, 1973. A classic in the field. In a very real sense, this edition and earlier editions set the tone for competitive texts, not all of which were able to meet the challenge. Last of the texts cited here, but not the least.

W. H. Sebrell et al.: *Food and Nutrition*. Time-Life, New York, 1967. A popularized account, better than most competitors.

P. S. Howe: *Basic Nutrition in Health and Disease*, 5th ed. W. B. Saunders, Philadelphia, 1971, and R. M. Deutsch: *The Family Guide to Better Food and Better Health*. Meredith, Des Moines, Iowa, 1971. Both of these interrelate nutrition and health, competently and in an interesting manner.

The field of nutrition is rife with popularizing publications and articles in the Sunday newspaper supplements; some of these are not useful, although to the uncritical reader they seem helpful indeed since they "expose" the alleged avarice and other nasty attributes of distant, unseen individuals who presumably have used their political and financial influence to cheat the public out of its rights. That this happens, there is no doubt, but many of the popularizing publications present the story as though such practices were rampant. Generally, the author identifies a few blatant errors (which may have been honestly or avariciously committed, magnifies these out of all proper proportion, presents them as typical, embellishes them further with bald assertions out of the author's own head, and purports to cite the whole mish-mash as frighteningly and factually true. One of the milder such is *Consumer Beware*, by B. T. Hunter, Simon and Schuster, New York, 1971. Like others of the same kind, this book can be easily identified as malformed by its shrill and assertive style, by the generally negative and catastrophic fatalism it induces in the reader. Compare the book by Hunter, for example, with another book written for the same avowed purpose—to help the consumer citizen learn about nutrition—*Foods without Fads*, by E. W. McHenry, Lippincott, Philadelphia, 1960. One look at the style of each book will serve to indicate which of the two is truly informative and which misleading. The same test should be applied to any book that appears to address nutritional topics; there are a lot of quacks around and they all, seemingly, write books on nutrition.

The following books are by authors like McHenry; they also present a balanced view: T. J. Waylert and R. S. Klein: *Applied Nutrition*. Macmillan, New York, 1965, composed and informative; M. B. Salmon: *Food Facts for Teenagers*. C. C. Thomas, Springfield, Illinois, 1965, addressed to the young with facts; and M. A. Benarde: *The Chemicals We Eat*. American Heritage Press, McGraw-Hill, New York 1971, slightly technical but readable.

Is there a difference between cake flour and bread flour? What makes the meringue "weep" on a lemon meringue pie? How can this be prevented? How can you make mayonaise or hollandaise sauce stay together? These and a host of similar questions have answers involving chemistry. Both the questions and answers are treated in three books: L. H. Meyer: *Food Chemistry*. Reinhold, New York, 1960; B. Lowe: *Experimental Cookery*, Wiley, New York, 1955; and *Food*, the 1959 U.S. Department of Agriculture Yearbook, Washington, D.C., 1959. The latter reference also contains all sorts of interesting and useful information, from "Food in Our Lives," a general discussion, to "Your Money's Worth," on food budgets, with lots of goodies in between.

B. T. Burton: *The Heinz Handbook of Nutrition*, 2nd ed. Blakiston, McGraw-Hill, New York, 1965. The well known H. J. Heinz Company has published sound nutritional literature for several years; among those offerings, this handbook has a well deserved reputation. Highly recommended, slightly out of date.

B. K. Watt and A. L. Merrill: *Composition of Foods*, U.S. Dept. of Agriculture, Washington, D.C., 1963. Exactly what it says in the title; an excellent and authoritative reference. Indispensable; look for a revised edition, to be published soon.

The following publications deal with the formidable problems associated with establishing sound nutritional habits in the general population. Long overlooked, this necessary matter is at last receiving some attention: J. Mayer: Toward a national nutrition policy. *Science*, **176**:237 (1972), an article reviewing progress since the 1969 White House Conference on Food, Nutrition, and Health; for a report on the corresponding 1974 conference, see C. Holden: *Science*, **184**:548 (1974). In the same context, look at the editorial statement from P. P. McCurdy: *Chem. Wk.*, Oct. 23, 1974, p. 5, and an accompanying article on p. 21 of the same issue to the effect that, with improved planning, twice the present world population can be adequately fed.

On the same topic, also see D. Hollingsworth and M. Russell (eds.): *Nutritional Problems in a Changing World.* Wiley, New York, 1973, the proceedings of a conference on nutrition now and in the future, wide ranging in content; and *Recommended Dietary Allowances*, 8th rev. ed. National Academy of Sciences Washington, D.C., 1974. Present knowledge of our nutritional needs is incomplete and research continues; this is the most recent brief summary.

N. S. Scrimshaw and V. R. Young: The requirements of human nutrition. *Sci. Amer.*, **235**:51 (Sept. 1976). All sorts of details on human nutrition, from what nutrients are in which foods to the developments of societies as a consequence of particular nutritional needs.

N. Roosevelt: Good Cooking. Harper, New York, 1959. Really a book of recipes (almost), without a formula in it, not even "H_2O"; but loaded with practical chemistry, applied where it does the most good (at least in the context of this listing). Written in a lively style; you'll like it if you try it.

E. Groth: *Science and the fluoridation controversy. Chemistry*, **49**:5 (May 1976). Facts, no matter how undisputable, are not the only basis for decision making. This well balanced article points out that in the fluoridation question, personal values are an important factor. These considerations strongly suggest the political and practical efficacy of alternatives to the fluoridation of public water supplies.

D. G. Crosby (coordinator): *Symposium on Natural Food Toxicants.* American Chemical Society, Washington, D.C., 1968. It is not well known that some foods contain toxic components through natural processes. After reading this one, you may wish you had remained in ignorance.

P. L. White and N. Selvey: *Nutritional Qualities of Fresh Fruits and Vegetables.* Futura, Mt. Kisco, N.Y., 1974. On the brighter side; the role of fruits and vegatables in our diet, from genetic engineering through harvesting processes, to storing, shipping, processing, and, eating.

A. Berg: *The Nutrition Factor.* Brookings Institution, Washington, D.C., 1973. Especially recommended to those who are concerned with the political and sociological consequences of nutritional policies and practice.

I. Raw and G. W. Holleman: Water-energy for life. *Chemistry*, **46**:6 (May 1973). Using the energy from sunlight, plants manufacture food by decomposing water (and also utilize carbon dioxide); animals recover the energy in food by "reuniting" the hydrogen and oxygen once again (also evolving carbon dioxide). An interesting summary of the details involved in the preceding sentence.

G. O. Kermode: Food additives. *Sci. Amer.*, **226**:15 (March 1972) and J. Z. Majtenyi: Food additives, food for thought. *Chemistry*, **47**:6 (May 1974). For thousands of years man has added nonfood substances to natural foods. Currently more than 2000 different substances are added (not all to the same food). These two articles survey the subject describing briefly in addition how the necessity and presumed safety of some of these food additives are determined.

R. Winter, *A Consumer's Dictionary of Food Additives*. Crown, New York, 1972. A reasonably complete list in alphabetical order of most of the food additives known to be used as of the date of publication.

T. E. Furia (ed.): *Handbook of Food Additives*, 2nd ed. Chemical Rubber, Cleveland, Ohio, 1972. This handbook would not fit into a small brief case. A complete listing of food additives, with additional information, more than most would want to know. Highly recommended as authoritative. A new edition can be expected soon, with more pages needed to include the most recent food additives.

Two publications from the Manufacturing Chemists Association, Washington D.C. are informative and, as might be expected, describe the advantages of and safeguards for the use of food additives; the titles are *Food Additives, Who Needs Them?* and *Food Additives, What They Are/How They are Used*. Both are anonymous. Legitimate concerns about food additives are indicated by the title of this article—B. F. Feingold: Behavioral disturbances, learning disabilities, and food additives. *Chemtech*, **5**:264 (1965).

Anon: *The Almanac of the Canning, Freezing, and Preserving Industries*. E. E. Judge and Son, Westminster, Md., 1977. Published annually, approximately every July 1. Among other things, this publication has an up to date section on food laws and regulations, including those which pertain to labeling and quality standards. Informative.

For down to earth practical information, how to buy food by the serving rather than by weight, and save money, how to understand nutritional labeling, the explanation of open code dating, comparisons of fresh, frozen, and canned foods, and a long list of similar helps and hints, see the *Shopper's Guide*, the 1974 Yearbook of the U.S. Department of Agriculture.

For insight into the technical details of food fortification, how it is really done, see *Technology of Fortification of Foods*, Food and Nutrition Board, National Research Council, National Academy of Sciences, Washington, D.C., 1976.

chapter 10

Polymers

In this chapter we will look at a variety of substances: nylon, starch, wood, dacron, cellophane, linen, cotton, protein, mylar, polyethylene, wool, silk, rubber. The common theme which unifies this variety is restricted flexibility.

Consider a bead necklace, open, not fastened at the ends (Figure 10–1). It is flexible; it can be arranged in several different ways, straight, curved at one end a little and curved sharply at the other, curved in the middle but straight at the ends, and so on. However, if each bead was shaped like the one in Figure

Figure 10–1.   Spherical beads on a string are flexible.

10–2 and then strung on a string, the flexibility would be restricted. It could still be curved here and there to some extent, but sharper curves would be impossible on a flat surface because the arms of adjacent beads would interfere with each other. To make sharper curves in this necklace, we would have to twist adjacent beads with respect to each other, so that the arms of those beads could overlap. We would have a kind of spiral, or helix, or zig zag, or combinations of these, and then the necklace could be curved sharply if we wished. If the two arms on each bead were different, say one small and one very large, this would restrict the flexibility in yet another way. If we had only one arm on each bead, or more than two, again the flexibility would be differently restricted. Or, we might have a set of necklaces, with beads all alike but with some peculiar shape, and every so often put some sticky glue (or chewing gum, maybe) on several different beads here and there. Then, if we mixed up all those necklaces in a pile, the tangle of necklaces would still be flexible to some extent, but the flexibility would be restricted in yet a different way than we have considered up to now.

RUBBER

With this picture in mind, we are ready to talk about our first polymer, rubber. Rubber molecules are very long, made up of repeating units. Each unit would look like Figure 10–3 if we laid it out flat. In Figure 10–3, the circles represent carbon atoms, the short lines not connected to anything on four of the carbon atoms represent bonds to hydrogen atoms (which are not shown to keep the clutter at a minimum), the dotted lines represent bonds to other similar units (we'll see how that goes in a minute), and the lines between the carbon atoms represent the carbon atom to carbon atom bonds. Notice that there is a double bond between two of the carbon atoms, C and Y, and single bonds otherwise.

Atom positions in rubber

The carbon atoms in Figure 10–3 are lettered, so that the bond angles can be easily discussed. First, the angle QZC is 109.5°; both angles, XCY and ZCY,

Figure 10–2.   A string of these beads would have restricted flexibility.

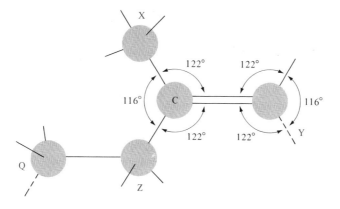

Figure 10–3.   One unit of a rubber molecule (schematic); note the angles and the identified carbon atoms, C, X, Y, Z, and Q (see text).

are 122°; and angle XCZ is 116°. You can think of carbon atom X as something like a top; it can turn on the axis, XC, spinning its three attached hydrogen atoms around. It is probably not spinning all the time, but it is loose in this sense. However, angle XCY stays the same all the time, or at least doesn't vary very much. Similarly, carbon atom Q can rotate on the QZ axis. Instead of the two hydrogen atoms on top, as pictured, they could be on the bottom, with the dotted bond indicating linkage to the adjacent unit pointing upward; or, the hydrogens could be sticking out the front of the page and the linkage sticking out the back, and so on. Of course, with all those other units linked on to carbon atom Q, one after another, that Q atom isn't going to do much, if any, rotating; it will stay like it is, however that might be. The point is that in one unit of a rubber molecule, carbon Q will be turned one way; in the next adjacent unit, some other orientation is likely, and still a different orientation in the next, and so on.

The same kind of statement applies to carbon atom Z, except that when it rotates on its ZC axis, it has to drag carbon atom Q too. Now, that Q atom has a whole bunch of exactly similar units attached to it. That dotted line on the left connects the Q carbon to a similar unit; on the "left" end of that one, there is another similar unit, and so on. So, it is pretty clear that carbon atom Z is not going to twist and turn as much as carbon atom X. Any way Z happens to be oriented is about where it is going to stay. Finally, atoms C and Y cannot rotate at all (well, hardly at all), they can be thought of as "locked" into position by the double bond between them.

So, a single unit of a big rubber molecule would be arranged as follows: Four atoms, X, C, Z, and Y are all in the same plane. C and Y cannot rotate, X can, but it doesn't make much difference to our story whether it does or not because only three hydrogen atoms are attached to it. But, because Z can rotate

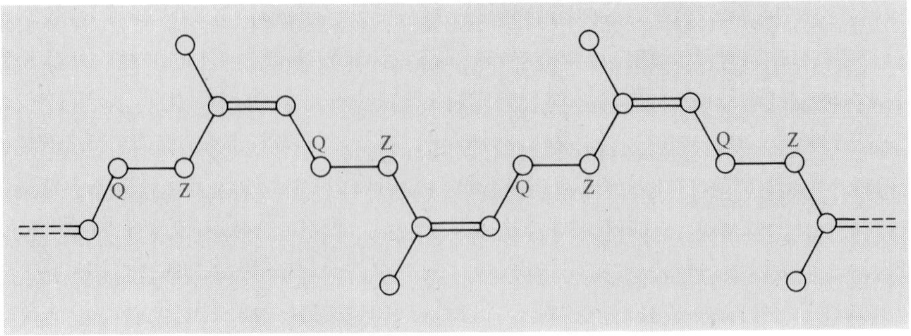

Figure 10–4.   Section of a rubber molecule (schematic).

about the ZC axis, carbon atom Q, and all the units that are attached to it, can be in a different plane than the plane occupied by the other four carbon atoms in that unit. And the same thing applies to all of the other "Q" atoms in all of the other units that are in that single rubber molecule. As a result, the necklace of units in our rubber molecule is very much convoluted indeed.

It would be helpful to show a picture of a rubber molecule, with all those units turned this way and that, except that the detail of such a picture would obscure the point we are considering. Instead, Figure 10–4 shows what a small segment of that necklacelike rubber molecule would look like if all of the carbon atoms were in one plane (hydrogen atoms are not shown). Notice that every other unit is upside down compared to its adjacent units, incidentally. In a real rubber molecule, compared to this picture, you can imagine that each of the Q and Z carbon atoms have been rotated on their respective axes this way and that.

Vulcanized
rubber

In vulcanized rubber (the only kind most people have seen) some of the double bonds are broken by the insertion of one or more sulfur atoms, when then form a bridge from one part of a rubber molecule to another, or from one rubber molecule to a different rubber molecule, thus making one big molecule out of two or more that were not as big before (Fig. 10–5. Hydrogen atoms are not shown in this diagrammatic sketch.)

The vulcanization process for rubber was discovered when Charles Goodyear dropped a mixture of rubber and sulfur on a hot stove in 1839. (Before that, people had mixed rubber and sulfur together, to make the rubber less sticky, but as far as is known no one thought of cooking the mixture.) When heat is applied, those double bonds are more likely to break apart, and the sulfur atom is thus enabled to bond to those carbon atoms. You can see that this is similar to the bits of glue on some of the beads in that tangled pile of necklaces. Without any fastening between one necklace and another in a tangled pile, the whole pile could be distorted by pushing or pulling or

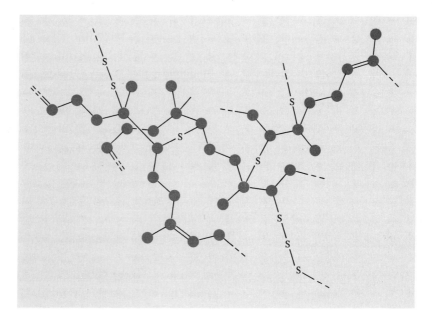

Figure 10–5. Section of a vulcanized rubber molecule (schematic).

squeezing it. With a few sticky places, it would not be as distortable; with several sticky places, it would be still less distortable; with a lot more, it would be almost solid.

Rubber bands have only a little sulfur mixed in before being heated. Rubber tires have more sulfur. A rubber tire is not as soft, not as distortable, as a rubber band. Other rubber articles, such as erasers, balls, innertubes, carpet backing, and so on, have varying amounts of sulfur (and other substances), depending upon the degree of softness that is desirable.

When a rubber band is stretched, we untangle the molecules. They tend to assume the shape illustrated in Figure 10–4, where all the carbon atoms were in the same plane. This tendency is inhibited both by the restrictions on the movement of those Q carbon atoms, restrictions on the rotation of carbon atoms Z on the ZC axis, and by the sulfur atom bridges. When we release a stretched rubber band, the Q atoms tend to go back into the position they had before we stretched it. We say that the rubber is elastic.

Synthetic rubber

Today there are several different kinds of rubber. The rubber we have described is natural rubber, obtained from the sap of living rubber trees. About 40% of the rubber used in commerce today is natural rubber. Various types of synthetic rubber make up the rest. All of these, natural rubber and the several kinds of synthetic rubber, are called polymers. Their molecules are made up of simpler unit parts, hundreds or thousands of them bonded together, like beads on a string with restrictions on their flexibility. The word, **polymer**, conveys some of this descriptive picture; poly of course comes from the Greek, meaning a lot of whatever it is, and mer from the same language, signifying part, or unit. There are many other examples of polymers, both natural and synthetic. We have already noted that proteins are composed of units bonded together in a long string or chain. Of course, in a protein molecule the units are not necessarily exactly the same, but they are similar in that they are all α-amino acid residues (α-amino acids with some water gone).

STARCH AND CELLULOSE

Starch and cellulose provide another example of natural polymers. In this case the linkage between the adjacent units is uniquely important. Let's try an example first. Imagine an equal number of human type male and female persons in a large room, like a gymnasium perhaps. Say we have about 100 of each type; ask five boys and five girls to form a line, holding hands, boys and girls alternatively. Have all the girls facing, say, north, and the boys the other way, south. Then take another five and five, similarly, in a line alongside the first line. And so on, until all those present are linked up in lines of ten persons each with the lines parallel to each other.

What would we observe next? A lot of conversations would take place from one line to the other, but the people in one line would not be talking to others in the same line very much because they are turned, every other one, the "wrong" way. In a sense, there would be a bonding, an attraction, from one line of people to the folks in another line. It would be difficult, we could say, to separate those lines of people.

Now, if we do the same thing but link them differently so that everybody in each line all faces, say, north, the results would be different. Those in any one line would not be talking much to people in another line, because those in each line would have their backs to the next line. It would be easier to separate one line from another. It would be difficult to break up a single line into separate people. All the folks in one line would be linked together, not only by holding hands but also by their conversations with their adjacent neighbors in the same line.

There are other ways, of course, to arrange these people. The point is that, even when we use the same units, we can alter the results we would observe merely by altering the linkage between the units.

The glucose molecule

Starch and cellulose are both made of glucose units, linked together. The linkage in starch and cellulose is different, although the units are substantially the same. Yet there is a lot of difference in their properties. We can east starch and it is nutritious. We can eat wood, sawdust, tooth picks, grass, all made of cellulose, but we would not be able to survive on it. A potato is mostly starch.

No one would build a house made of potatoes nailed together. Wood is mostly cellulose; many people live in wooden houses. The linkage between glucose units makes the difference.

Let's look at the details of the linkage. It is not possible to show these as clearly as we would like because the details are three dimensional and the pages in this book are two dimensional. You have already seen other two-dimensional representations of glucose in Chapters 5 and 9. Here, in order to illuminate the linkage details, a different representation will be used.

We can best visualize this by thinking of a three-leg lounging chair (Figure 10–6a), a slanted back and a slanted leg rest which forms the front leg of the chair. The seat of the chair is square. Next, remove the two back legs but imagine that the chair remains in the same position (Figure 10–6b). Then, remove the padding from the back, seat, and leg rest, leaving only the frame (Figure 10–6c). At five of the corners, put a carbon atom with an oxygen atom at the sixth corner (Figure 10–6d). The frame now represents shared pair electron bonds, six of them, between adjacent atoms. Finally, add other bonds, as shown in Figure 10–6e. They are positioned at five of the corners and lead to the other atoms that complete the picture. This is β-D-glucose. By allowing line intersections to represent carbon atoms, as we have done before, the structural arrangement may be made a bit clearer (Figure 10–6f).

Figure 10–6g is another form, called α-D-glucose, which we will consider later. Note that the only difference is the altered arrangement on the carbon shown on the right end.

There is considerable restriction in the bond angles in the glucose unit whether it is α- or β-glucose. All the angles are fixed. Around the carbon atoms, the angles are 109.5°. For the oxygen atoms, the bond angles are approximately 100°, varying slightly depending upon whether it is the oxygen atom in the "frame," in the ring, or one of the oxygen atoms outside the ring. In the glucose structure there is not much rotation of atoms on the bond lines, except for the carbon atom shown on the left of the structure, appended to the ring. That single carbon atom can rotate, swinging its attached hydrogens and oxygen and the hydrogen connected to the oxygen, round and round. As.it happens, this is of secondary importance for our purposes.

Cellulose

When one β-D-glucose combined with another, and with another, and so on, the linkage is obtained by the loss of a hydrogen atom from the oxygen shown on the right and the loss of a hydrogen and an oxygen from the carbon shown on the left. The residual glucose unit (Figure 10–7) linked to other similar units to form a cellulose molecule. A cellulose molecule consists of thousands of glucose units. The linkage from one unit to the next is the key point. Notice that every other glucose unit is linked "upside down" compared to the units on either side. This linkage from one unit to the next generates a linear shape for the total molecule. A segment of such a molecule is shown in Figure 10–8.

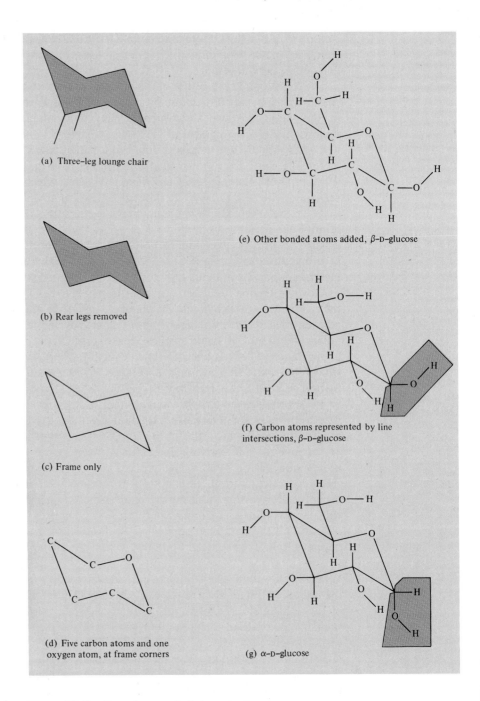

(a) Three-leg lounge chair

(b) Rear legs removed

(c) Frame only

(d) Five carbon atoms and one oxygen atom, at frame corners

(e) Other bonded atoms added, β-D-glucose

(f) Carbon atoms represented by line intersections, β-D-glucose

(g) α-D-glucose

Figure 10–6.   From lounge chair to α-D-glucose.

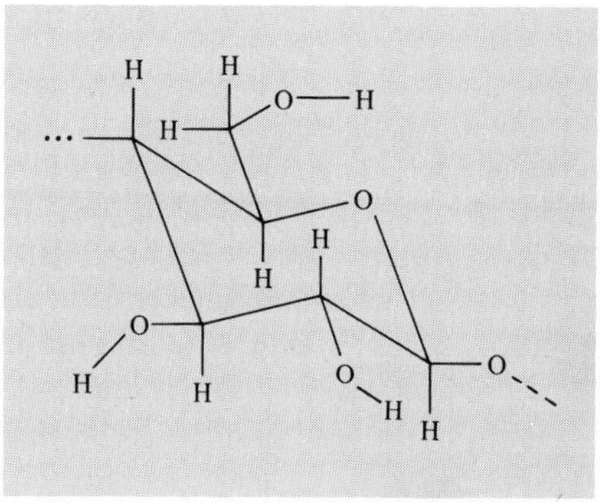

Figure 10–7. Residual glucose unit, β-D-glucose.

Alongside of this more or less linear cellulose molecule, you should imagine another cellulose molecule, and still another along side that one, and so on, for several parallel molecules, like matches lined up parallel in a full box of matches.

Notice in Figure 10–8 that a linear cellulose molecule will have oxygen–hydrogen groups sticking out at various angles and directions. So will the adjacent linear cellulose molecules. As we know, the oxygen–hydrogen bond is polar, very much so, with the oxygen distinctly negatively charged and the hydrogen positively charged. (Our first example of this was the water molecule, where we called the plus to minus attraction between water molecules the hydrogen bond.) The situation is the same in cellulose as in water. Hydrogen bonds exist between two adjacent cellulose molecules, several hydrogen bonds, holding these two molecules together. And the molecule adjacent to the first two is similarly held, and so on, to form a compact bundle of linear molecules. No wonder wood is as strong as it is. There are of course other factors involved in the strength of wood, but clearly one of the important factors is the linear

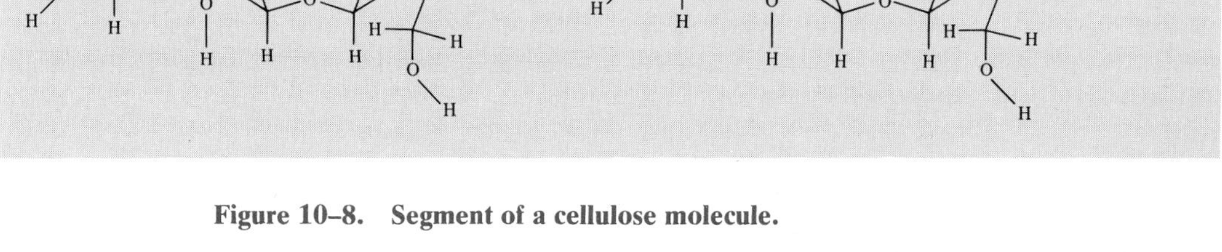

Figure 10–8. Segment of a cellulose molecule.

shape of a cellulose polymer molecule and the intramolecular hydrogen bonds that can be formed, thousands of them, between all those adjacent molecules.

Starch

By a seemingly simple change, from the β to the α structure, a markedly different polymer molecule results. In α-D-glucose molecules (shown in Figure 10–9a again for easy reference) the position of the carbon–oxygen–hydrogen bonds on the right end carbon is different from the β arrangement. Figure 10–9b shows the α-D-glucose residue, the same as the molecule, less one oxygen and two hydrogens. When these units are joined, one possible set-up looks like Figure 10–10.

Figure 10–10 appears to be curved. In three dimensions it would look helical. But before we get to that, notice that this time there is no alternating up, down, up, down arrangement. All the units are oriented similarly. To get at the helix your attention is called to the bridging oxygens, between adjacent rings. Rotation is possible around those bridging oxygen bonds. If all the units rotate slightly, and in the same direction, the results produce the helical structure for the molecule as a whole.

By a slight rotation of each successive unit, hydrogen bonds can be established between one unit in the molecule and another unit in the *same* molecule. In cellulose, the hydrogen bonding was from one molecule to another. In this arrangement, with the α linkage, the hydrogen bonding is within the same molecule. The result is called starch. There is no fibrous structure as there was in cellulose. Different molecules are not bonded (very much) to each other, but each unit of the molecule, each unit within the molecule, is bonded to other units. The whole molecule has a helical shape, like

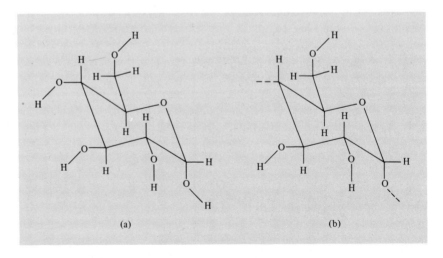

(a)   (b)

Figure 10–9.   (a) α-D-Glucose and (b) residual glucose unit.

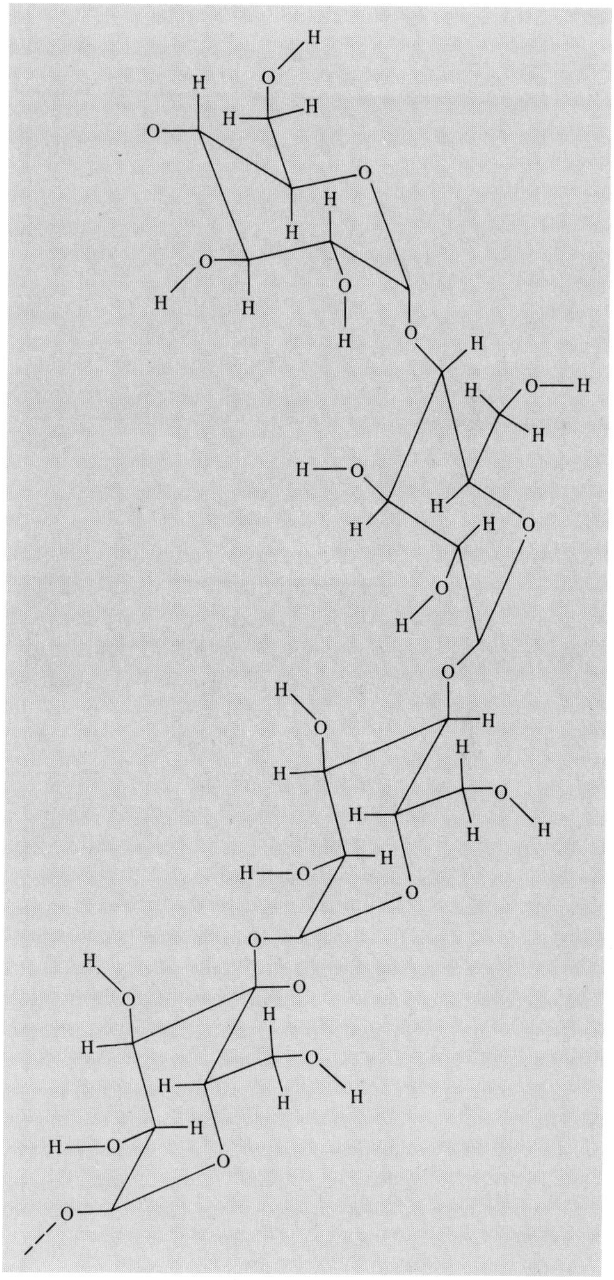

**Figure 10–10.
Segment of a starch
molecule.**

a "spiral" staircase, with a hole down the center (where the central pole of the staircase would be).

As it happens, the hole is just the right size to fit iodine molecules. A whole bunch of iodine molecules can fit into this hole in a starch molecule, one behind the other. The electrons on the iodine molecules, which are then very close to neighboring iodine molecules, and those to their neighbors, and so on, get a little bit confused. The electrons zip around, helter skelter, among all of the trapped iodine molecules. In this condition the color of the trapped iodine molecules is a deep blue. In some homes tincture of iodine is still used as an antiseptic for minor cuts; this is simply a solution of iodine molecules in alcohol. A drop of tincture of iodine on a raw potato will produce a deep blue stain (don't eat the potato).

There are several forms of starch. The discussion up to this point describes amylose or potato starch. Amylopectin, another form of starch, from other plants, consists of shorter chains of α-D-glucose units, and therefore shorter helices, but with other short chains of glucose units branching off from the helix, here and there.

Photosynthesis   In the preceding chapter we learned that when glucose, from starch, reacts with oxygen in the body, energy is released in a series of steps involving AⓅⓅ and AⓅⓅⓅ, especially, and forming carbon dioxide and water as the ultimate products. Now that we know more about the starch molecule, it is worth going over some of the details to see how energy from the Sun and carbon dioxide and water, interact to form starch and oxygen. The process is called **photosynthesis**.

The complete details of the photosynthetic process are not yet known, although enough is known to more than fill a book of this size. Briefly, then, in photosynthesis light energy from the Sun is absorbed by an unknown molecule (probably, it is a fairly complicated structure) and eventually an electron is loosened from some molecular species present within the green leaves of a plant. Energy is required to take an electron away from an atom or molecule, as we know; in this case the energy absorbed from the Sun can be thought of as residing in this loosened electron.

Of course, this event happens several times each instant in any green leaf that is exposed to sunlight. As the several loosened electrons are transferred from one molecule to another and to another, and so on, within the cells of the plant, they lose this energy, which is ultimately captured in two ways. In one, an AⓅⓅⓅ molecule is formed when some of the energy is captured by an AⓅⓅ and Ⓟ species. In the other, the energy is used to break up a water molecule into oxygen, hydrogen ions, and an electron.

The oxygen ultimately escapes from the green leaf, becomes part of the atmosphere, and we humans and other animals breathe it. The electron that has lost its energy ultimately replaces the electron that was given up by the not yet known molecule, and this molecule is ready once again to be recycled in the process. The hydrogen ions are used in the formation of reduced nicotine adenine dinucleotide phosphate, a fairly complicated molecule usually called NADPH, for short. That is, NADPH is an energy carrier, similar to A℗℗℗. The energy held in the chemical bonds of NADPH also comes from sunlight in another absorption process in which chlorophyll is probably the initial absorber and NADPH is the ultimate receiver at the end of a series of stepwise reactions.

We can summarize at this point. Sunlight is absorbed in two separate processes. After a series of reactions, oxygen is released to the atmosphere and two kinds of energy carrying species are formed within the plant cell, A℗℗℗ and NADPH. Water has been consumed.

In the total photosynthetic process, carbon dioxide is also consumed. In a nutshell, carbon dioxide reacts with NADPH to form glucose, water, and NADP.

$$6\,CO_2 + 24\,NADPH \rightarrow (CH_2O)_6 + 6\,H_2O + 24\,NADP$$

The NADP is returned to be recycled. Actually, the equation showing the formation of glucose [as $(CH_2O)_6$] is much more complicated. It takes place in a series of steps and requires energy; some of the energy is supplied by the NADPH and the rest by several A℗℗℗ molecules. The process is cyclic, as we would have guessed by now, anyway.

The synthesis of starch It will help to focus on a five-carbon molecule, ribulose-1,5-phosphate, called RuDP (Figure 10–11). Briefly, aided by enzymes as catalysts and using the chemical bond energy supplied by NADPH and A℗℗℗, the RuDP reacts with CO_2 to form two three-carbon molecules; these react further, as we shall see. Every three times this series of reactions takes place, we get six molecules that each contain three carbon atoms. Five of these are used to make three RuDP molecules, and the remaining three-carbon molecule is equivalent to

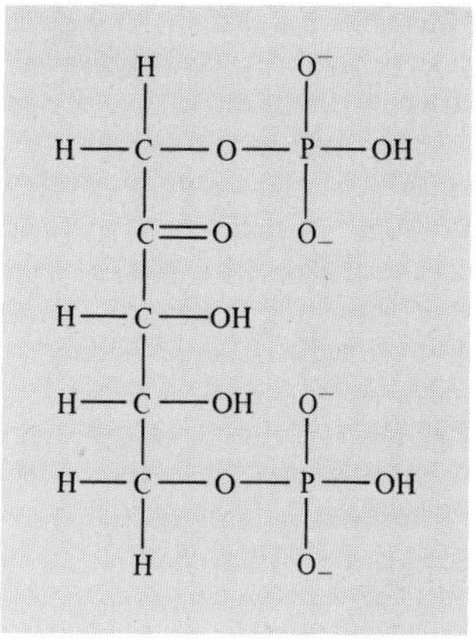

Figure 10–11.   Ribulose-1,5-phosphate (RuDP).

half of a glucose molecule, which, of course, contains six carbon atoms. So, when the series of reactions is repeated six times, we get one glucose molecule in addition to six RuDP molecules.

Let's look at some of the details in the series of reactions. RuDP reacts with carbon dioxide and water to form two 3-phosphoglyceric acid (PGA)

RuDP $+ CO_2 + H_2O \longrightarrow 2$ PGA

molecules. Notice that the PGA is identical to 3-phosphoglycerate, discussed in Chapter 9, except for one hydrogen ion, which is an incidental difference. In the metabolic processes, 3-phosphoglycerate often lacks that hydrogen ion, and in photosynthesis the PGA often has that "extra" hydrogen ion.

PGA $+ A\textcircled{P}\textcircled{P}\textcircled{P} \longrightarrow$ DPGA $+ A\textcircled{P}\textcircled{P}$

Each PGA reacts with $A\textcircled{P}\textcircled{P}\textcircled{P}$ to form diphosphoglyceric acid, (DPGA). DPGA reacts with NADPH to form glyceraldehyde-3-phosphate,

DPGA   +NADPH ⟶   GALP   + (P) + NADP

called GALP for reasons that are not obvious, and other products. Then, in a series of reactions, which we need not detail here, five GALP molecules become three ribulose-5-phosphate molecules. They are usually called Ru5P. Finally, Ru5P reacts with A(P)(P)(P) to form RuDP and (A)(P)(P).

Notice that for each three RuDP molecules that we start with we get six PGA molecules and that we need only five of these to regenerate three other RuDP molecules, for recycling over and over again. This leaves us with one PGA molecule. In the living plant tissue, these PGA molecules are the basis for the formation of a variety of other molecular products, such as rubber, turpentine, vegetable oils, proteins, the colored pigments of many flower petals, the noxious exudate that causes poison ivy rash in many people, and glucose, from which starch and cellulose are derived. In a typical plant, about half of the PGA "extra" molecules are converted either to starch or to cellulose and the other half is used in the formation of all the other products specific to

GALP   Ru5P

Ru5P   RuDP

that particular plant. The details are not yet well understood for all of these different possibilities, although we probably know more about the conversion of PGA into starch and cellulose than any other series of reactions.

Very briefly, for each series of reactions, chemical bond energy is needed and is supplied by either NADPH or by A(P)(P)(P), or both. Each reaction step in any series of reactions involves enzymes as catalysts, and the production of one product, or another instead, is controlled within the plant tissue by other mechanisms that provide or deny the enzymes involved in a particular series of such reaction steps.

MANUFAC-TURED NATURAL POLYMERS

As we know, the strength of wood, its tendency to split, the fact that it swells when wet, and other properties, can be attributed in large part to the fibrous nature of the cellulose molecule and the presence of thousands of O—H groups on each long molecule. Cotton and linen are cellulosic fibers, and their properties can be largely attributed to the same structure.

Rayon

Rayon is also a cellulosic fiber, and cellophane is a cellulosic sheet material. To make rayon, cotton or wood is cooked in large vats with a solution containing hydroxide ions, OH^-. After sufficient cooking, the cellulose in the wood (or cotton) dissolves (or, perhaps better, disperses) in the aqueous alkaline liquid.

This mixture is then forced through a spinnerette that looks very much like a shower head except that the holes are much smaller. The spinnerette however is submerged in an acid bath. The hydronium ions, H_3O^+, in the acid bath react with the hydroxide ions in the squirted out alkaline liquid, forming water. The cellulose is no longer soluble and becomes a solid, but in the form of a filament. The filament produced in this way is used to make rayon yarn from which rayon fabric is made by a mechanical process that is, in itself, very interesting. You can get an idea of the small size of the holes in the spinnerette by unwinding a thread of rayon into its component filaments.

Cellophane

Cellophane is made in the same way except that, instead of a spinnerette, the mixture is extruded through a long and very narrow slit into an acid bath. In cellophane the cellulose molecules are arranged every which way, which helps to account for the fact that cellophane is relatively easily torn apart. In the spinnerette, at least to some extent, the cellulose molecules tend to line up parallel to each other as they are squirted out.

SYNTHETIC POLYMERS

We could say that cotton and linen are natural fibers and that rayon is a manufactured natural fiber. It is appropriate to consider wholly synthetic polymers next. In Chapter 8, the figure summarizing the utilization of sources of energy included a notation for the use of petroleum (although that is not specifically indicated in the figure) as a synthetic raw material for the manufacture of fibers, plastics, and paints. We shall now consider this particular use. Synthetic fibers, plastics, and paint materials are polymers. In each instance, the polymer molecule consists essentially of a carbon to carbon skeleton on which other atoms, mostly hydrogen, are affixed.

The carbon and hydrogen atoms come from petroleum. They could be obtained from other sources, such as cellulose, coal, or even limestone and water. However, the manufacturing processes would have to be changed radically at some increased cost. As a result, the cost of the synthetic polymers would also increase. It is certain that continued use of what amounts to a small percent of our total petroleum consumption for the manufacture of synthetic polymers has only a small effect upon the rate of depletion of this valuable resource. By foreseeing the ultimate end of this supply of carbon and hydrogen atoms well in advance, it is more likely that the transition to other sources can be made at lower cost. There are some who argue that petroleum should be used only for chemical processes and never as a fuel, this prolonging the eventual day of reckoning. Few agree with this radical position, for many reasons.

Polyethylene

Now, let's get to our first synthetic polymer, polyethylene. An ethylene molecule is shown in Figure 10–12. In the presence of another molecule that

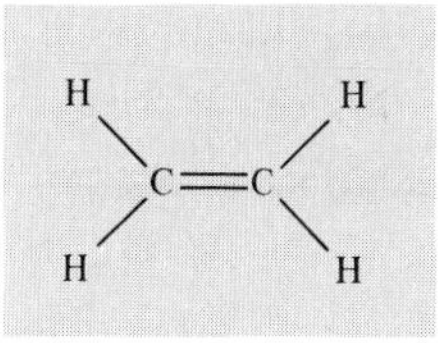

Figure 10–12.   Ethylene.

tends to take electrons readily (an oxidizing agent), one of the electrons in the double bond between the two carbon atoms is removed. Together with an electron from the oxidizing agent an electron pair bond is formed with one of the carbon atoms.

In this equation, Q· represents the oxidizing agent; notice that its lone electron is represented by a dot on the right. Any species with one (or more) unpaired electron is called a **free radical**. A free radical will oxidize, take an electron from, anything handy if it can get away with it. Notice that the species on the right of our equation is a free radical. Its lone electron is the other electron that was formerly in the double bond of ethylene.

So, this new free radical reacts with another ethylene.

And this product is a free radical, too.

The polymer molecule grows and grows and grows. Eventually, some other reaction becomes more likely to happen. For example, two free radicals will meet (as they happen to grow toward each other, and we get a much larger, molecule, in one step (and two less free radicals). This is called a **termination** reaction.

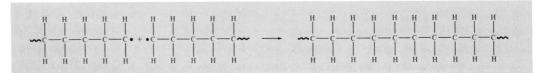

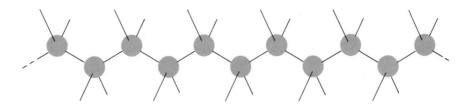

Figure 10–13.   Section of a polyethylene molecule (schematic).

Other kinds of terminations are also possible. Typically, in a single polyethylene molecule, there will be more than 2000 carbon atoms. The molecule is quite long and narrow, something like a long piece of thread. A short section of a polyethylene molecule can be visualized schematically, as in Figure 10–13. The balls represent the carbon atoms, the lines between them represent the bonds, and the short lines pointing upward and downward, more or less, represent the bonds between a carbon atom and a hydrogen atom. To minimize clutter, hydrogen atoms are not shown.

The angle between any two bonds on the same carbon atom is 109.5°, and all bonds can rotate. This has no effect if we consider the carbon to hydrogen bonds, but it has a considerable influence with respect to the carbon to carbon bonds. The molecule can and, in the liquid state, does take almost any twisted and tangled shape we could think of, subject to the restriction that all bond angles on a single carbon atom are 109.5° and that no two atoms can be in the same place at the same time. The liquid state of a lot of polyethylene molecules all in the same container resembles a tangle of very long, very active writhing snakes, all very much tangled up with each other.

However, as the liquid is cooled, it solidifies, and this is a different story. That is, in any tangle of writhing snakes, there will be some snake bodies side by side, parallel, from time to time here and there in the tangled mess. If the snakes are actually polyethylene molecules, as shown in Figure 10–13, then at those places where the molecules are parallel and side by side, they will fit together. As the liquid polymer molecules cool, they tend to move around less writhingly. They tend to stay together wherever they happen to be side by side, and parallel. In the solid state of the polymer, many molecules are side by side, parallel to each other here and there, as shown in Figure 10–14. Solid polyethylene can be formed by cooling the liquid in molds to make many common and useful objects. Also the liquid can be squirted from a spinnerette, and as it cools fibers are formed. The fibers are called "olefin" fibers. You may have heard of them.

Acrylic polymers   You know acrylic fibers by their trade names such as Orlon or Acrilan. Their synthesis is the same story all over again, with only a slight difference. We start

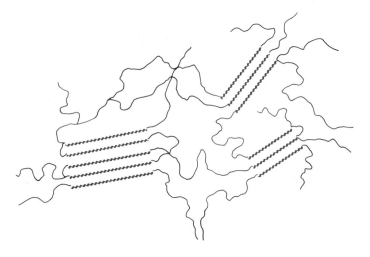

Figure 10–14. Tangle of polyethylene molecules showing microcrystalline regions.

with acrylonitrile (Figure 10–15). The build-up of a polymer molecule, **poly-acrylonitrile**, is begun in the same was as polyethylene and continues similarly. A section of the polyacrylnitrile molecule might be imagined to look like Figure 10–16.

The polyacrylonitrile molecule is identical to polyethylene except for the nitrile group, C bonded to N, which replaces one of the H atoms on every other carbon atom. Now, a nitrile group is much larger than a hydrogen atom, so the flexibility of the polyacrylonitrile molecule is less than that of a polyethylene molecule. Additionally polyacrylonitrile cannot be dyed readily, and this would limit its usefulness as a fiber. To permit dyes to "take," we need some oxygen atoms, which are slightly negatively charged, in the polymer molecule to which dye molecules are likely to adhere. (Or else we must devise some special dyes.) Oxygen atoms are incorporated into the polymer molecule by mixing in a little vinyl acetate before polymerization begins. Figure 10–17 shows a schematic form of vinyl acetate. As the polymerization of the acrylonitrile proceeds, every once in a while, at random, a vinyl acetate is incorporated. So, at random in the polymer molecule (on the average about every 16 or

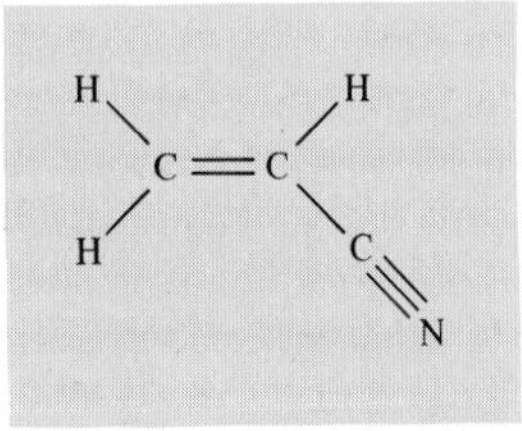

Figure 10–15. Acrylonitrile.

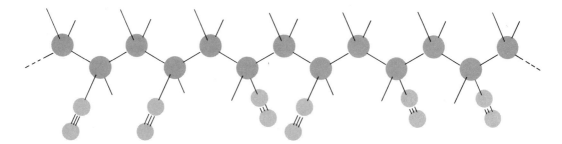

Figure 10–16. Section of a polyacrylonitrile molecule (schematic).

17 carbon atoms), there is a carbon atom with some oxygen atoms, and other atoms also, attached (Figure 10–18). This combination polymer can be used in fibers, squirted out through a spinnerette, of course.

Poly(vinyl chloride)   One of the economically important polymers is poly(vinyl chloride). Approximately 1% of the total United States annual gross national product involves this polymer; approximately 2 million people are employed in the poly(vinyl chloride) industry. The variety of uses for this polymer includes floor covering, garden hose, waterproof fabrics, and phonograph records. Vinyl chloride has the schematic structure shown in Figure 10–19.

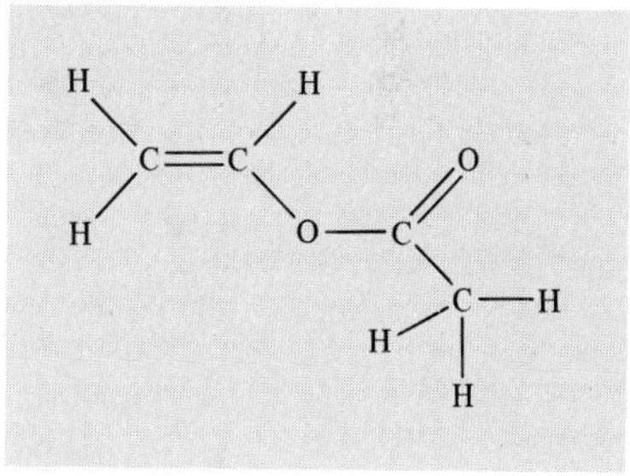

Figure 10–17.   Vinyl acetate.

Figure 10–18.   Section of a mixed polymer molecule, polyacrylonitrile with some vinyl acetate.

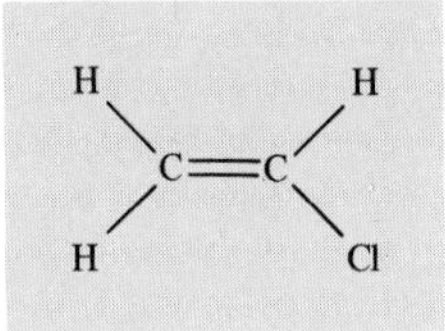

Figure 10–19. Vinyl chloride.

We would expect the polymer molecule of poly(vinyl chloride) to look something like Figure 10–20; notice that the large chlorine atoms (the larger balls) are randomly disposed on alternate carbon atoms. We call this disposition **atactic**, and will discuss it briefly later. For now, you can see that, as the polymer molecule might otherwise twist and bend, the chlorine atoms would get in the way, at least to some extent. As we might expect, poly(vinyl chloride) would be stiff and rigid, not a soft and pliable polymer.

To soften the polymer and make it suitable for many different purposes, lubricating molecules, called **plasticizers**, are mechanically mixed into the solid polymer. One such plasticizer is dioctyl phthalate (Figure 10–21). The plasticizer molecules get between and among the polymer molecules and render the whole polymerized mass flexible. Unfortunately the plasticizer slowly migrates to the surface and may evaporate or be gradually rubbed off in use. Eventually, a useful object made of poly(vinyl chloride) becomes stiff and useless. Poly(vinyl chloride) is used as upholstery covering material in many automobiles. In this case, the slowly evaporating plasticizer may deposit on the cool surface of the windows inside the closed passenger compartment. It must be scrubbed off, with some vigor and effort now and then in order to see clearly out of the windows.

Polypropylene Polypropylene exhibits some interesting properties. The fundamental molecule is propylene (Figure 10–22). And, as we might expect, a section of the polymer molecule looks like Figure 10–23. However, this artist's sketch is misleading. The CH_3 groups, shown attached to every other carbon atom, are too large to allow the polymer molecule to be stretched out flat as it is shown. If

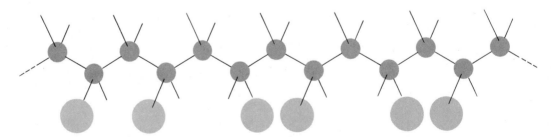

Figure 10–20. Section of a poly(vinyl chloride) molecule (schematic).

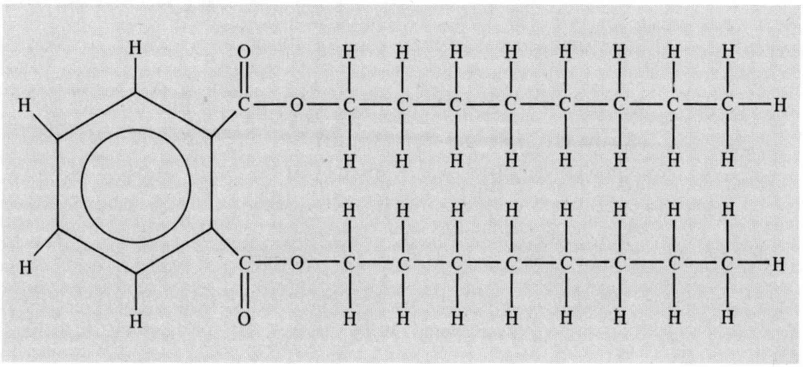

Figure 10–21. Dioctyl phthalate.

it were stretched out flat, those CH_3 groups would interfere with each other. So, the polymer molecule is twisted a little, at each succeeding CH_3 group.

The whole polymer looks like a helix (circular stair case), although this time there is no hole that amounts to anything down through the center (as was the case for starch). Solid polypropylene consists of parallel helices, one fitting snugly against another. When the bundle of parallel helices is pulled or stretched, the helices unwind a little and the material stretches. When the tension is released, the material shrinks slightly. In the form of fibers, polypropylene ought to make an excellent rope, and it does. It is used extensively

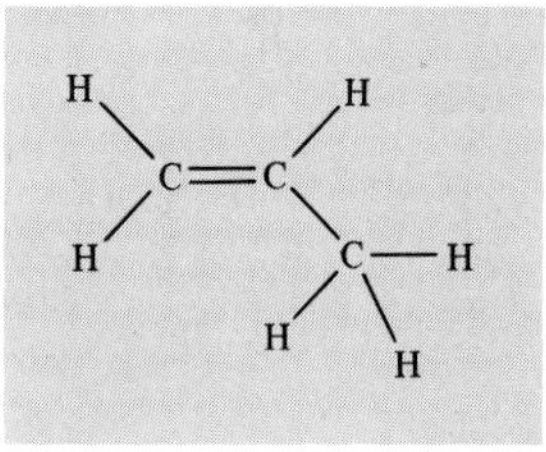

Figure 10–22. Propylene.

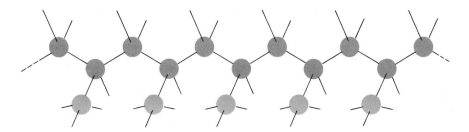

Figure 10–23. Section of a polypropylene molecule (schematic).

for marine "line" (rope) because it is not only a strong, slightly elastic, fibrous material but it is also resistant to attack from molds when wet. Finally, it is less dense than water and therefore less likely to be lost if tossed overboard.

Polypropylene fiber is called **olefin** fiber as polyethylene fiber is. It is spun from spinnerettes, and the fibers newly extruded from the holes in the spinnerette are pulled as they are formed to cause the helices to line up parallel and snug with each other. In addition to its use as rope, polypropylene fiber is used for carpeting and other similar uses.

Isotactic and atactic polymers

We have really been discussing **isotactic** polypropylene molecules thus far. Isotactic is a word with Greek roots, iso signifying the same, and tactic conveying the idea of position. In isotactic polypropylene, all of the CH_3 groups are in the same position relative to each other.

In **atactic** polypropylene, the CH_3 groups are randomly positioned with respect to each other. Atactic polypropylene molecules have no regularized structure, and the material made from this form of polypropylene is tough but has little tensile strength or elasticity. Atactic polypropylene is used to make automobile battery cases, steering wheels, dentures, washing machine agitators, and other similar objects.

Teflon

We have by no means exhausted the possibilities for polymers related to polyethylene, but one more specific example will be sufficient here. If all of the hydrogen atoms in ethylene are replaced with fluorine atoms, we have perfluoroethylene (Figure 10–24). And polyperfluoroethylene looks like Figure 10–25. However, this polymer has larger fluorine atoms in place of the small hydrogen atoms in polyethylene. The size of the fluorine atoms tends to keep the molecule from bending around every which way.

Further, as we learned in Chapter 5, fluorine is very electronegative. Once it is bonded (to carbon in this instance) it acts as though it has a satisfactory share of electrons and shows very little tendency to do anything about any other atom's electrons. As a result, polyperfluoroethylene molecules are inert; they do not react (unless you really force them to); they do not even attract other species. You could fry an egg on a sheet of polyperfluoroethylene, and the egg would not stick to the polymer.

As you have guessed, the trade name of polyperfluoroethylene is Teflon. One of the problems in manufacturing Teflon-coated cooking ware is to get the Teflon to stick to the metal of the pan.

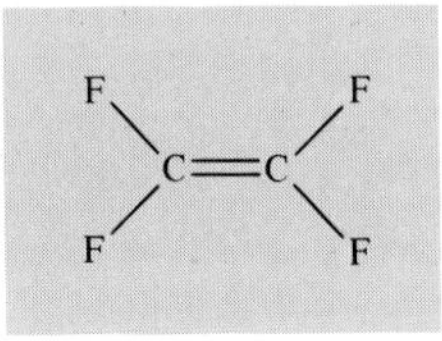

Figure 10–24.   Perfluoroethylene.

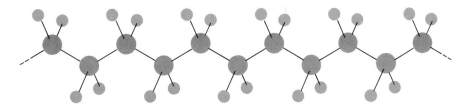

Figure 10–25.   Section of a polyperfluoroethylene molecule (schematic).

Copolymers

Finally, to generalize, recall the incorporation of vinyl acetate with acrylonitrile in a mixed polymerization process to produce a polymer molecule that had desirable properties. Such polymers are called copolymers, and there are many other examples. Acrylonitrile (Figure 10–26a) butadiene (Figure 10–26b) and styrene (Figure 10–26c) are shown. Their copolymer is called ABS, a very tough resilent material used for helmets and other similar products that have special shapes and must be tough and sturdy.

When we copolymerize two of these, styrene and butadiene, we get a copolymer that is rubbery, for much the same reasons that real rubber is rubbery, and the copolymer is known as SBR or GRS synthetic rubber. Dynel, a synthetic fiber used mostly for wigs, is another copolymer, of vinylchloride and acrylonitrile.

This short excursion has barely touched the surface of the subject of polyethylene-type polymers; there are many other types.

Nylon

An example of one other type is nylon. Nylon is closely related to proteins. A protein polymer molecule, as we know, can be thought of as composed of α-amino acids. These acids have an NH_2 group and a COOH group, at least one of each on each molecule. When an NH_2 group on one molecule interacts with a COOH group on another molecule, the two are bonded together (water

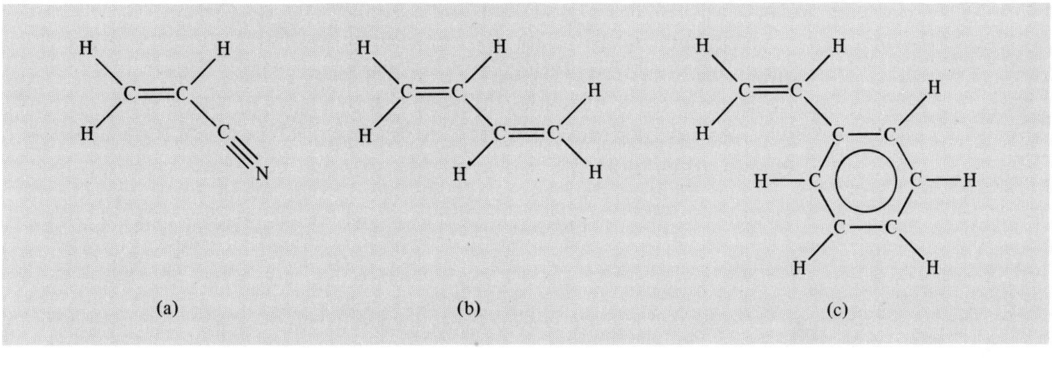

Figure 10–26.   (a) Acrylonitrile, (b) butadiene, (c) styrene.

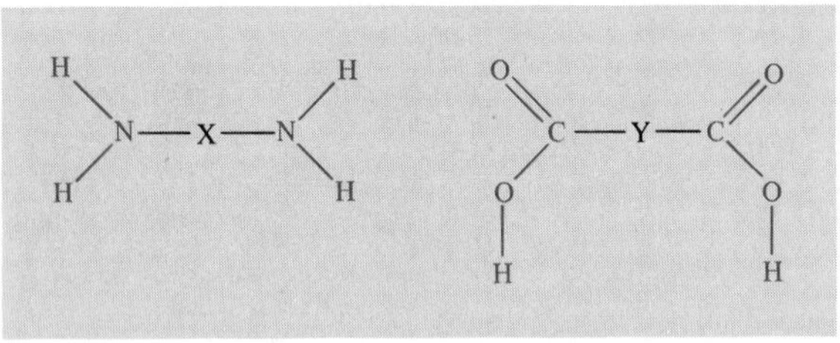

Figure 10–27. Two different bifunctional molecular species.

is lost). Then a third molecule can react with the unreacted COOH or NH_2 group on the bonded molecule, COOH reacting with NH_2 or NH_2 with COOH, which ever is available. The process continues until a large α-amino acid polymer is formed, which we call a protein.

Silk, wool, and other animal hairs and furs are examples of protein fibers. Nylon is an example of a closely related polymer. For example, think of a molecule with two NH_2 groups, one at each end, and of another molecule with two COOH groups, one at each end (Figure 10–27). The X and Y in our imaginary molecules represent a chain of carbon atoms, with other atoms attached to each one. Now, one NH_2 group can react with one COOH group.

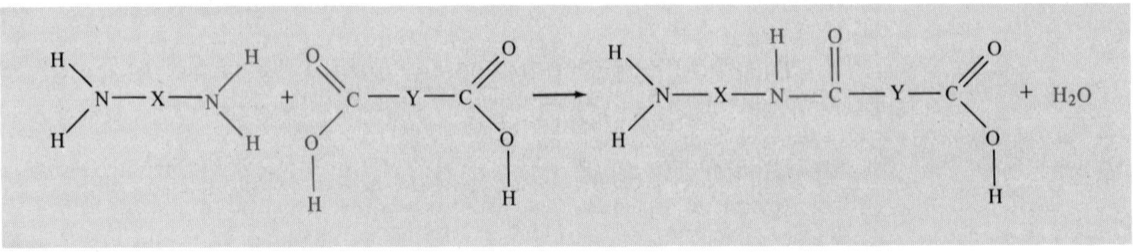

The product has an NH_2 group at one end and COOH group at the other. Another of our original molecules can react with this paired, or **dimer**, molecule, to form a triple combination, a **trimer**.

We have to pick one or the other end of the dimer to think about, so think about the end with the NH_2 group. Therefore, for the addition of the third molecule, we must pick the one with the two COOH groups, one at each end. (If we had selected the other end, with the COOH group, then we would now need the other molecule with an NH_2 group at each end.)

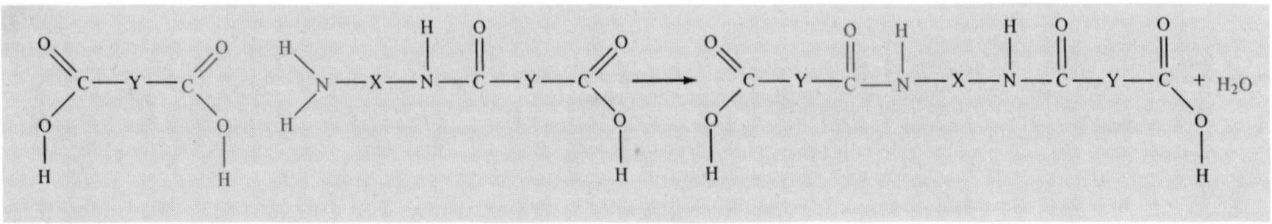

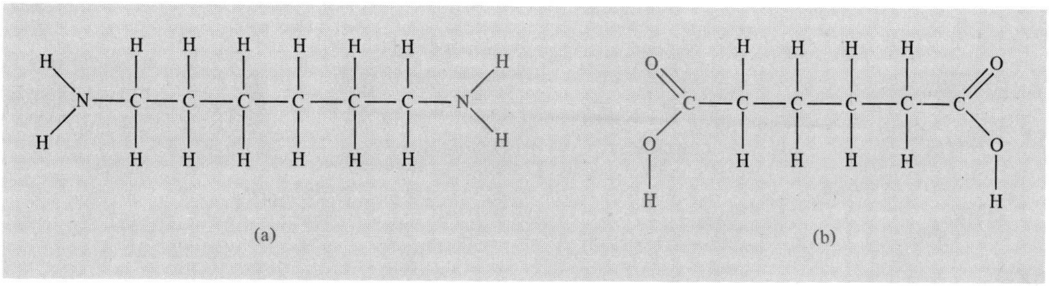

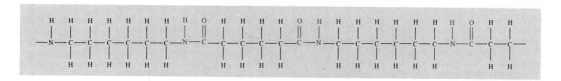

Figure 10–28. (a) Hexamethylene diamine and (b) adipic acid.

And so it goes, the polymer molecule grows by the reaction of an NH_2 or COOH group at one end or the other with single molecules that have two COOH or two NH_2 groups, respectively, at both of their ends. (Of course both ends of the molecule can react at the same time also.)

Nylon 66 is formed from hexamethylene diamine (Figure 10–28a) and adipic acid (Figure 10–28b). A section of a Nylon 66 polymer molecule would look like Figure 10–29 schematically. Notice the oxygen atoms and the nitrogen atoms, both slightly negatively charged, and the hydrogen atoms attached to the nitrogen atoms; these hydrogen atoms are slightly positively charged. The Nylon 66 polymer molecule as a whole is flexible, and can form hydrogen bonds within itself and with other Nylon 66 polymer molecules that happen to be nearby. As the Nylon 66 is spun from a spinnerette, it is stretched to about four times its initial length. This stretching tends to line up the Nylon 66 polymer molecules parallel to each other, forming hydrogen bonds between adjacent molecules. The result is a very strong fiber.

Polyester polymers

As we know, molecules of fat contain the linkage of carbon and oxygen atoms shown in Figure 30. This linkage is called an **ester link**. It is the basis of polyester polymers. To make Dacron, a polyester fiber, for example, we begin with two different molecules, terephthalic acid (Figure 10–31a) and ethylene glycol (Figure 10–31b). Notice the similarity between ethylene glycol, with two carbon atoms and two OH groups, and glycerine (the basis of a fat molecule), which has three carbon atoms and three OH groups. Notice also the relation to

Figure 10–29. Section of Nylon 66 polymer.

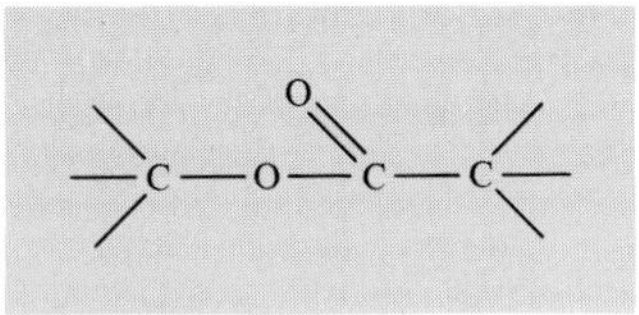

Figure 10–30. Linkage of carbon and oxygen molecules characteristic of fats (ester link).

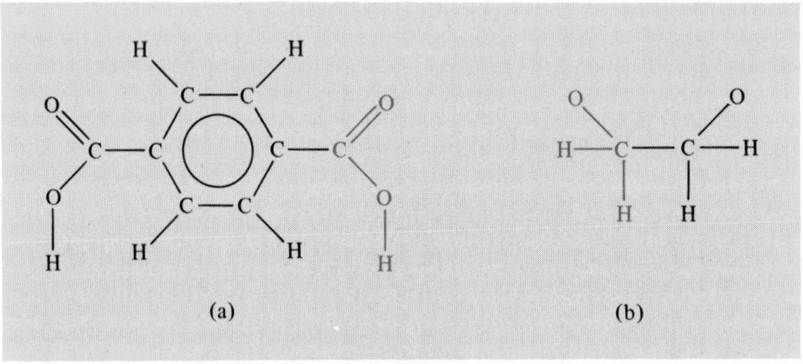

Figure 10–31.   (a) Terephthalic acid and (b) ethylene glycol.

the fundamental reactions with nylon. These also began with two different molecules, each of which had similar groups on its two ends.

Terephthalic acid reacts with ethylene glycol to form an ester link.

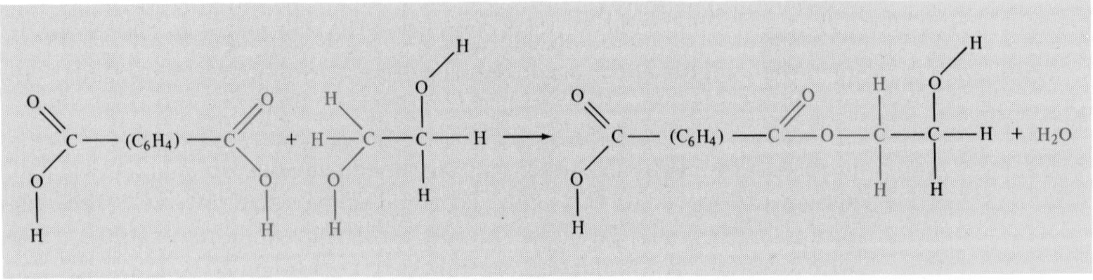

And either another terephthalic acid molecule or another ethylene glycol molecule can react with the product of first reaction. We will pick an ethylene glycol molecule.

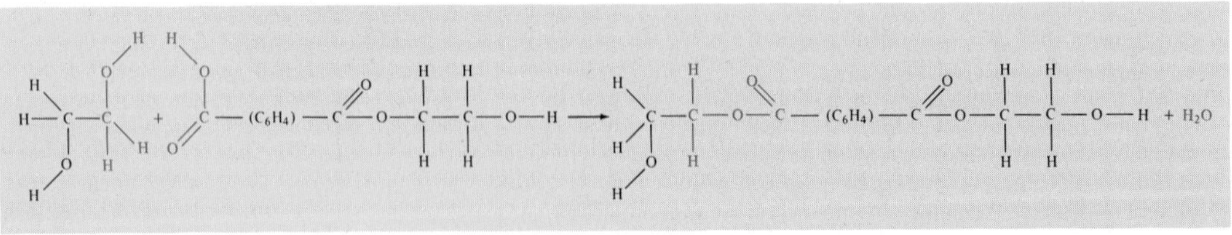

And so it goes; the molecule grows; a polymer forms. The fiber spun from a spinnerette is called Dacron; the same polymer, extruded from a narrow slit, is called Mylar.

Summary of fiber molecules

As we have seen, the properties of a polymer depend upon the composition and upon the structure, or shape, of the polymer molecule. In fact, by selecting the appropriate starting molecules, it is possible to construct polymer molecules with a variety of desirable properties. Table 10–1 summarizes the properties of several fibers that are of concern to most consumers.

TABLE 10–1.   Qualities of Untreated Fibers

Fiber	Wrinkle Resistance	Moisture Absorption	Strength	Pilling Resistance	Heat Resistance	Resists Ageing in Sunlight	Laundering Stability
Acetate	Ac	Md	Pr	Ex	Ac	Ac	Ac
Acrylic	Ac	Lo	Ac	Ac	Ac	Ex	Ac
Cotton	Pr	Md	Ac	Ex	Ex	Ac	Ac
Linen	Pr	Md	Ac	Ex	Ex	Ac	Ac
Nylon	Ac	Md	Ex	Pr	Ac	Pr	Ac
Olefin	Pr	Lo	Ac	Ac	Pr	Ac	Ex
Polyester	Ex	Lo	Ac	Pr	Ac	Ex	Ex
Rayon	Pr	Md	Ac	Ex	Ac	Ex	Pr
Silk	Ac	Md	Ac	Ac	Ac	Pr	Ac
Wool	Ac	Hi	Ac	Ac	Ac	Ac	Pr

Ex=excellent; Hi=high; Ac=acceptable; Md=medium; Lo=low; Pr=poor.

The fiber called acetate in Table 10–1 has not been described in this chapter. Briefly, this material is derived from cellulose by replacing some of the hydrogen atoms of some of the OH groups in a cellulose polymer molecule with an acetate group (Figure 10–32).

Pilling, as used in the table, refers to a condition in which the ends of fibers in a fabric that are near each other tend to tangle into small balls or pills. All fibers will form pills. Generally, the stronger the fiber, the less the tendency for the pills to break off, an interesting example of an undesirable side effect caused by a property, fiber strength, which otherwise is generally desirable.

Notice that many qualities of a fabric or material constructed from fiber are missing from the table Thus, insulating ability, weight (or density), feeling of bulk (or "hand"), and other similar properties depend largely upon the construction of the fabric rather than upon the construction of the polymer molecules. Further, some of the properties of the fiber in the fabric can be altered by finishes applied (usually) to the fabric. Thus, some wrinkle resistance can be added, as it were, to cotton fabrics; wool fabrics can be treated to have

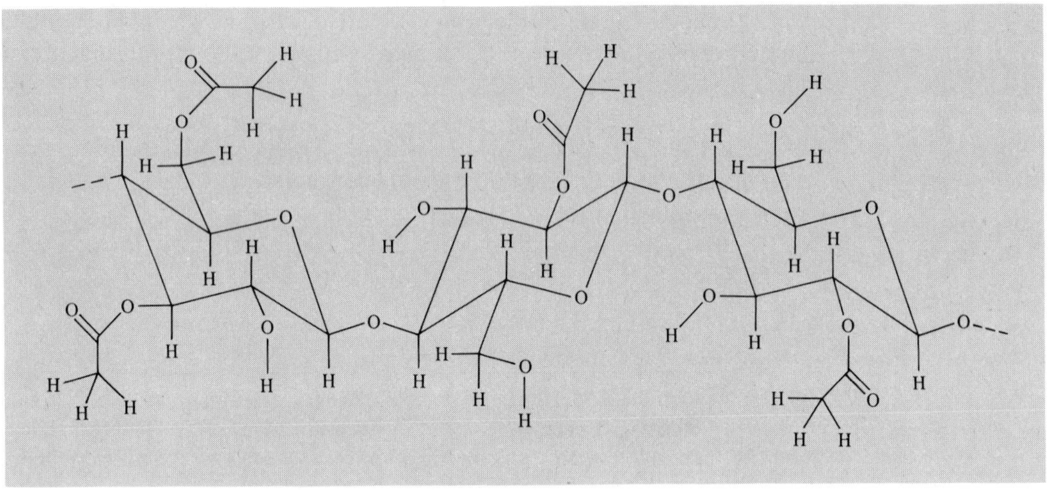

Figure 10–32. Section of acetate (fiber) polymer molecule.

lower moisture absorption. The chemistry of these finishing treatments is interesting but cannot be treated here.

Our emphasis thus far has been upon polymer molecules that were long, narrow, generally flexible to at least some extent, and generally capable of fitting together in parallel arrangements at least here and there. Such polymer molecules can be used as the basis of fibers. Here and there in our discussion, we saw examples of uses of these polymers where the molecules were not parallel to any extent but tangled in a kind of network.

Network polymers

There is a kind of polymer molecule that is deliberately designed to form a network as part of the molecular structure; that is, a network of bonded atoms, different from a tangle of long polymer molecules. As our first example of a

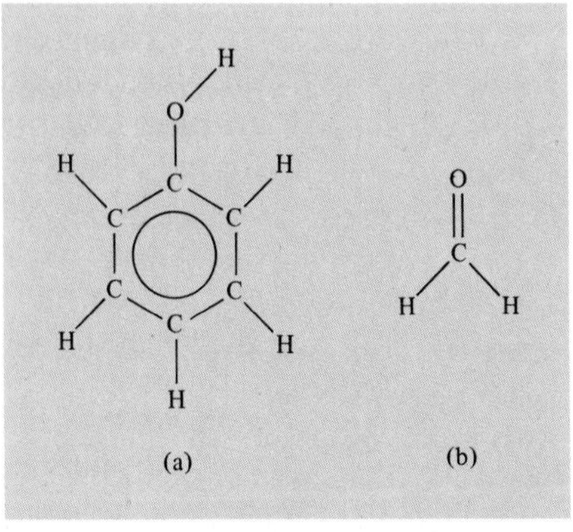

Figure 10–33. (a) Phenol and (b) formaldehyde.

network polymer we will consider the first synthetic polymer ever made, **phenol formaldehyde** polymer, known today under many commercial names, such as Bakelite phenolic and Durez. This polymer is used in a variety of ways, as telephone casings, brake linings, clutch facings, for glues and other adhesives, structural parts in electronic devices, and many others. Figure 10–33a is a schematic of phenol and Figure 10–33b of formaldehyde.

One formaldehyde molecule can react with two phenol molecules.

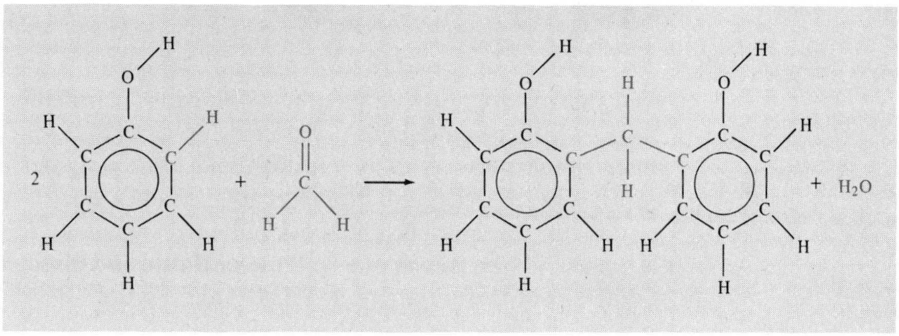

When several such reactions take place, a network polymer is formed; Figure 10–34 shows a section of that resulting molecule.

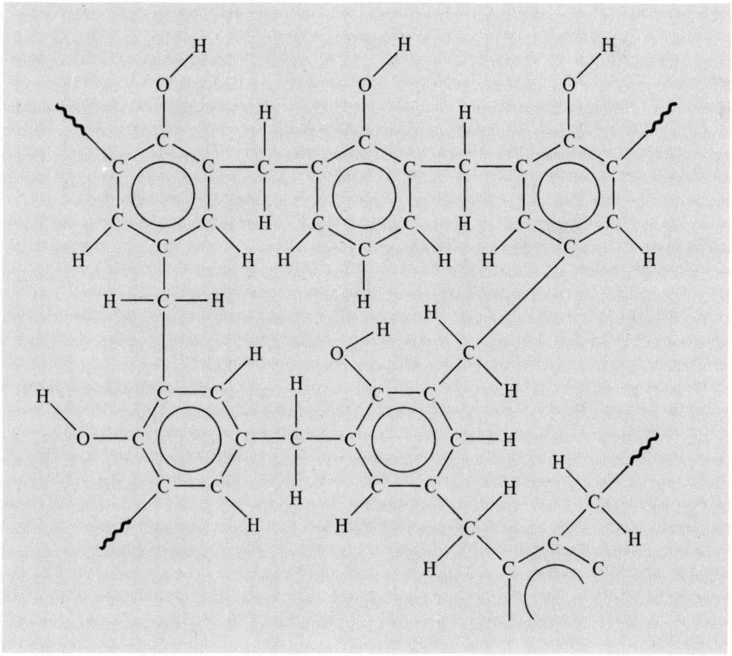

Figure 10–34. Section of phenol-formaldehyde polymer molecule.

Urea reacts with formaldehyde similarly.

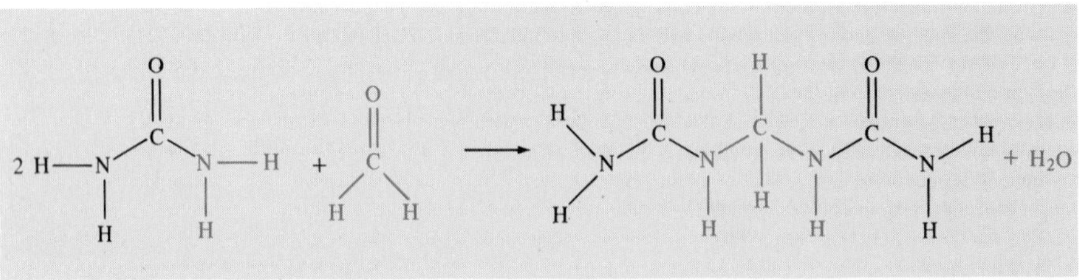

And can form a network polymer, a portion of which is illustrated in Figure 10–35. **Urea formaldehyde** polymer is known as Plascon or Beetle. It is used in much the same ways as phenol formaldehyde polymer, and, in addition for buttons, bottle caps, and "unbreakable" dishware.

Our third and last network polymer, **melamine formaldehyde**, is called Melmac and has uses similar to those of Bettle and Plascon. Figure 10–36 gives a diagram of the melamine molecule, and Figure 10–37 shows the malamine formaldehyde polymer molecule network.

PROTEINS

We begin this chapter with a look at the natural polymers, rubber, cellulose, and starch, and considered some of the details by which these are produced in plants as a result of photosynthesis. It is appropriate to conclude this chapter

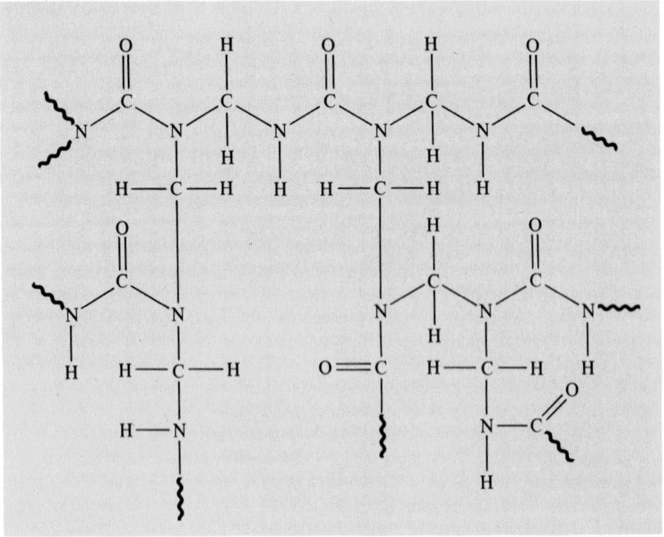

Figure 10–35. Section of urea-formaldehyde polymer molecule.

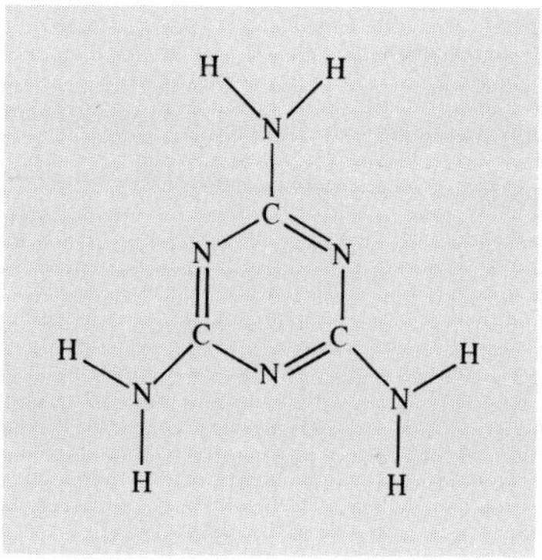

Figure 10–36.   Melamine.

with some of the details of protein synthesis; this will also serve as an introduction to the following chapter.

We will construct a protein molecule in two ways, descriptively and schematically. To begin the schematic construction, recall that there are many different α-amino acids. Six of these, mentioned specifically in Chapter 9, are serine, leucine, histidine, valine, proline, and methionine. In Figure 10–38 we

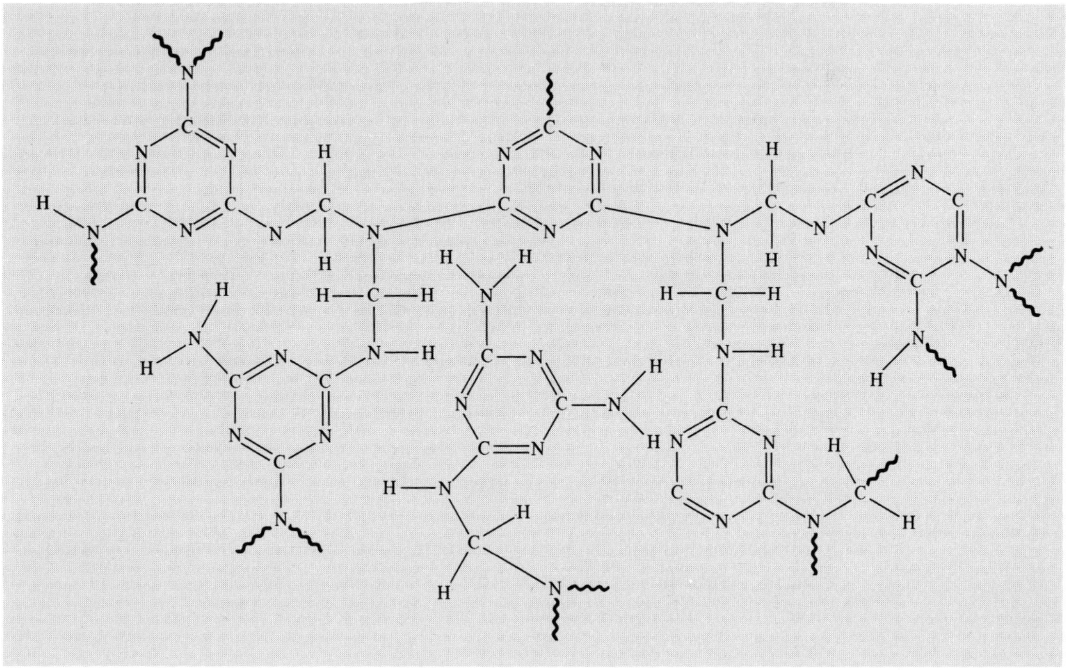

Figure 10–37.   Section of melamine-formaldehyde polymer molecule.

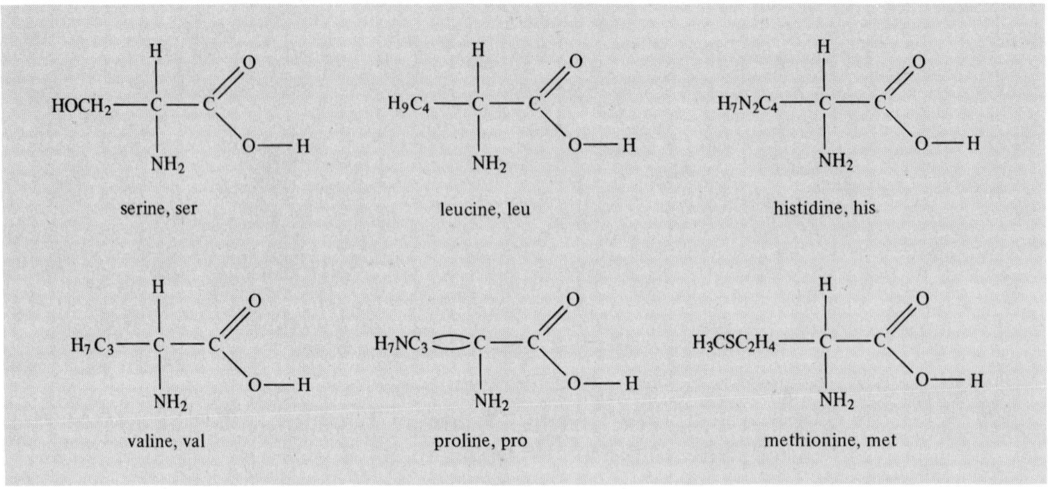

Figure 10–38.   Six α-amino acids, with three letter abbreviations.

draw their schematic structures again, a little differently than before. Note also the three letter abbreviation for each.

Construction of the α helix

Since a protein molecule is *any* combination of a large number of α-amino acids, we could build a very large number of different protein molecules from merely these six α-amino acids. We will build one with the following sequence: val-ser-leu-his-his-pro-met-val-pro-ser-ser-his-leu-val-val-ser-val-leu-val-met-ser. (Of course, this is a "small" protein molecule, not really worthy of the name, but it will do as our example.)

Let us continue with the instructions. Refer to Figure 10–39 and cut it from this book. Cut apart the three strips of the figure and reassemble the pieces into one long strip. Join the ends with paste or transparent tape, fitting the points of the a, a arrows and the points of the b, b arrows together, as in Figure 10–40a. Next, curl the left end so that the symbol $\overset{O}{\underset{||}{}}$ nearest the left end (at the top edge) is placed *over* the same special marker $\overset{O}{\underset{||}{}}$ symbol near the middle of the first strip-piece, at the bottom edge, as in Figure 10–40b. Fasten the curl permanently in this position with a spot of paste or small piece of transparent tape. You have just made a symbolic hydrogen bond between the hydrogen atom and that oxygen atom. Continue the curl with the long strip, making symbolic hydrogen bonds as you go. Eventually, you will have a paper tube with symbols on it that represents an α helix. The α helix is made up of α-amino acid residues in a long chain, curled into a helix and held in the curled condition by hydrogen bonds. Your model should look like Figure 10–41 when it is finished; it represents a section of a protein molecule.

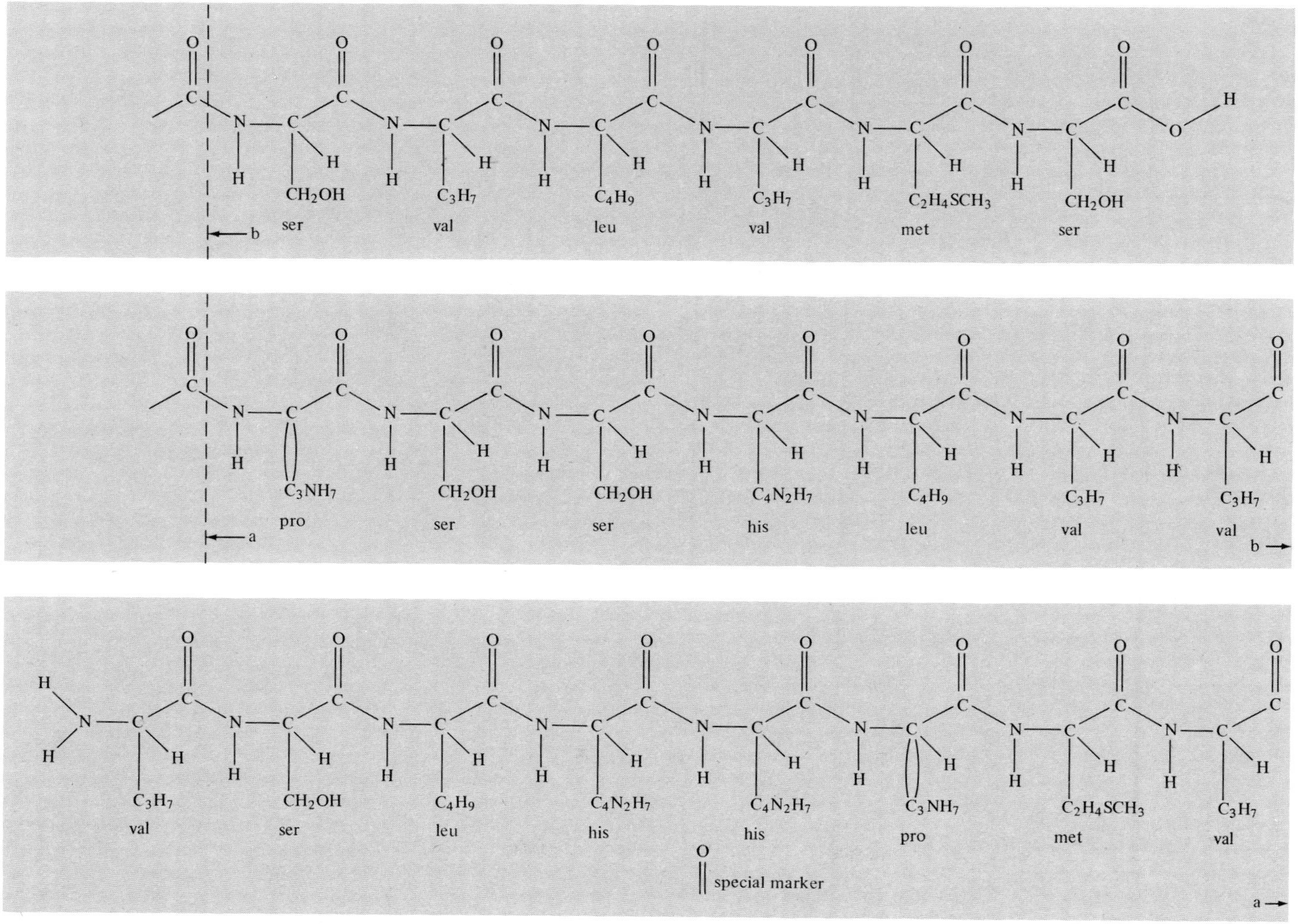

Figure 10–39. Template for protein model.

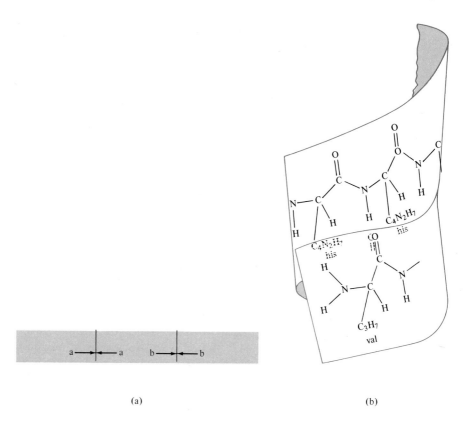

(a)   (b)

Figure 10–40. Construction details for protein model.

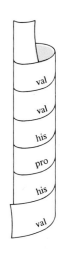

Figure 10–41. Finished model of protein structure.

Protein molecules in the form of an α helix are found in human hair, fur, and wool. Roughly, several helices are themselves coiled together, in a sort of superhelix, each helix held to nearby neighbors by still more hydrogen bonds. Other arrangements are possible, for example a single long helix can itself be curved in various directions or at sharp or shallow angles to form a uniquely shaped "tangle." Instead of forming a helix at all, the chain of α-amino acid residues can form a flat or pleated sheet in which adjacent rows are held together by hydrogen bonds. Silk, for example, has this kind of structure.

Synthesis of proteins in the body

Let us review what has happened. On the printed pages of this book, we have used a 26-letter code, a few punctuation marks such as commas and periods, special symbols such as capital letters and small letters, and peculiar curved and straight marks in the illustrations to provide instructions on the construction of a schematic protein molecule. It was a unique protein molecule, with a unique sequence of six different α-amino acids.

Our bodies must also construct proteins, real molecules, many different kinds, each one unique. If this is to be done, then somewhere inside our bodies there must be a set of instructions, one set for each different protein molecule we need. (Actually, we have several duplicate sets of instructions for each different protein molecule. In various cells in our bodies, duplicate protein molecules are often being constructed simultaneously.) Each cell that has the job of synthesizing proteins contains a set of instructions in a special three-letter kind of code language that governs the construction of each needed protein. This code exists in two forms. One form is in a double helix molecule called **DNA, deoxyribonucleic acid**. The code in the DNA molecule enables the cell to manufacture another molecule, a long chain something like a protein (in a loose way). This other molecule contains the second code; it is called **messenger ribonucleic acid**, or **mRNA**. There is a different mRNA molecule for each different protein molecule that a cell is able to synthesize.

The process goes something like this. From the DNA some mRNA molecules are produced. These mRNA molecules attach themselves to a glob-shaped molecule, called a **ribosome** (see Figure 10–42a) inside the cell. The ribosome moves along the mRNA and reads it, we can say, identifying what kinds of α-amino acids are needed, one after the other, to make a particular protein molecule. As the ribosome moves along the mRNA, another ribosome falls in line, behind the first one, and does the same thing. Behind the second, there is a third ribosome, and so on. After a while, the picture is something like Figure 10–42b. Before long, the mRNA is loaded with ribosomes, each one reading the code and making the same protein molecules, all alike. The short wiggly line you can see coming out of each ribosome represents the protein molecule it is producing. In Figure 10–43, the ribosomes move from left to right, so the protein molecules coming from the ribosomes on the right end of

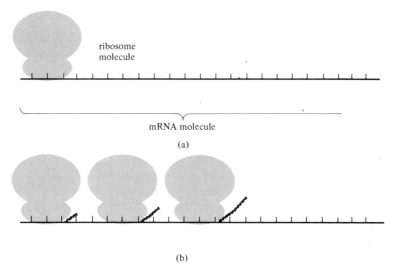

Figure 10–42. **Ribosome molecule attached to mRNA molecule (schematic). (a) One ribosome attached. (b) Additional ribosomes attached (note that the first one has moved to the right).**

the whole business are almost finished. In fact, you can see one that is finished, just at the edge on the right end.

Let us look at this in more detail. mRNA is a long, long molecule but not a protein. It is made up of four different molecule residues, linked together one after another. The four different residues are **uracil** (U), **cytosine** (C), **adenine** (A), and **guanine** (G). Although it would be interesting to discuss the details of these four residues and how they are held together in a long, long chain, we will not take the time to do so. Instead, Figure 10–44 presents a symbolic picture of part of a mRNA molecule with one ribosome on it (the other ribosomes are not drawn, to make the picture less cluttered).

This ribosome is "reading" the code of "letter combinations," three letters at a time, on the mRNA molecule. GGC is the code for glycine, the

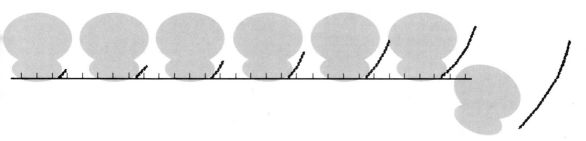

Figure 10–43. **Ribosomes moving along mRNA.**

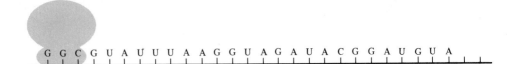

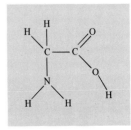

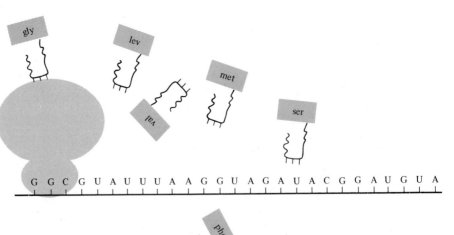

G G C G U A U U U A A G G U A G A U A C G G A U G U A

Figure 10–44. Ribosome "reading" code on mRNA.

H
H
C
C
O
O
N
H
H
H

Figure 10–45. Glycine.

simplest α-amino acid (Figure 10–45). The next three letters to be read are GUA, which is the code for valine; UUU is the code for phenylalinine, another of the α-amino acids, and so on.

Now, notice Figure 10–46. The new molecules are **transfer RNA** molecules, **tRNA** for short. They are not the same as mRNA; for one thing, they are shorter and bent differently. There are at least 20 different tRNA molecules, one for each α-amino acid. In Figure 10–46 you can see that one

gly
lev
met
val
ser
G G C G U A U U U A A G G U A G A U A C G G A U G U A
phe

Figure 10–46. tRNA molecules, with attached α-amino acids, in vicinity of ribosome and mRNA.

tRNA molecule has attached itself to a serine molecule, another to a valine molecule, and so on.

At the moment, the ribosome is reading the code GGC. As you can see in Figure 10–47a, the tRNA with the glycine attached to it is now fastening onto the ribosome. Whenever the ribosome reads GGC, it will "allow" only a

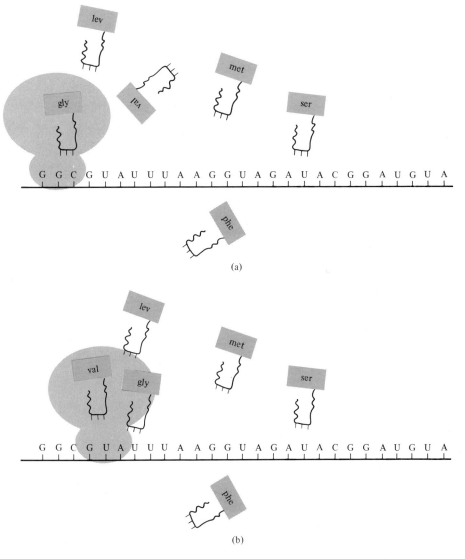

Figure 10–47.   Interactions of ribosome, mRNA, tRNA, and α-amino acids, building a protein. (a) First step, (b) second step.

tRNA with a glycine to become attached to it. Since the next three letters are GUA, the next tRNA to be involved will be one that has a valine attached to it. And so on.

Now two tRNA molecules, one with a glycine, and one with a valine, are next to each other on the ribosome molecule (Figure 10–47b). Next, the two α-amino acids, glycine and valine, are bonded together, with the loss of a water molecule. The first tRNA is released and will eventually be attached to another glycine molecule and deliver it to a ribosome when the mRNA code so indicates. Shortly, the same will happen to the tRNA that is unique for valine. And so on.

In Figure 10–48 you can see what is happening after several α-amino acids have been linked together to form one end of a protein molecule. Notice that the wiggly, loose end of the molecule is already beginning to curl up to form an α helix.

This is the way protein molecules are made within the cells of an animal or plant. Each living being is born with all the coded information that is needed to make the necessary proteins (unless the code is defective). That coded information is contained in the DNA molecules. From the DNA molecules all sorts of instructions are obtained. One set of instructions involves mRNA molecules. Many other sets involve many other molecules that we have not discussed in this chapter nor even in this book.

Many details of the process remain unknown. But it is now possible to produce new forms of some bacteria, although there are serious questions about the consequences of doing so. For example, if it were possible to produce a bacteria that could convert grass into petroleum this might not be a wise way to solve a petroleum shortage. If those bacteria happened to be able to live in the rumen of a cow, it is conceivable that, inside the cow, oil would be formed instead of glucose. The cow would die, of course; and eventually we might lose our sources of dairy products. Going much further in our conjectures, we can in

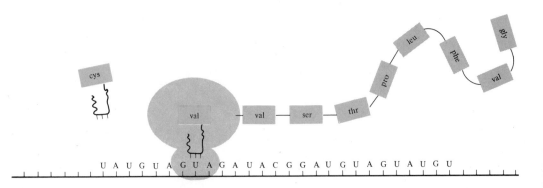

Figure 10–48. Building a protein, several steps later.

principle forsee the time when the characteristics of humans could be controlled in advance of their conception. (We are a long way from this; at present). Clearly, the future possibility of altering the characteristics of living beings by deliberate design is portentous. Careful and wise decisions involving moral, ethical, ecological, and philosophical considerations, as well as the scientific aspects, will be required.

Very briefly, many people believe that they were made by God in his image and likeness, and therefore men should not be allowed to alter the characteristics of other men. They are less concerned with such changes in animals because they believe God has given man charge over the animals. Others will not accept this, arguing that to tamper even with animals is a violation of mans' stewardship. Still others argue from an agnostic or athestic position that any alteration that is desirable and necessary may be made, once we have the knowledge needed. Some point out that it is too dangerous to attempt to acquire this knowledge; we must necessarily begin to acquire it by working with simple organisms, such as bacteria, and this risks altering a bacterial strain from relative innocence into a markedly virulent new kind of living being that might wipe out the human race. Those who recognize each person's right to establish his own well thought out position assert that, since man has obviously been given an intellect by his Creator, he is obliged to use it well. Because the efforts of our intellect have brought us this far, it is up to us to use our knowledge, seeking God's guidance as we do so, and to strive to solve the myriad of problems that this challenge will surely bring upon us.

SUGGESTIONS FOR DISCUSSION

1. What does the term, "restricted flexibility" mean to you? Use examples from this chapter to illustrate your concept. Can you think of other examples from your life that are flexible yet restrictive in nature, such as laws or rules, the wishes of your friends, your needs, the niceties and luxuries you particularly like? How are these similar or dissimilar to the chemical restricted flexibility described in this chapter?

2. Another theme in this chapter could be termed "connected and repetitive." Again, what does this mean to you? Cite both chemical examples and examples from your own life to illustrate your discussion.

3. We have looked at several different polymers in this chapter. Identify at least one that was of interest to you and at least one that did not interest you. What characteristics of those polymers you have identified caused your selection for each extreme?

4. Select any polymer mentioned in this chapter (not necessarily one of those you have identified in response to the preceding question) and, with the help of an encyclopedia or other general reference source or sources, spend about an hour reading about that selected polymer. From that study return to this chapter and find a paragraph or two that you can now rewrite in a better fashion. Revise that selected passage.

5. Termites, cows, sheep, and horses can extract nutritious values from cellulose (wood for termites and grass for the other animals). See if you can guess or find out why.

6. Can you summarize, say, 80% of the details of photosynthesis that are given in this chapter into one fairly long (if necessary) paragraph? If so, do so. If not, justify your inability to do so in an equally long paragraph.

7. Briefly summarize at least two arguments favoring the restriction of the utilization of petroleum solely for the synthesis of polymers. Briefly summarize at least two arguments favoring the elimination of petroleum as a raw material in the synthesis of polymers. Finally, make a decision of your own in this matter and supply at least two more arguments to support your choice or choices.

8. The monomer, single molecule that is, vinyl chloride, is known to cause cancer in some animals other than man and is strongly suspected to be a carcinogen (cancer producer) in man when exposed to it perhaps for a long or short period in the manufacture of the polymer. Considering the economic importance of this polymer and in particular the fact that the jobs of many workers depend upon the manufacture of polyvinylchloride, would you or would you not order the cessation of polyvinylchloride manufacture if it was in your power to do so? Justify your decision; or suggest other alternatives, and justify your choice of one of those.

9. Just for fun, and in 25 words or less, describe a helix to someone who does not know what it is but without using the words "circular stairway" or "spiral".

10. Refer to the description of protein synthesis. In your opinion and assuming that the description (which actually is theoretical not factual) is at least close to the truth, use that the details of protein synthesis support one of the following propositions: (1) Life just happened to be evolved. (2) Life was developed or created by a divine all powerful Creator. (3) Life was developed by a Creator in a special way for each different species without evolution. (4) Some other similar proposition. Then, find at least one short argument that tends to disprove your contention.

11. Review the last two paragraphs of this chapter. Make a tentative decision that it is proper or improper for man to attempt to alter the instructional code governing the characteristics of plants, animals, or men. Justify your decision, think of an argument against your decision, and finally be prepared to offer a rebuttal in favor of your original contention.

12. In the fine arts, particularly architecture, there is a well known (sometimes disputed) dictum, "form follows function." Much of the content of this chapter argues for a corresponding molecular-chemical dictum, "function (properties) follows (depend upon) form (molecular structure)." Respond to these two dictii by agreeing or disagreeing with either or both and account for your opinion.

chapter 11

Metals and Life

by C. A. McAuliffe

In the two preceding chapters we have emphasized the utilization of energy via its storage as chemical bond energy. Along the way, we also noticed a few other details involving nutrition and common sense and polymers from nature and artificial polymers (to mention only the high spots). In our summary thus far, however, it was necessary to skip over and detour around several other high spots; otherwise the discussion would have become too complicated to be interesting. But before leaving the subject of chemistry and life altogether, we ought to pause long enough to examine at least one more facet, the involvement of metals in living processes.

Up to now, except for a brief mention in Chapter 9, our attention to molecules involved in life has been confined to molecules containing carbon, hydrogen, oxygen, nitrogen, and occasionally sulfur. It is appropriate now to call your attention to some other elements and to the roles they play in your life, even now, as you read these words. It is a pleasure to include in this book this chapter, the contribution of a respected colleague, a bioinorganic chemist, C. A. McAuliffe.

Two decades ago biochemists and medical researchers were aware that about one third of all enzymes contain metal atoms. However, the part that those metal atoms play in the reactions of enzymes has only recently been realized. Now we know enough to say confidently that the metal atom is frequently the very center of activity. Before we examine some examples of the importance of metal ions in enzymes, there is a detail to attend to. Biochemical nomenclature can be quite complicated, and we ought to note these definitions and statements:

1. **Enzymes** are biological catalysts, that is, they speed up the rate at which substances react in the body.
2. Enzymes that contain metals are called **metalloenzymes**.
3. All metalloenzymes consist of two important parts, the apoenzyme and a small prosthetic group.
   (a) The **apoenzyme** is the bulk of the molecule and consists of protein.
   (b) The **prosthetic group** is the metal atom and its nearest atom neighbors.
4. In a biological reaction an enzyme interacts with another molecule (the **coenzyme**); after the reaction is over, the coenzyme departs, chemically altered.

THE ROLE OF METALLOENZYMES

We are aware, from the implications in the preceding chapters or from other sources, that there are thousands of different, complicated, biochemical reactions that occur in the body every second. These reactions are extraordinarily efficient. If we could understand the principles of body chemistry and apply these to industrial processes, we could conserve both energy and material. Recently it has become clear that it is the metal atom in metalloenzymes that controls the activity of the enzyme to a great extent. Chemists understand the ways in which metals control such activity as the effect of a metal ion on the structure and shape of the rest of the metalloenzyme. As you will see, the rest of this chapter is an extended discussion of these two statements.

Nitrogen fixation by nitrogenase

We shall begin by briefly reconsidering nitrogen fixation, a topic already mentioned in Chapter 7. In the plant world a metalloenzyme called nitrogenase turns atmospheric nitrogen into nitrogen compounds for plant growth. It is through this enzymatic fixation of nitrogen that atmospheric nitrogen is made available to plants and, subsequently, to man and other animals as a constituent of all protein molecules. Let us briefly review the details of the Haber process, the artificial fixation of nitrogen.

$$N_2 + 3\,H_2 \rightarrow 2\,NH_3$$

Compared to natural nitrogen fixation, the Haber process involves high temperatures and pressures or electric discharges to force the reaction. This is because molecular nitrogen, $N \equiv N$, is very unresponsive to chemical reaction. The very large amount of energy needed to separate the molecule into atoms, so they can react further, accounts for the relative inertness of nitrogen molecules.

However, nitrogenase enzyme acts at ordinary conditions of temperature and pressure, much lower than the 400–600°C of the Haber process! Nitrogenase is known to contain the elements molybdenum and iron. It is thought that nitrogen initially binds to one of these elements while the other metal provides electrons for the reduction of nitrogen to ammonia.

There is an immense store of nitrogen in the atmosphere (the air is 80% nitrogen), so the eventual elucidation of the way in which nitrogenase works will not only provide cheaper nitrogen fertilizer (thus helping to alleviate the ever-growing gap between food production and increase in population) but should also provide a valuable insight into the mechanism of reactions involving the nitrogen molecule. Possibly a new branch of synthetic chemistry will be opened.

Selectivity of living organisms

One of the characteristics of living matter is that it is selective with respect to its environment. An example of this selectivity is the quite striking difference between the abundance of available elements on the Earth and those in living matter. About 99% of the Earth's crust is composed of the elements oxygen, silicon, aluminum, sodium, calcium, magnesium, potassium, and iron. Living organisms, on the other hand, are 99% hydrogen, carbon, nitrogen, and oxygen. Only in the case of oxygen is there any parallel between abundance in the Earth's crust and in biological material. Stated otherwise, it must be concluded that living matter prefers to be made up of the lighter elements rather than heavier elements!

In an earlier chapter we discussed the cosmos. It is interesting to note that in concentrating lighter elements in its material makeup living matter is very much like the cosmos. The stars are essentially composed of the lighter elements. Because of the very weak gravitational pull of our Earth the lighter

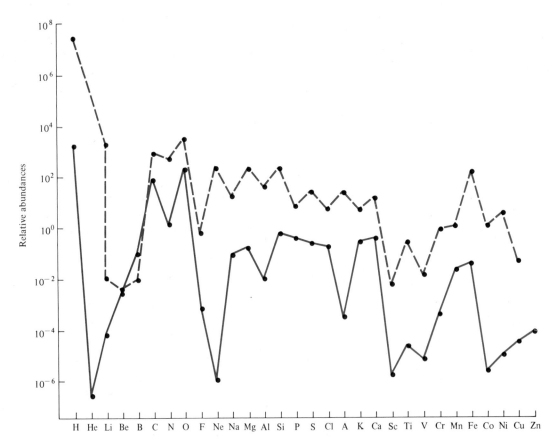

Figure 11–1.   Comparison of abundance of light elements in the cosmos (– – – –) and in living matter on Earth (————).

elements have left us—they have been "distilled off" into space—and we are left with an uncharacteristic amount of the heavier elements. The cosmos as a whole, however, is predominantly composed of light elements. Figure 11–1 illustrates the parallel that exists between the 31 lightest elements of the cosmos and their distribution in biological systems. The interesting question thus exists: is this striking parallel entirely a coincidence or is the distribution of the elements in biological systems a "photograph," or replica, of the chemical composition of the Earth when life began to evolve?

THE ESSENTIAL ELEMENTS

Only about 20 elements have been shown beyond any doubt to be essential to life (17 elements are now known to be essential to man); they are listed in Table 11–1. About ten more are thought to be essential, but it is an extremely difficult

TABLE 11–1.   Elements Used in Living Systems and Their Role

Element	Symbol	Atomic Number	Concentration in g/70-kg (154-lb) person	Role
Hydrogen	H	1	6,580 ⎫	Structural components
Carbon	C	6	12,590 ⎪	of all biological
Nitrogen	N	7	1,815 ⎬	material
Oxygen	O	8	43,550 ⎭	
Sodium	Na	11	70	Na^+, sodium ion, is an important constituent of fluids outside cells
Magnesium	Mg	12	42	Active center of many enzymes
Phosphorus	P	15	680	Involved in biochemical reactions providing energy for life processes
Sulfur	S	16	100	Found in proteins and other important substances
Chlorine	Cl	17	115	Important negative ion
Potassium	K	19	250	Found inside all cells; involved in electrical activity in nerves and in muscle contraction
Calcium	Ca	20	1,700	Important constituent of membranes; involved in muscle activity
Boron	B	5	*	Important for healthy plant growth
Silicon	S	14	*	Found in lower life forms
Vanadium	V	23	*	Found in certain sea creatures
Manganese	Mn	25	1	Vital part of many enzymes
Iron	Fe	26	7	Transports oxygen in man
Cobalt	Co	27	1	Center of vitamin B_{12}
Copper	Cu	29	1	Important in many enzymes, and oxygen transport in marine life
Zinc	Zn	30	1	Vital part of many enzymes
Molybdenum	Mo	42	1	Vital part of a few enzymes

* These elements have not been shown to be essential to human life.

task to *prove* that an element is essential. Moreover, some elements (hydrogen, carbon, nitrogen, oxygen, sodium, magnesium, phosphorus, sulfur, potassium, and chlorine) have been shown to be vital to all living matter (plants and animals), some (iron, copper, manganese, zinc) have been shown to be vital to certain species and may thus be essential to all, but the universal importance of such elements as boron, silicon, vanadium, cobalt, and molybdenum has not yet been proved beyond all doubt. This is because the problem of showing an element to be absolutely necessary to any biological system entails isolating that system without *any* trace of that particular element.

For instance, the nutritional necessity of some ultratrace elements was not even suspected until their absolute absence in certain soils caused disease and abnormalities. The loss of kinkiness of sheep's wool, detrimental effects on the nervous system, and anemia in the sheep were cured by addition of trace amounts of copper to their diet in an area of Australia where the soil had no copper whatsoever. The diseases known as "heart-rot" in beets, "cracked stems" in celery, and "internal cork" in apples are caused by a deficiency of boron in soils in which these plants are grown. The addition of 0.1 part per billion (ppb) parts of soil will cure the conditions. But, note carefully, 1 ppb additional of boron will poison the plants and cause other disorders or kill them. Obviously, the link between boron in soil and the health of these plants is delicate.

ENERGY FROM THE SUN: HOW LIVING ORGANISMS HARNESS IT

The whole chemistry of living creatures can be classified simply into two main processes: (1) the harnessing of solar energy into driving chemical reactions that produce oxygen and reduced organic compounds from carbon dioxide and water (discussed in Chapter 10) and (2) the oxidation of these reduced organic compounds with the production of carbon dioxide, water, and energy (discussed in Chapter 9). For both these processes metal ions and organic compounds called porphyrins are essential.

The porphyrins

Figure 11–2 illustrates **porphin** itself, the simplest of this class of compounds. It is shown schematically, with a carbon atom understood to be present at every corner where two or more lines join. The dot pairs on the nitrogen symbols

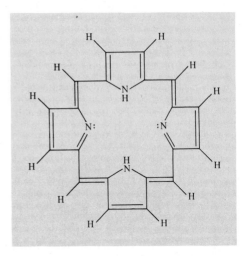

Figure 11–2. Porphin.

represent pairs of electrons. Notice the "hole" in the center of the molecular diagram. This hole is more than diagrammatic; there is a real hole in the porphin molecule. A metal ion, such as Mg^{2+} or Fe^{2+}, fits nicely in the hole, displacing the two hydrogens. Such a metal ion, symbolized by M, is bonded by four pairs of electrons, one pair from each nitrogen. The arrows drawn between the nitrogens and the metal ion signify that both electrons in the shared pair come from the nitrogen atom.

Actually, it is not porphin that is important but the derivatives of porphin. Porphin molecules that contain organic groups substituted for some of the outermost hydrogens around the complex ring are called **porphyrins**. Porphyrins are rigid molecules, and the ring complex is exactly flat. Porphyrins also react with metal ions, to form **metalloporphyrins**.

The metallo-
porphyrins

The most important magnesium metalloporphyrin is **chlorophyll** (Figure 11–3). The role of chlorophyll in "fixing" radiant solar energy in chemical bonds in living plants was mentioned in Chapter 10. The absorption of light energy enables the green leaved plants to produce sugars from carbon dioxide and water, as we know. The overall reaction can be written:

$$6\ CO_2 + 6\ H_2O \xrightarrow{\text{energy}} C_6H_{12}O_6 + 6\ O_2$$
$$\text{glucose}$$

In point of fact, in green plants there are **photosynthetic** systems, known as chlorophyll a and chlorophyll b, both very similar. As you know, the photosynthetic process is complicated. Many other parts of the process also involve metal atoms (manganese, iron, and copper are known to be involved).

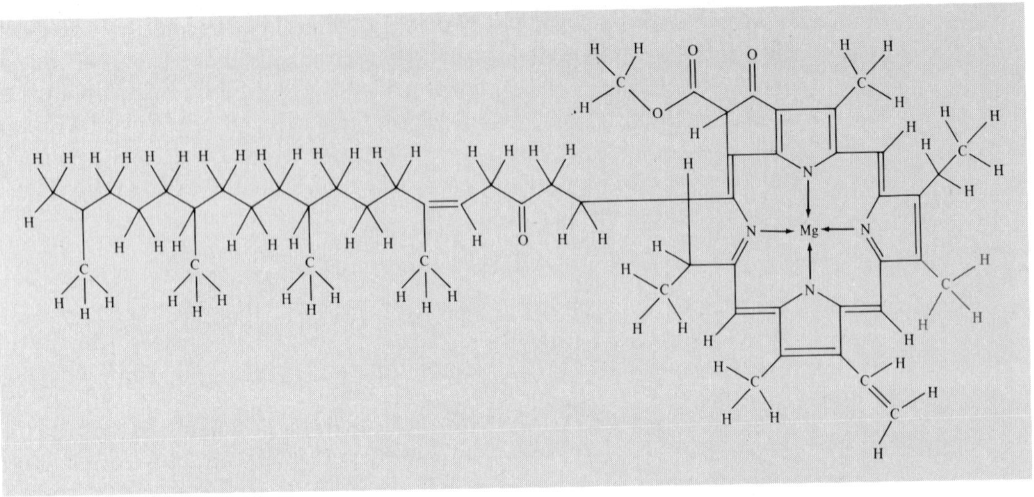

Figure 11–3. Chlorophyll a. Note the methyl group, 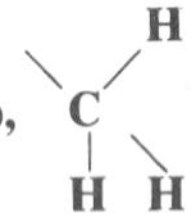 in color. In chlorophyll b, the structure is identical except that the methyl group is replaced with a formyl group, H.

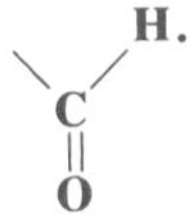

In the current search for alternative sources of energy to replace oil, it is quite obvious that much research must be done to mimic the action of photosynthesis, that is for man to make an artificial photosynthetic device able to convert incident solar radiation into electricity. Such a device is quite likely to incorporate a metalloporphyrin as its center, although recent work on a solar energy device has utilized a blue-green algae to decompose water into hydrogen and oxygen. This algae contains the nitrogenase enzyme (with molybdenum and iron as the active metals.) This enzyme is not a porphyrin.

Were man to design a molecule capable of absorbing solar energy, it is quite unlikely that he would be able to think up a system better than that of chlorophyll. The radiant energy which comes to the Earth from the Sun can be conveniently divided into three types: ultraviolet (absorbed in the Earth's atmosphere by oxygen and ozone), visible (that which corresponds to the sensitivity of the human eye and can be absorbed by metalloporphyrins), and infrared (absorbed by carbon dioxide and water). Of this "spectrum" of radiation from the Sun it is fortunate that by far the most intense is the visible part; as this visible radiation is not appreciably absorbed by the atmosphere, it is almost all available for photosynthesis and at the service of man to produce organic molecules (and store energy). The porphyrin molecule, because of its double bonds, can absorb this energy.

Porphyrins alone would absorb the visible light and then immediately reradiate this energy (through fluorescence). However, when a metal atom is fixed in the center of the porphyrin molecule, the visible light is absorbed, we say that the molecule becomes excited. In this case, it stays excited for a small, finite amount of time. During this short time interval the absorbed energy is used for photosynthesis, and a small amount of remaining energy is then reradiated (through phosphorescence). We thus note that it is a bioinorganic molecule containing carbon, hydrogen, nitrogen, oxygen, and a metal, that makes the Sun's energy available to mankind via photosynthesis.

The cytochromes

We know from the equation describing the photosynthetic reaction that oxygen is always produced. In early evolutionary times this oxygen undoubtedly ripped apart (oxidized) the organic sugars that were the other product of photosynthesis, thus effectively putting a block on further evolutionary development. Fortunately, early organisms evolved a counterattack through molecules known as **cytochromes**. These are iron-containing porphyrins, also called a **prosthetic** or **heme** group, surrounded by protein apoenzyme. It is a cytochrome that allows the oxidation of the sugars in the reverse of the photosynthetic reaction

$$6\,O_2 + C_6H_{12}O_6 \rightarrow 6\,CO_2 + 6\,H_2O$$

but also allows the energy from this process to be *stored* in the organism for future use. Thus we see that not only is the metal-containing molecule chlorophyll used in trapping solar energy but yet another molecule that depends for its effectiveness upon the inclusion of a metal, cytochrome, is used in living organisms to make that energy available to the organism.

RESPIRATION

Let us move to a different topic. We take respiration—breathing—very much for granted. We draw in air, hopefully fresh air, which is one fifth oxygen and four fifths nitrogen; it is pulled into our lungs and then we breathe out again. Simple, isn't it? Well, it is not so simple, and once again a metal atom, iron, is at the center of things, making life go on.

We need molecules to pick up oxygen from the lungs and transport it (these are **hemoglobin** molecules) round the body where other molecules (**myoglobin** molecules) store it until it is needed for biological reactions. It is hemoglobin that gives the red color to human blood. It seems that only molecules that contain metal atoms can transport oxygen in biological systems. In crabs, for example, another iron-containing molecule, hemocyanin, is responsible for this process. The blue color of fresh crab blood is due to hemocyanin. There are some very primitive sea creatures, the ascidians, that use a vanadium-containing molecule for breathing purposes.

Hemoglobin and myoglobin

Let us look more carefully at hemoglobin and myoglobin. Myoglobin is a much smaller molecule than hemoglobin (myoglobin has a molecular weight of 17,500 whereas hemoglobin has a molecular weight of 64,450). It is mainly made up of protein, the apoenzyme. This protein twists and winds itself around the most important part of the molecule, the prosthetic or heme group where the iron atom sits inside a porphyrin ring quite like magnesium in the chlorophyll molecule. The chemical structure of the heme group is shown diagrammatically in Figure 11–4.

We can consider the heme molecule to be a flat disc for simplification. This heme disc fits neatly into the apoenzyme protein chain that circles and weaves around it. The protein chain is joined to the prosthetic group, the iron atom in the porphyrin ring, through part of an amino acid called histidine (indicated as "his" in Figure 11–5). Now then, directly opposite the histidine, on the other side of the iron atom, is the site at which the oxygen molecule is held onto the iron in the myoglobin molecule. As we might expect, myoglobin exists in high concentrations in muscle tissue where it is ready to release oxygen when physical work is demanded. Myoglobin gets its oxygen from hemoglobin, as we shall see.

Like myoglobin, hemoglobin has been studied by x-ray crystallographers. Hemoglobin contains four heme units in each hemoglobin molecule. Each of these units can form a bond to one oxygen molecule, but the uptake of

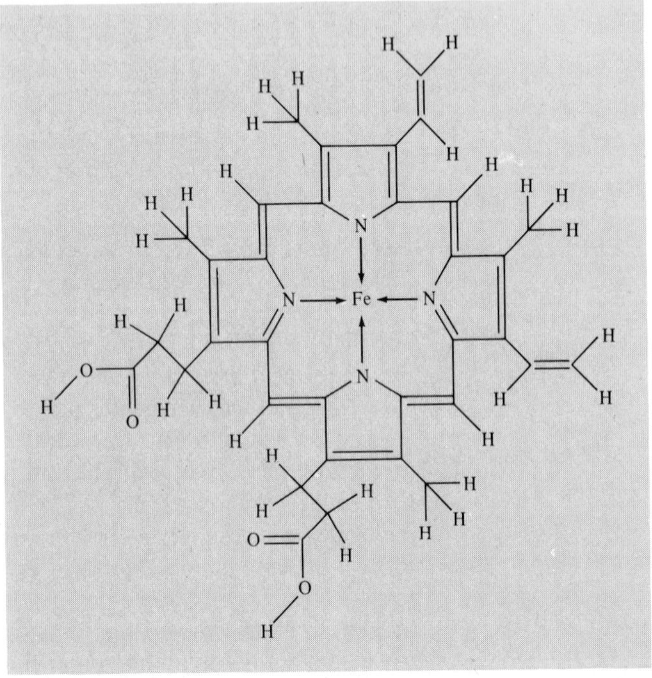

**Figure 11–4.
Heme group.**

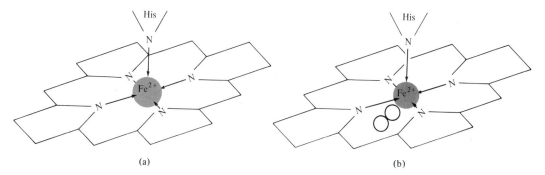

(a)   (b)

Figure 11–5. **(a) Diagrammatic perspective view showing the flat heme group with Fe²⁺ ion in center and a histidine residue (his) "above" the ion. The histidine is part of the apoenzyme, not otherwise indicated. Note that the Fe²⁺ ion is "above" the plane of the heme group. (b) A similar view, but suggesting an oxygen molecule coordinated (bonded) with the ion, shown "underneath" the ion. Note that the ion is now slightly smaller in diameter and is now centered in the plane of the heme group.**

oxygen by these four heme units does not happen simultaneously. In fact, nature has performed a very elegant task in designing hemoglobin as an oxygen carrier. Once one molecule of oxygen is bound to one heme unit, the whole structure and properties of hemoglobin begin to change. The result of the change is that the next oxygen molecule is bound more readily than the first. Similarly, the third oxygen is bound still more tightly, and the fourth the tightest of all.

All four heme units have oxygen coordinated (bound) to them in the lungs where there is a high pressure of oxygen. The **oxyhemoglobin** (as the fully oxygenated hemoglobin is called) is then carried by the blood round the body to the tissues where it enters areas of low oxygen pressure. Here the oxyhemoglobin releases the oxygen, which then binds to myoglobin. Note, here again, the elegant designs of nature. Remember that after one oxygen molecule became bound to **deoxyhemoglobin** (hemoglobin which has no oxygen coordinated) it was easier for subsequent oxygens to coordinate. In an analogous manner, once one oxygen molecule has been lost from oxyhemoglobin, it becomes increasingly easier for the second, the third, and then the fourth to be lost. If you think carefully about the significance of this phenomenon you will understand that hemoglobin is an exceptionally efficient machine.*

The Nobel prizewinners, Perutz and Kendrew, have contributed greatly to our knowledge of the workings of hemoglobin by their x-ray crystallographic work extending over a decade. As oxygen begins to bind to deoxyhemoglobin

---

* The uptake and release of oxygen seems to paraphrase Scripture: to him that hath shall be given and to him that hath not shall be taken away. The more oxygen the hemoglobin has the more it can have (up to four molecules); and once oxygen begins to leave the subsequent loss of the remainder is enhanced.

the whole molecule undergoes quite drastic structural changes. The heme units are placed quite separately in the molecule as a whole, and the effects of oxygen binding on one heme site are in some marvellous way transmitted through the interconnecting protein apoenzyme to the other prosthetic groups. In some way it would appear that hemoglobin "breathes" (that is, changes its shape back and forth) as oxygen is picked up and released! Even changes around the iron atom in the heme units have been detected. The unbound iron atom is larger (and also has different magnetic properties) than the bound iron atom in heme. In deoxyhemoglobin the iron atom is too large for the hole in the porphyrin ring and sticks up above it. Once oxygen is bound, the radius of the iron atom decreases and the iron "falls" into the center of the ring. (See Figure 11–5.) In this way, also, it may be thought that even the heme units "breathe" as they go from deoxy- to oxy- and back again.

SODIUM AND POTASSIUM

All animal and plant cells are encased by a membrane. Its function is extremely important and is dependent on metal ions. This membrane is permeable (that is, it allows certain types of matter to pass through it), but it is a very selective permeable membrane and this allows for quite different concentrations of sodium and potassium ions inside and outside the cell.

Sodium and potassium are alkali metals—the word *al-quily* is Arabic for plant ashes and derives from the fact that these metal ions are left in the ash after desert plants are burned. Although the relative abundances of sodium and potassium in the Earth's crust are similar, plants contain about ten times as much potassium as sodium. Since our bodies obtain their supply of these minerals from plants (directly or indirectly), it is easy to see why we have to supplement the amount of sodium by addition of sodium chloride (salt) to our food. Also, since growing plants take up so much potassium from the soil, the need for potassium-containing fertilizers is easily understood. (See the discussion on clays in Chapter 12 for further details.)

The ion pump

Sodium is found on the outside of cell membranes and potassium inside the cells. Because there is a high concentration of potassium ions (K^+) inside cells and a low concentration outside cells (and vice versa for sodium ions, Na^+), common sense predicts both that K^+ would diffuse out (that is, move through the cell wall membrane) into the fluid outside the cell and Na^+ would diffuse in the opposite direction, into the cell. This type of diffusion does occur. However, nature has designed an elegant trick to overcome this tendency and keep the concentrations of K^+ (inside) and Na^+ (outside) high; the mechanism is called the **ion pump**. A phosphoprotein (protein with a phosphate group) is involved in selectively "grabbing" K^+ from outside the cell and carrying it back through the cell membrane and for doing the reverse for Na^+. Just how this is done has yet to be fully understood. One theory suggests the sequence shown in Figure 11–6.

Thus, although there is a continuous migration of K$^+$, at high concentration inside the cell, to a lower concentration outside the cell, the process described in Figure 11–6 continuously renews the concentration of K$^+$ inside the cell. Perhaps a similar or perhaps a distinctly different set of details act conversely with respect to Na$^+$, keeping the concentration higher outside the cell than inside. Whatever the real details might be, both processes require the expenditure of energy, probably supplied by phosphorus-containing energy carriers such as NADPH and A℗℗℗.

(a) Outside the cell membrane an energy carrier we have seen before in Chapter 10, NADPH, releases a hydrogen ion, H$^+$, and an electron, e^-, becoming NADP, which is recycled somewhere else back into NADPH..

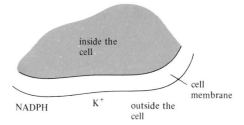

(b) The H$^+$ remains in the fluid outside the membrane, but the electron is "grabbed" by constituents in the membrane and transported across to the inside of the cell.

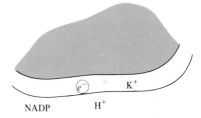

(c) Simultaneously, a potassium ion, K$^+$, is removed from the fluid outside the membrane and transported by some mechanism through the membrane and into the inside of the cell.

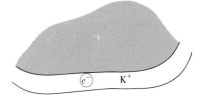

(d) The two processes, electron transport through the membrane and potassium ion transport through the membrane, are thought to occur side by side, as it were, in separate but adjacent "channels" within the membrane.

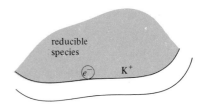

Figure 11–6.   Possible mechanism for the ion pump according to steps (a), (b), (c) and (d).

Functions　　We have seen that the localized Na^+ and K^+ concentrations are kept high, but what function do these ions perform in the body? Na^+ depresses the activity of the enzymes that cause a muscle to expand; thus, the muscle contracts. K^+ is involved especially in heart muscle contractions. Both ions are used in the transmission of electrical nerve impulses. Sodium salts play other roles in the body. Sodium chloride provides hydrochloric acid for gastric tract juices, and sodium hydrogen carbonate (the inexpensive pharmacological agent in many expensive patent remedies for acid stomach) helps maintain the delicate acid-base balance of body fluids. The acid-base balance is very important in the transport of dissolved carbon dioxide, which the body exhales, as mentioned in Chapter 7. One salt of sodium should be mentioned because of its detrimental effect, sodium urate, the sodium salt of uric acid. This is a very insoluble compound. When it is deposited in cartilage, it produces the agonizing condition known as gout.

One more metal　　In addition to sodium, potassium, iron, and magnesium, the biologically important metals mentioned in this chapter thus far, there are several others. We could review Chapter 9 at this point and describe the details for all the rest of the metals mentioned there. For each, we should mention the several functions in living tissue. But as you have guessed, a great deal of this information is not yet known. Further, it is necessary to keep this chapter in reasonable bounds so we will look at one more metal, zinc, and at one of its functions.

　　　　Our last metalloenzyme is carboxypeptidase, a zinc-containing enzyme. This enzyme is involved in the digestion of protein. As we know, a protein molecule is hydrolyzed, broken up into its constituent α-amino acids, in the digestion process. This reaction is catalyzed by carboxypeptidase in a regular manner. The α-amino acid at the end of a protein molecule is removed.

This leaves a shorter molecule, but it still has an α-amino acid on its end, so it happens again.

And again and again, until eventually there is only one α-amino acid left. Notice that this enzyme attacks the α-amino acid that has the COOH group (or COO⁻, where the proton has gone to an NH₂) exposed; it does not react with the other end of the protein, which has an exposed amino group, NH₂. Carboxypeptidase is secreted by the pancreas.

METALS AS POISONS

The concentrations of metal ions in the body are particularly controlled by certain proteins and hormones. If this balance is disturbed it is clear that disorder must arise. Trace amounts of certain metals are needed for enzyme activation, but the presence of greater quantities often has a poisonous effect. Conversely, a deficiency of any metal may result in reduced enzyme activity. We all know of the "tired blood" condition remedied by treatment with iron tablets, but the converse exists among the Bantu tribes in South Africa. The Bantu brew their beer in iron pots, and the consequent build-up of iron in their bodies is known as siderosis.

Usually metal poisoning of this type is treated by the use of chelating agents. (Chelate comes from the Greek word for claw, that is, a chelating agent grabs a metal ion in its "claw.") A chelating agent will bind more strongly to the metal than does the body protein. A typical use of such an agent is in Wilson's disease, caused by copper poisoning. The condition is characterized by tremors, stiff joints, difficulty in swallowing, and nervous disorders. The normal body content of copper in adults is 0.10 to 0.15 g, most of it being deposited in the locus caeruleus in the brain stem. In newly born babies, there is a particularly high concentration (2% of the total weight) of copper in the liver. This, however, disappears after a few months once the baby's metabolism has synthesized the copper containing protein ceruplasmin.

In Wilson's disease copper concentrations up to 100 times normal are frequently observed. The excess copper is deposited in the brain, liver, and kidneys, with the distressing results already noted. If the excess copper is removed, the symptoms disappear. A highly effective treatment is dosage of the patient with 1 to 2 g daily of penicillamine (Figure 11–7a). This substance

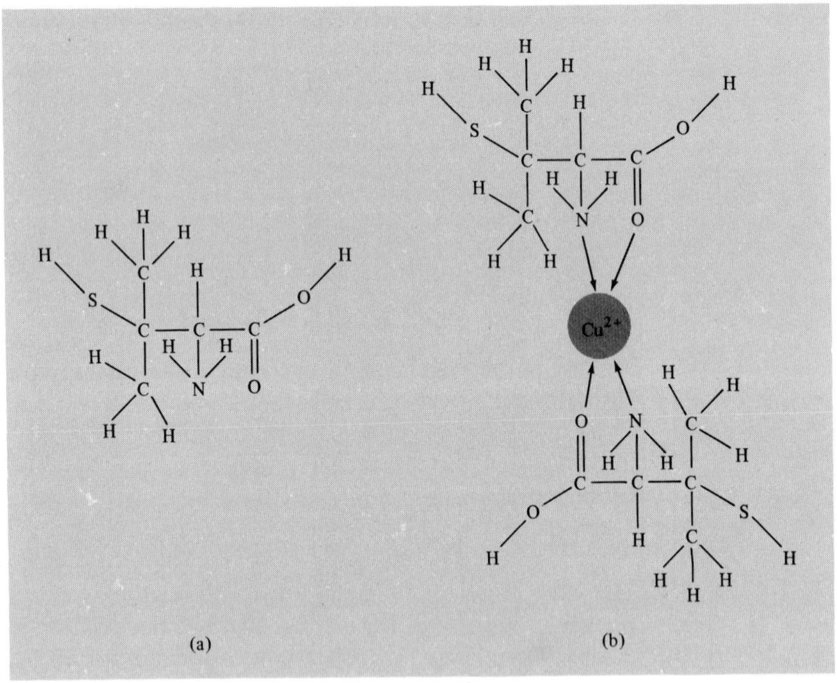

Figure 11–7. (a) Penicillamine. (b) Copper ion chelated by two molecules of penicillamine.

chelates the copper in the body, as is shown diagrammatically in Figure 11–7b; the arrows, as before, symbolize shared pairs of electrons from the oxygen and nitrogen atoms of the penicillamine. Fortunately, not all of the copper in the body can be removed in this way.

FOR FURTHER INFORMATION

With this chapter we have completed our three-chapter examination of chemistry and life processes. It is evident that many interesting aspects of this topic have been omitted and many sophisticated and complex portions of the subject have been dismissed very lightly. Hopefully, some students will now be excited enough about the subjects presented to wish to learn more. The references cited will at least get you started, although a complete and annotated listing is not possible (it alone would fill this book). The citations are divided into two parts, the first rather closely related to this chapter and the second dealing with general topics on chemistry and life.

References dealing primarily with the content of this chapter:

E. Frieden: The chemical elements of life. *Sci. Amer.*, **225**:52 (July 1972). A broader look at the topic of the title, covering more elements, in less depth, than this chapter.

M. F. Perutz: *Sci. Amer.*, **211**:64 (Nov. 1964); D. C. Phillips: *Sci. Amer.*, **215**:78 (Nov. 1966). Two articles, on the structure and function of myoglobin and lysozyme.

For more details on lysozyme, which is very widely distributed in nature (it was first found in nasal mucus, now obtained for laboratory studies from egg white) and is probably the most intensely studied enzyme today, see D. M. Chipman and M. Sharon: *Science*, **165**:454 (1969).

If you would like a book about enzymes, *Catalysis and Enzyme Action* a paperback (so it's not too large) by Myron L. Bender and Lewis J. Brubacher, McGraw-Hill, New York, 1973, would be just about right.

H. Neurath: Protein digesting enzymes. *Sci. Amer.*, **211**:68 (Dec. 1964). More about enzymes such as carboxypeptidase. For enzymes in general, and why we think they act the way they do, see J. P. Changeaux: *Sci. Amer.*, **212**:36 (April 1965).

The structure of cell membranes was merely mentioned in this chapter, for more details, try C. F. Fox: *Sci. Amer.*, **226**:30 (Feb. 1972). For a more detailed treatment, see S. J. Singer and G. L. Nicholson: *Science*, **175**:720 (1972). Briefly stated here, the membrane of a cell probably is composed of a solution, fats and proteins mixed with each other in an orderly way. The question dealt with is what are the details of this order?

Another emphasis on order seems to be the key in studying the structure of DNA. For this, look at one or more of these three. F. H. C. Crick: *Sci. Amer.*, **197**:188 (Sept. 1957); R. L. Sinsheimer: *Sci. Amer.*, **207**:109 (July 1962); and R. A. Deering: *Sci. Amer.*, **207**:135 (Dec. 1962).

The monthly publication, *Scientific American*, has been cited several times. Among the many delightful characteristics of most of the articles published, the clarity and esthetic quality of the illustrations is notable. This is particularly important when we deal with complicated molecules that have remarkable shapes. One of the outstanding articles in this respect (from a large group which is itself outstanding) is R. E. Dickerson: The structure and history of an ancient protein, *Sci. Amer.*, **226**:58 (April 1972). After you look at that, enjoy R. E. Dickerson and I. Geis: *The Structure and Action of Proteins*, Harper and Row, New York, 1969. It is loaded with interesting pictures of protein molecules, mostly enzymes. Then, try the stereo supplement to this book, by the same authors and publisher; fifty-five stereo drawings of nine protein molecules, a real treat.

M. Perutz: Haemoglobin: the molecular lung. *New Scientist and Science Journal*, June 17, 1971, p. 676. The personal story by the discoverer of the way hemoglobin is and acts; delightful and informative.

These two have the same title, Biological nitrogen fixation, but are otherwise different, the second citation being more extensive: G. N. Schrauzer: *Chemistry*, **50**:13 (March 1977), describes the use of chemical techniques to study enzymes. W. J. Brill: *Sci. Amer.*, **236**:68 (March 1977), emphasizes the interaction between bacteria and the roots of leguminous plants.

And finally, for this section of the references, here are several that treat the subject of bioinorganic chemistry: W. F. Fallwell: Chemical frontiers in biology. *Chem. Eng. News*, April 6, 1970, p. 58; R. J. P. Williams:

Endeavour, **26**:96 (1967); *Bioinorganic Chemistry.* American Chemical Society Publication No. 100, Washington, D.C.; F. R. Jevons: *The Biochemical Approach to Life.* Basic Books, New York, 1968; and M. N. Hughes: *The Inorganic Chemistry of Biological Processes.* Wiley-Interscience, New York and London, 1972.

These references are more general in nature:

M. Calvin: *Chemical Evolution.* Oxford University Press, Oxford, 1969. If you found Chapter 1 interesting, and would like to see how that story is related to the matters discussed in these three chapters on chemistry and life, read this book. On the same topic, see S. W. Fox et al.: Chemical origins of cells. *Chem. Eng. News,* June 22, 1970, p. 80.

For a short, readable (but take notes while you read) account of biochemical principles and details, see the chapter titled, Biochemical Processes in R. T. Morrison and R. N. Boyd: *Organic Chemistry,* 3rd ed. Allyn and Bacon, Boston, 1973.

The references that follow are all textbooks, some more formidable than the others, some fairly brief, some not. Instead of annotating each, I suggest you find out for yourself by selecting perhaps two or three from your library stacks, giving each one a ten minute skim, and then dipping more deeply into only one.

L. Stryer: *Biochemistry.* Freeman, San Francisco, 1975.

R. J. Light: *A Brief Introduction to Biochemistry.* W. A. Benjamin, Menlo Park, California, 1968.

A. L. Neal: *Chemistry and Biochemistry, a Comprehensive Introduction.* McGraw-Hill, New York, 1971.

E. Baldwin: *The Nature of Biochemistry.* Cambridge University Press, Cambridge, 1962.

K. Harrison: *A Guide-book to Biochemistry.* Cambridge University Press, Cambridge, 1968.

J. S. Fruton: *Molecules and Life.* Wiley-Interscience, New York, 1972.

A. L. Lehninger: *Biochemistry: The Molecular Basis of Cell Structure and Function,* 2nd ed., Worth, New York, 1975.

E. E. Cohn and P. K. Stumpf: *Outlines of Biochemistry,* 3rd ed., Wiley-Interscience, New York, 1972.

M. Yudkin and R. Offord: *Comprehensible Biochemistry.* Longman Group, London, 1973.

R. F. Steiner: *The Chemical Foundations of Molecular Biology.* Van Nostrand, Princeton, 1965.

In addition to these several other books on the same general topic could have been identified. You will find them in the same location on the shelves in your library stacks; pick one or more of those for a quick, or deep, dip, too. However, one demands an annotation. T. S. Hall: *Ideas of Life and Matter,* 2 vols. University of Chicago Press, Chicago, 1969. For those who like history, philosophy, and biology mixed with chemistry, or would like to find out if they do. An extended essay, really, on the central problem of biology: What is life? Or, what is it about inanimate matter which, when combined in certain ways makes life possible? Good questions! Although he does not provide complete answers, Hall's comments are well worth your attention.

chapter 12

Geology, Chemistry, and People

Each of us is aware of the Earth, its scenic vistas, the hills and valleys, the rivers and streams, the plains. Climb a mountain, walk on a beach, work in a garden, look at a building, drive on a road, hammer a nail; directly or indirectly we are involved with our Earth in these activities. Spend a quarter, write with a pencil, watch a TV picture, look in a mirror, eat with a spoon, drink from a glass, hold a china cup, walk on a sidewalk, use a tape recorder, fill a dental cavity, slice bread, turn on a faucet, enjoy a sculptured form, listen to a phonograph record;

in these and in almost every other human activity we are appreciating or using the minerals and rocks from the Earth's crust.

The Earth exists and therefore we can exist. Our life depends upon air and water, and also upon what we take from the crust of the Earth. We call our planet Mother Earth for good reason. Wherever we go we remain attached to our planet; even the astronauts must return home. Humans are curious, we need to know how things happen. An appreciation of our Earth is wholesome and satisfying.

THE CHANGING EARTH

The Earth is billions of years old, but contrary to our intuitive feelings, its present condition is transient. Before our eyes the Earth is continually changing. The average rate of wearing down of any continent is almost 4 cm (about $1\frac{1}{2}$ in.) in 1000 years. In terms of a human lifetime this is not much, but in the period since the Earth began, the continents could have been worn down to sea level more than 200 times.

The tectonic plates

Yet the continents still exist. Some processes must be at work to build them up while they are being worn down. Mountain ranges are forming now. Not only is there a wearing down and a building up, there is a sideways movement as well. Think of the continents as large log rafts with the individual logs fastened together, each raft frozen in an ice floe. As the ice floes move, so do the continents. Where the floes touch each other, there is a constant grinding, earthquakes and volcanoes in our terms. The Pacific Ocean is narrowing as we in the United States move ever closer to Asia, and the Atlantic Ocean is widening. We are locked in a great tectonic plate that moves westward, slowly and inevitably, 2 or 3 cm each year. This rate is about 0.0025 mm/hr; the tip of the hour hand on a wrist watch moves about 5 mm/hr, 2000 times faster.

Along a more or less north-south line in the middle of the Atlantic Ocean, from an up-welling beneath the ocean floor, our tectonic plate is continuously formed. As it grows along this edge, the rest of it, with us on it, moves westward. Our tectonic plate bumps into the tectonic plate associated with Asia, just off the border of California, and slides under the Asian plate, down into the interior of the Earth. A little east of that location, under the western part of the United States, deep in the Earth, the under-sliding of our tectonic plate causes some commotion, which we see indirectly as the magnificent mountains in western United States. In South America, the Andes mountain range is the result of the corresponding tectonic dynamics.

Many years ago, the land mass we call India was far south of its present location. As it moved, locked into its tectonic plate, toward Asia, there was an eventual collision that lasted for thousands of years. India finally came to a semihalt after wrinkling up the surface of south-central Asia into what we know as the Himalayas.

We humans think of the Earth as static, unchanging, and in our terms it is. But in terms of its total history, the Earth is as dynamic as we are, almost alive, ever changing. Let us briefly review some of that early development which we noted in the first chapter of this book.

The creation of the Earth

Over a period of time, which probably lasted for a few million years, small chunks of matter and large chunks were attracted each to the other by gravitational force. Eventually a large body of stuff accumulated. This building up process was not tranquil. Imagine, for instance, what might happen if, out of the sky, a boulder the size of a house should appear headed straight for where you were. When it hit, the ground would shake, and at the point of impact, probably, the stones and pebbles would be melted.

Initially, when the very first two pieces that were to form our Earth collided with each other, the collision was less dramatic. As more and more chunks were added and their total mass increased, the consecutive impacts of the succeeding chunks that hit became steadily more and more violent. With each impact, that is, some energy was added to the growing, forming planet. After a few million years of this, things got pretty hot. And they still kept on coming; things got hotter yet!

As nearly as we can guess now, those early chrondites (the name of those chunks) were composed of elements in about the same proportion as now exists in the Sun, except that the lighter elements such as hydrogen, helium, and others were present in lesser proportion. At first, while the temperature was still fairly low, the lighter elements that remained were trapped by the ever continuing accretion of more stuff on top of them. Later, when the temperature was much higher, some of the trapped lighter elements did escape, but not all. There is still some hydrogen (with oxygen, in the form of water) in the interior of the Earth, for example.

The six elements which then comprised the major portion of the total were (from most to least) iron, oxygen, silicon, magnesium, nickel, and sulfur. Oxygen and sulfur will combine with the metals, to form iron oxide, silicon oxide, magnesium oxide, nickel oxide, and the corresponding sulfides. Of these the oxides and sulfides of silicon and magnesium are the most stable. As it happened, there was not enough oxygen to form oxides with all of the metals (not enough sulfur either), so the predominant oxides that did form were the stable ones, silicon oxide and magnesium oxide (and the same more or less for the sulfides). Turning this around, we can say that although there was some iron oxide and nickel oxide (and iron and nickel sulfide), most of the iron and nickel remained in the uncombined state, as metals.

At this point in the development of the Earth, we have a hot, liquid, mixture of mostly iron, nickel, silicon oxide, and magnesium oxide. (To keep the story brief, we will omit further details about the sulfides,) Now, as every one knows, if you mix oil and water together, or try to, you get a mix of bits of oil and bits of water each separate from the other. Oil and water will not dissolve in each other. Eventually, if we let the mixture stand, the water settles to the bottom and the oil floats on top. The same sort of thing occurred in the early history of our Earth. Liquid iron and nickel will dissolve in each other. Liquid silicon oxide and magnesium oxide will dissolve in each other. The other metal oxides, present in much lesser amounts, and other compounds also present, will in general dissolve in liquid silicon and magnesium oxide but not in iron–nickel. The other metals present in trace amounts would tend to dissolve in the iron–nickel liquid mixture. Eventually, since the iron–nickel mixture was the denser of the two liquids (as water is more dense than oil), it settled to the "bottom,") to the center of the Earth, and the oxide liquid mixture floated, on top.

The "solid" Earth

The picture at this point is a hotter than red hot, liquid, two-part sphere, about 4000 miles in radius, with an iron–nickel core, and a metal oxide layer around that core. From time to time, chrondites appear in the distance, are attracted by gravity, and plunge into the cauldron. It must, indeed, have been a remarkable sight! As time passed, of course things became a little calmer, and the molten sphere cooled off, little by little.

The metal oxide layer became solid, at first bit by bit, here and there. As it became solid, new compounds were formed. Compounds that were stable at higher temperatures were changed into compounds that are stable at lower temperatures. (We see this happening today in the weathering of some rocks to form clay. Clay is a more stable compound at the temperature of the Earth's surface than feldspar, the kind of mineral from which clay is formed.) As the process of solidification continued, however, pockets that were still liquid remained. Meanwhile, deep in the interior, in the very center, the core was under great pressure (more than a million times greater than the pressure we experience from the atmosphere on the surface of the Earth). Under this pressure, the center of the core, though then at a temperature probably exceeding 6000°C, was solid. It was surrounded by a very hot mixture, which was also composed mostly of iron and nickel, that was liquid because it was under less pressure. This fluid outer core is still a liquid, billions of years later, and moves or circulates slowly even today. It is the motion of that hot liquid that is believed to be responsible for the Earth's magnetic field, the north and south magnetic poles.

On top of the outer core the metal oxide mixture exists, even today, with regions where it is still a liquid, or at least a mushy, fluidlike solid. The whole covering over the core is called the **mantle**. Today it is believed to be more than 90% silicon oxide and magnesium oxide, along with some iron oxide; with 4% or less, each, of aluminum oxide, calcium oxide, and sodium oxide, and less than 0.6% of all other oxides including less than 0.1% water (hydrogen oxide). The Earth's crust was formed from this ancient mantle. Bit by bit, some of the liquid portions of the mantle were squeezed, this way and that, some of it upward, toward the surface. This "magma" squirted out here and there in what must have been a remarkable pyrotechnic display, eventually cooled and solidified, and formed the crust on which we live.

The layers of the Earth

The crust of the Earth is thinner than an eggshell in proportion to the size of the Earth. The total mass of the crust is less than 1% of the total mass of the Earth. On the average, the crust ends 17 km beneath us although it is quite variable in thickness, and the mantle begins. (For comparison, this is about 10 miles below us.) For the next 400 km (250 miles) below that, more or less, we would encounter the "upper mantle," where there are still regions of liquid or mushy solid, from which we still get volcanic action. Then 1000 km further down, and on to a distance of about 2900 km, is the "lower mantle," with a "transition zone" between 400 and 1000 km thick. Some of the heat still in the interior came from the impact energy of those chrondites. Some came from the falling of the molten iron–nickel mixture to the center of the Earth. The rest came from, and today is still generated from, the energy of radioactive decay processes, largely from radioactive isotopes of potassium, uranium, and thorium.

These names for different regions of the mantle, although important, really only reflect how little we know about the inside of the Earth. That is, about all we know is that the interior is layered, something like an onion, and that each layer is different in some way from the other layers. The innermost layer, the solid iron–nickel core under great pressure, is a sphere with a radius of about 1300 km.

As is obvious, we know more about the crust than we do about any other named part. It is awesome to realize that everything we see or use on this Earth comes directly or indirectly from the Sun's energy and from a fraction of 1% of the Earth. Surely in the years to come we will be able to use wisely much more of our Earth than is now possible. What we know now about the Earth is derived from a limited variety of information. From meteorites we can make guesses about those early chrondites; by setting off explosives at known distances we can measure how long it takes for the detonation to be detected somewhere else. From minerals in the crust and from planned experiments in laboratories that reproduce those minerals, or other species that are related to natural minerals, we can guess how the minerals might have been formed. For example, mix equal amounts, about half a teaspoon of each, of baking soda and salt and notice that in the mixture of these solids, it is difficult to tell one from the other. Next, dissolve the mixture in about half a glass of water and pour some of the solution into a shallow saucer. Let it stand undisturbed until all the water has evaporated. In the process, as you will then see, a partial separation of the salt and soda has been achieved. More sophisticated experiments at higher temperatures and pressures are more revealing, of course.

At some places on the Earth's surface, the force of gravity is above or below average; probably, a large body of material very dense or very light, lies under the surface at such a spot. Accurate measurements of these gravitational anomalies can reliably suggest what it is underneath, and perhaps even how it might have been formed there. Similarly, at some places there are magnetic anomalies, where a compass needle points almost any direction except more or less to the north and south. Again, careful measurements are helpful in estimating the reasons. Volcanoes, hot springs, geysers, and other eruptions of hot material from under the surface help to provide information about what is underneath.

The liquid, or mushy solid, regions in the upper mantle are composed of **magma**, the name given to a liquid region there. This magma is under high pressure and occasionally now (more often, long ago) finds its way upward. As it flows up, coming in contact with the cooler rock as it moves, the magma cools slightly. Some of the substances that were soluble in very hot magma are not now soluble, and they precipitate, leaving the rest of the liquid magma with a changed composition. The process continues as the magma is forced upward. Eventually, if it breaks through the surface, we may see a terrifying display of molten lava (magma of altered composition which has reached the surface of the Earth). When the lava has cooled sufficiently and solidified, it can be examined and analyzed, which gives us some clues about what the magma itself might be like, deep, deep in the interior.

Some information about the inside of the Earth can be theorized about from the slight wobbling of the Earth as it rotates on its axis (like a defective golf ball spinning in flight). When the Earth shakes during and after an earthquake (the Earth rings much like a bell in this circumstance), measurements of the vibrations at several points near to and far from the earthquake center reveal details about the inner structure of the Earth. It is from these seismological data in particular that we conclude that the Earth has a layered onionlike structure. Different layers, transmit earthquake vibrations, or waves, differently.

We cannot learn much about the interior of the Earth by making a hole, from the surface downward, at least not yet. The deepest drilled hole thus far is 8 km (about 5 miles), and the deepest mine shaft is about 3.5 km. As far as we know the only substance available direct from the upper mantle is diamond. Long ago, and in only a few locations, diamond bearing portions of the upper mantle were forced upward under high pressure. Today in a diamond mine, when a diamond is found in the digging, it pops out of the surrounding material like a watermelon seed when pressed between thumb and forefinger.

Composition of the Earth

From all of this, we think that, by weight, the Earth as a whole is 34.6% iron, yet the crust is only 5.0% iron. As a whole, the Earth contains 15.2% silicon, but the crust is 27.7% silicon. On the other hand, the whole Earth is

12.7% magnesium, but we find only 2.1% magnesium in the crust. Similarly, the total per cent of nickel is 2.4 and of sulfur 1.9, yet less than 1% of the crust is composed of these elements. There is a mere fraction of a per cent of sodium and of potassium considering the whole Earth, but 2.8% of the crust is sodium and 2.6% is potassium. These kinds of comparisons show very clearly that, as the Earth developed, several different chemical and physical processes occurred which eventually formed a crust that was quite different from the Earth as a whole. Even though the crust is a very small fraction of the Earth, it is an important, and different, fraction for us.

Composition of the crust

On and immediately under the continents, the crust is mostly granite. Under a relatively thin layer of sediment, the ocean floor is mostly basalt. These two, granite and basalt, constitute the major part of the crust.

Granite is a mixture of several different minerals, orthoclase feldspar, $K(AlSi_3O_8)$; quartz, SiO_2; plagioclase feldspars, such as $Ca(AlSi_3O_8)_2$; micas, such as $KAl_2(AlSi_3O_{10})(OH)_2$; and a variety of other minor constituents. In almost any chunk of granite rock, it is easy to see many of these components since each has a different color or different texture, and each occurs in crystallites large enough to be distinguished by the unaided eye.

Feldspars and micas are not single compounds, as is quartz. For example plagioclase feldspars are themselves mixtures with the general formula $M(AlSi_3O_8)_2$, where M may represent calcium, Ca, or two sodiums, Na_2, or any mixture in between, with a little calcium and a lot of sodium, or the converse, or about half and half, or any other composition.

Basalt is a mixture also, largely composed of plagioclase feldspars; pyroxenes, $MSiO_3$, where M in this case represents Fe, Mg, or Ca; and olivine $(Mg,Fe)_2SiO_4$. The parentheses around Mg,Fe and the comma separating the two symbols signifies one or the other, magnesium or iron, but not both, for each SiO_4 unit. Olivine is a mixture of Mg_2SiO_4 and Fe_2SiO_4.

MINERALS AND THEIR STRUCTURE

Each component of granite and basalt is a silicate. In silicates, the silicon is in an oxidation state of +4, and each silicon atom is surrounded by four oxygen atoms tetrahedrally arranged. Often, any given oxygen atom is shared by two silicon atoms, and forms a corner of the tetrahedron for both silicon atoms. To see this clearly, examine the illustrations and construct your own models using several balls of clay all the same size. (Marbles can be used also, except that it is necessary to also use a little glue, or maybe chewing gum, to get the marbles to stick to each other. Small balls made from bread, by crushing and rolling a few slices into spheres, could also be used.)

The structure of silicates

Put three balls together, as in Figure 12–1a. In the middle of the three, put a much smaller ball, or imagine that you have done so, as in Figure 12–1b. Now,

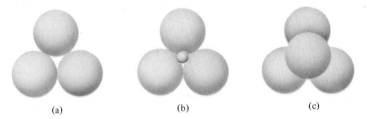

(a)　　　　　　　　(b)　　　　　　　　(c)

Figure 12–1.　Steps in the construction of a model of an SiO₄ tetrahedron.

on top of the small ball, put a fourth ball the same size as the first three, as in Figure 12–1c. Notice that the small ball, in the center of the four, must be very much smaller than the four; if it is too large, it will prevent the four from touching each other.

You have constructed an SiO_4 tetrahedron; the four oxygens are represented by the four equally sized balls, and the much smaller silicon atom is represented by the smaller ball in the center. Since silicon has an oxidation state of +4 and each oxygen has an oxidation state of −2, the whole SiO_4 unit can be thought of as an ion with an overall −4 charge $[4 \times (-2) + 4 = -4]$ or SiO_4^{4-}.

A pedagogical caution is in order. Before we get finished, this is going to get complicated. The fundamental learning problem you will meet as you read further is to convert a flat two-dimensional picture on the printed page into a mental three-dimensional image. It is almost impossible to do this without the help of real three-dimensional models. If you have not yet obtained the clay, or bread, or marbles and chewing gum, you are urged to do so before reading further.

Next, use five balls all the same size, and put them together as in Figure 12–2a. Now, if you have some different colored clay, put two very small balls in the center positions of the three and three (one counted twice) balls, as in Figure 12–2b, and then put another large ball, two total, on top of each small ball (Figure 12–2c). You have constructed an Si_2O_7 unit. Given the +4 oxidation state of silicon (+8, total) and the −2 oxidation state of oxygen (−14, total) the ionic charge can be calculated to be −6, or $Si_2O_7^{6-}$.

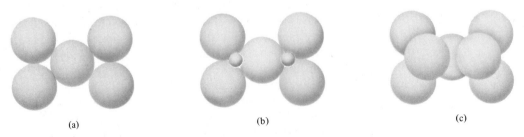

(a)　　　　　　　　　　(b)　　　　　　　　　　(c)

Figure 12–2.　Construction steps for an Si₂O₇ model.

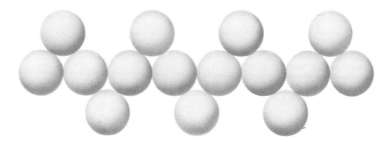

Figure 12–3. Construction of a model showing several adjacent tetrahedra sharing oxygens; first step.

We can make a similar polymeric structure, with several tetrahedra sharing oxygens instead of two sharing one oxygen. Begin with 22 or more balls and arrange 15 of them as in Figure 12–3. In each of the three-ball centers we put small, differently colored balls. As before, these represent silicon atoms. On top of each "silicon atom" we put one regular sized ball representing the fourth oxygen to be associated with each silicon. Notice that each silicon shares two of its oxygens and that the shared oxygens are counted twice, once as belonging to one silicon and once as belonging to the other silicon on the other side. (The silicons and oxygens on each end, of course, should be thought of as though the chain went on and on, forming a very long polymer of SiO_3 units.) We can assign a charge of -2 to each SiO_3 unit $[3 \times (-2) + 4]$.

The pyroxenes

This gives us our first real example, the pyroxenes. A pyroxene is an SiO_3 polymer, with one Fe^{2+}, Mg^{2+}, or Ca^{2+} for each SiO_3^{2-}. The SiO_3 polymer chains are lined up parallel to each other, and sandwiched in between are positively charged ions that hold adjacent chains together. In Figure 12–4, these ions are represented by M^{2+}. The positive ions may all be of the same type or each of the three kinds, Fe^{2+}, Mg^{2+}, Ca^{2+}, may be present, or two may be present and one completely absent.

The micas

To construct an example of a mica would require a very large number of balls to represent the oxygens, as you can see from Figure 12–5. In this arrangement, three oxygens from each tetrahedron are shared with adjacent tetrahedra, thus producing an infinite sheet. Notice that all of the labeled "basal oxygens" are shared and that none of the "apical oxygens" are shared. In the mica structure, one out of every four silicons is replaced by an aluminum, so instead of the unit formula $Si_4O_{10}^{4-}$, we have $AlSi_3O_{10}^{5-}$. (The oxidation state of aluminum is $+3$.)

The next stage in the mica structure requires two infinite sheets, one over the other, with the apical oxygens of both sheets pointing inward. Think of a sandwich made of two slices of bread, both buttered on one side. Put the sandwich together by touching buttered side to buttered side. The bread slices

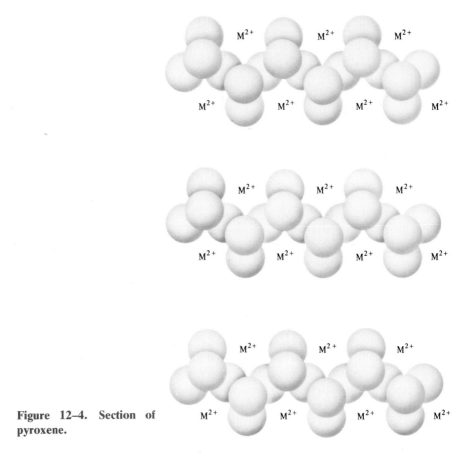

Figure 12–4. Section of pyroxene.

represent the basal oxygens plane; the buttered surfaces represent the apical oxygens plane. Mixed in with the apical oxygens, in the butter so to speak, are more aluminum ions, Al^{3+}, two for each $AlSi_3O_{10}^{5-}$ unit. In addition, here and there in the holes in the infinite sheet are some hydroxy ions, OH^-, two for each $AlSi_3O_{10}^{5-}$ unit.

Now take two more infinite sheets, and put them together like the first two, apical oxygens pointing inward. Then put that sandwich on top of the first (like a Dagwood sandwich) and in between the two sandwiches put some potassium ions, K^+, one for each $AlSi_3O_{10}^{5-}$ unit. The result, if repeated several times, to make a super-Dagwood sandwich, would represent the structure of one kind of mica, $KAl_2(AlSi_3O_{10})(OH)_2$.

Quartz

In quartz, all four oxygens are shared. One way to visualize this is to begin with our infinite sheet, except that half the apical oxygens are on the other side of the basal oxygen plane. This is shown in Figure 12–6. Half the apical oxygens are

apical
oxygens

basal
oxygens

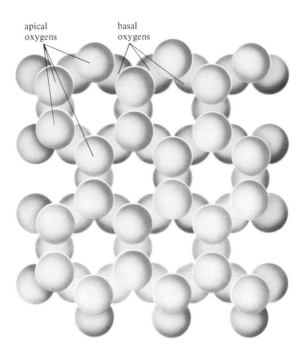

**Figure 12–5. Section of
mica layer showing several
tetrahedra sharing oxygens.**

now barely visible because they are on the other side, mostly hidden by the three nearest basal oxygens.

Next, think of another infinite sheet, like the first, with half of its apical oxygens on one side and half on the other. Fit it to the first sheet, like two slices of bread for a sandwich, so that the apical oxygens on one fit into the vacant places on the other. Then, get a third sheet like the other two, and make a three-slices-of-bread-like sandwich, fitting the apical oxygens on one side of that third sheet into the vacant places on the adjacent sheet. That is, think of each infinite sheet with half of its apical oxygens on one side and half on the other. And, on either side you want to think of, where there would have been oxygens (if all the apical oxygens had been all on the same side) there are now many vacant places. As we build up the multilayered sandwich, the apical oxygens that are on one side of a sheet fit into the vacancies on the sheet next to it. Of course, the converse applies; the vacant places on the first sheet are filled by the oxygens on the second. Any two sheets mutually fit together, each filling the other's vacant places on the sides that touch. The resulting structure would be a form of quartz.

Feldspars

Orthoclase feldspar is almost like quartz. Merely substitute an aluminum for every fourth silicon, thus forming a unit with the formula $AlSi_3O_8^-$. And for each such unit put one potassium ion in one of the handy dandy spaces between

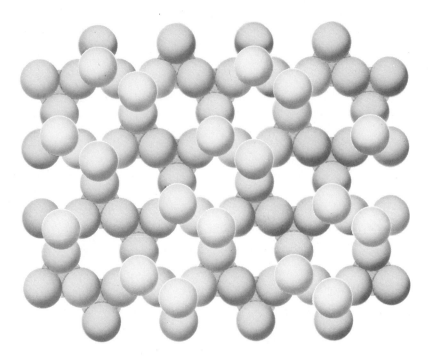

Figure 12–6.   Visualization of layered two-dimensional structures, for conceptualization of three-dimensional quartz structure.

or among the other atoms. The formula for orthoclase feldspar is, then, $K(AlSi_3O_8)$, in which the parentheses emphasize the unit of this polymer.

Plagioclase feldspar is almost the same. In the original quartz structure, substitute one aluminum for every fourth silicon. But instead of adding potassium, add either sodium ions, Na^+, or add one calcium ion, Ca^{2+} instead of two sodium ions. The formula for plagioclase feldspar can be written $(Ca,Na_2)(AlSi_3O_8)_2$.

Olivine

Olivine, the last of the components named thus far and not yet described in detail, is the easiest to construct, as long as we don't get too fussy about details. Make several separate tetrahedra out of sets of four each clay (or other) balls. In olivine, there is no oxygen atom sharing between tetrahedra. As usual, each tetrahedron represents four oxygens and, in the center, a silicon, SiO_4^{4-}.

Select any one tetrahedron, and next to it, upside down, put another tetrahedron. Then another and another and so on, one "right side up" the next to it "upside down" alternating. They should be arranged like a checkerboard more or less, not like a single up, down, up, down, up . . . chain. On top of that, make another layer, like a checkerboard, and so on. In the spaces between the oxygens, every so often, there is either a magnesium ion, Mg^{2+}, or an iron ion,

Fe^{2+}, such that, for each independent SiO_4^{4-} unit, we have two ions, both the same or each different, either way. The formula for olivine is, then, $(Mg,Fe)_2SiO_4$.

Freeing the elements

Now, the point of all this is very closely related to the preceding chapter. There, we noted the importance of a number of different elements in life processes. Except for carbon, hydrogen, and oxygen, the other elements that are essential for life came from the rocks in the crust of the Earth, either from the granite or from the basalt, or from both. Fortunately for us, the silicate structure, with independent SiO_4^{4-} units or with more complicated polymeric units, holds the other elemental ions, such as potassium, magnesium, iron, and many others, loosely enough so that they can become available to living tissue. The processes that have loosened these substances required thousands of years, or more. As a result, not only were those essential for life made available but many other minerals, some of them not silicate-based, were also formed. Many of these form the basis for industrial operations, such as the manufacture of steel, or aluminum, or cement, to name a few.

ROCK AND ITS FORMATION
Igneous rock

A brief discussion of modern theories of mineral formation is interesting. Think of the crust as being originally composed of igneous rock. Originally, this rock was liquid and hot. It solidified upon cooling to form the igneous rock of the crust. The original liquid magma generally had a high percentage of iron and magnesium silicates. This was either extruded to the surface in what we would today call a volcano or it formed an intrusive pluton, a mass of magma forced near the surface that did not burst through. As this liquid cooled, some of the dissolved substances were no longer soluble at the lower temperature and they precipitated as solids within the rest of the liquid. These solids tended to sink to the bottom of the liquid.

Generally, the first solids to precipitate were the apatites, or phosphate-containing rocks. Today, these form the basis of the phosphate fertilizer industry. Next to precipitate as the liquid cooled further were the magnetites, rocks rich in iron oxide. Following this, other metal oxides of chromium and titanium. And finally, of course the silicates solidified, such as the pyroxenes and olivine. As a result, natural processes of solidification have provided us with a separation. If we need phosphate rock for fertilizer or iron oxide for the manufacture of steel, we find these separated from each other at least to some extent. Were they not naturally separated, it would be much more difficult to obtain either fertilizer or steel from the raw materials.

Other effects also assisted in the differentiation. As the liquid mass cooled, it sometimes developed cracks in the solidified parts. The remaining liquid either was forced into these cracks or, if two adjacent blocks of solid were squeezed together at one end of a crack, the crack was closed off. Later very hot water was sometimes forced in through small holes in the constriction (so small as to prevent the viscous molten rock from entering, but big enough to let water pass). The hot water could dissolve some of the material from the walls of the crack, enlarging the space, and carrying off the most water soluble components to another region, where they were later deposited. Emeralds were formed in this way, as were some of the micas and tin oxide deposits, called cassiterite, for example.

Sedimentary rock

Exposed igneous rock is subject to weathering and other attritional processes. The rock that forms is called sedimentary rock. Although only about 5% of the crust today is sedimentary rock, it makes up 75% of the exposed land area, and ocean floor. So, even though the sedimentary cover over the igneous rock is very thin, most of the rock we see is sedimentary.

Sedimentation processes are still going on. Each day, all the rivers of the world carry 10 million tons of sediment from the continents into the oceans. One fifth of that total is carried by one river, the Mississippi. Sedimentary rocks contain essentially all of our oil deposits, natural gas, and coal. The phosphate rock that is now accessible is largely sedimentary, as is all the salt in salt mines. Sedimentary rocks are used directly as building stones, especially limestone and sandstone; indirectly they are the source of our cement, bricks, glass, and plaster.

Sedimentary rocks are formed by three major processes: detrital, biological, and chemical. In the detrital process, small or large chunks of rock are carried from where they were before to somewhere else. The transporting agents are wind, streams, surface flowing water, and ocean currents. The largest chunks of rock are probably carried by glaciers; the smallest are so fine that they are not distinguishable under an ordinary microscopic examination. The smallest particles are the clay particles. As they collect in one place and their weight builds up, they form shales. Sand particles form sandstone, of course, in the same way.

Limestone is the best known example of a sedimentary rock of biological origin. It is calcium carbonate, $CaCO_3$, and is formed as shells or "skeletons" by tiny organisms in the ocean. As these die their hard parts drift down, eventually accumulating on the ocean floor. The material piles up, is compacted, and becomes limestone. Other biological sedimentary rocks include chert, or flint, formed from silicon dioxide as a product of other tiny organisms in the ocean.

Gypsum, calcium sulfate dihydrate, $CaSO_4 \cdot 2H_2O$, is an example of a chemical sedimentation. Calcium sulfate is slightly soluble in water. However, if a large body of water containing this dissolved substance is trapped by an upwelling process so as to form a large lake, the water eventually evaporates and the dissolved material remains behind. Borax is another substance formed by the same process; its formula is $Na_2B_4O_7$, and its chemical name is sodium tetraborate. Common salt is formed in the same way; in this case the deposits of sodium chloride, have been covered later with an overburden, and we must dig beneath the present surface to obtain the salt.

Recently, nodules ranging in size from that of a grain of wheat or smaller to that of a potato have been found on the ocean floor in some areas. The processes that formed the nodules are not yet known but are presumed to be a chemical sedimentation. The nodules contain manganese, nickel, copper, cobalt, iron, and other metals. At current prices, the total value of the nickel and copper in the nodules is estimated to be more than $1 trillion.

Metamorphic rock

The third major class of rock is metamorphic rock. We can get a clue about the process of formation from the Greek roots of the name; meta suggests change and morph suggests form. What ever these rocks might have been before, igneous or sedimentary, their form has been drastically changed. The agent of change is not weathering, as it was for sedimentary rocks. Weathering is a fairly mild agent. For metamorphic rocks, the agent of change is high temperature or high pressure, or both, often in the presence of water. At high temperatures and pressures water is a very potent agent for change; it will dissolve almost anything, particularly silicate structured rocks. Limestone becomes marble under those high pressures and temperatures (no water is involved). Sandstone becomes quartzite; shale becomes slate.

There are three kinds of metamorphic processes, contact, regional, and metasomatic. In **contact** metamorphic processes, magma from the mantle is forced upward and intrudes into the crustal layer. The magma is very hot of course, and exerts a high pressure. As a result, the crustal rocks next to the intrusion are heated and compressed; they change their form and composition. Gases from the magma react with the crustal rock, causing further changes.

Regional metamorphism involves the burial of sedimentary or igneous rocks under a thick overburden; what was on top of the crust is now deeply enclosed and subject to high temperatures and pressures. The conditions are not as severe as they are in contact metamorphism, but the pressures and temperatures are high enough to cause the same general kinds of changes. This kind of metamorphism is the most common; many of the rocks we see are regional metamorphic rocks. The thick layer of overburden has long since been removed by weathering and erosion. Only sedimentary rocks are more common than regional metamorphic rocks.

Roughly translating the Greek roots, **metasomatism** suggests a change in the body. In rocks this refers to a change in the chemical composition, a significant replacement or movement of some of the constituents. For example, basalt contains olivine, Mg_2SiO_4, and pyroxene, $MgSiO_3$. At the high temperatures and pressures involved, these two minerals will react with water when it is present to form a new mineral, serpentine, $Mg_3Si_2O_5(OH)_4$. The equation for this reaction is

$$Mg_2SiO_4 + MgSiO_3 + H_2O \rightarrow Mg_3Si_2O_5(OH)_4$$

In most metasomatic processes, water or carbon dioxide is involved. We can illustrate the phenomenon of a slow metasomatic process by watching a much more rapid reaction. Gold can be dissolved in salt water in the presence of oxygen.

If you wish to see this happen, obtain a few cents worth of scrap gold leaf from a friendly jeweler. A piece or pieces not larger than about 50 mm^2 (a bit smaller than your little finger nail) will do. First dissolve as much salt as you can in about half a glassfull of water; be patient, salt dissolves rather slowly in water. Pour the solution into another glass, leaving the undissolved salt behind. Then put the gold leaf into the salt solution in the second glass and stir vigorously for several minutes. If you wish, stir for a while, rest yourself, then stir again; it will take a lot of stirring. Eventually, the gold will disappear completely. The equation for the reaction is

$$H_2O + 2\,Au + 4\,Cl^- + \tfrac{1}{2}O_2 \rightarrow 2\,AuCl_2^- + 2\,OH^-$$

The gold, Au, forms a complex ion, $AuCl_2^-$, which is invisible in the water solution. To get the gold back in metallic form, drop a small piece of raw (untreated) cotton batting into the solution and wait several days. The gold will precipitate onto the cotton. However, with the small amount of gold suggested and the fact that it will tend to deposit widely dispersed over the cotton fibers, the layer of gold will be so thin that it will be difficult to detect.

TYPES OF CRYSTAL STRUCTURES

We can illustrate another geologic process in the form of a puzzle. Figure 12–7 consists of a square-bounded figure, approximately 150 mm on each side. Eight pennies in a line extend over a distance of approximately 150 mm. So, by lining up 64 pennies, as shown by the circles, we can see that they fill the square nicely. The puzzle is to put four more pennies, with all 68 fitting nicely, into the square-bounded figure. No pennies may be on top of each other, no part of any penny may extend beyond, or even on, the boundary lines. (If you want the easy answer, it is printed upside down in mirror image form at the end of this chapter. Or look at Chapter 5; a portion of Chapter 5 is closely related to this section of this chapter.)

The pennies represent a two-dimensional crystal structure, in what we might call a checker-board or square array, with lots of empty space between any four pennies. The new arrangement of pennies you found could be called "close packed," with less empty space between any three pennies. In a hypothetical geological process with the pennies in the square array, we can

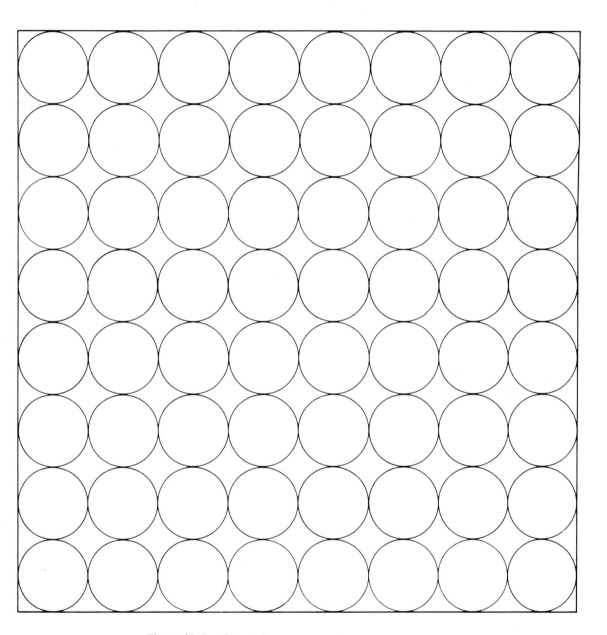

Figure 12–7. Square for penny puzzle.

imagine that by putting pressure (plus a little jiggling, maybe) on all four sides, to shrink the square, the sixty-four pennies would tend to assume a close packed array, occupying less area. The geological processes that occur at high pressures, and in three dimensions, are analogous in many cases. The components of a crystalline mineral shift to form new arrangements with smaller empty spaces among the atoms in the mineral.

The crystal structure of minerals is a topic of interest in itself, apart from the geochemical processes involved. Thus, except for a few amorphous substances such as opal, most natural gems and synthetic gems are crystalline. Snowflakes are crystals; the salt in a salt cellar is crystalline.

In Chapter 5 we learned that there were several different crystal structures possible, mostly depending upon the relative sizes of the ions in that crystal. To see this in more detail it will be helpful to use several coins. For example, how many nickels can you fit around a dime, with all coins flat, and each nickel touching the dime, no two coins on top of each other?

Five nickels will fit around a dime in this manner. Next, try a penny; how many nickels can fit around a penny? Although the penny is larger than the dime, we can still fit only five nickels around it. But a quarter, which is bigger yet, can have six nickels fitted around it. And a half-dollar will accommodate seven nickels. The larger the central coin, the more nickels (or other coins) that can be arranged around it.

Tetrahedra

Now, let's get back to that SiO_4^{4-} ion and build the same model we had in Figure 12–1 from clay balls (all the same size) or marbles and a little glue, or other available material. First, put three balls together as in Figure 12–8a. Then, put a fourth ball on top of the three, to make a tetrahedron, as in Figure 12–8b. Look "inside" the structure and notice the space in the center. We could put a smaller ball in that space. That smaller ball would then have four balls around it, much like the nickel coins around the penny. As long as the outside balls are from about $2\frac{1}{2}$ to about $4\frac{1}{2}$ times bigger than the central ball, we can fit four balls around the central ball in a tetrahedral structure.

Figure 12–8. A tetrahedron (b). Note the central space in the triangular arrangement in (a); there is a corresponding open space in the tetrahedron.

(a)　　　　　(b)

Octahedra

If the central ball is a bit larger, we can get six outer balls around it. To see this, arrange four balls to make a square array as in Figure 12–9a. Then, on top, in the center, put one more ball (Figure 12–9b). And now, underneath, in the

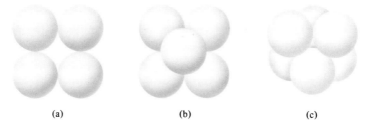

(a)   (b)   (c)

Figure 12–9.   Constructing an octahedral arrangement. Note the larger central space (compare Figure 12–8a) in the square (a) and the correspondingly larger space inside the octahedron (c).

center, put the sixth ball (Figure 12–9c); the structure is called an octahedron. Look "inside," and notice that the central space is larger than it was for the tetrahedral arrangement.

The central ball in the small space of the tetrahedron had four balls outside. We say that the **coordination number** of the small ball in the center of the tetrahedron is 4. For the octahedron, with six outside balls, we say that the coordination number of that slightly larger ball in the center is 6. A still larger ball would have a coordination number of 8. (We have skipped coordination numbers of 5 and 7 because there is no way to put 5 or 7 balls symmetrically around a central ball. In addition, in the crystal structures found in nature, coordination numbers of 5 and 7 are very rare. Coordination numbers of 2, 3, 9, 10, and 12 are also rare. There are no known examples of coordination number 11, or of 13 or higher.)

To see a coordination number of 8, first put four balls in a square, as in Figure 12–10a. Then, on top of those four, put four more balls, as in Figure 4–10b. Notice that the space in the center is the largest yet. You may find it helpful at this point to review the discussion on crystal structure in Chapter 5, where the major emphasis was upon the balls, the ions, around the central space which may or may not have held a smaller centrally located ion.

Figure 12–10.   Construction of a simple cubic arrangement. Note the much larger space in the center of the cube (b).

(a)   (b)

Fluorite structure

We can look at two examples of the crystal structure of natural minerals. First, fluorite, CaF_2, which is composed of calcium ions, Ca^{2+} and fluoride ions, F^-, with twice as many fluoride ions as calcium ions.

Begin with nine balls, all the same size, arranged to form a nine-ball square (Figure 12–11a). Then at the two places marked x (Figure 12–11b), put two smaller balls, one at each x, about three fourths as large as each of the nine balls. Follow this by covering the whole thing with nine more of the larger balls (Figure 12–11c). Put two more smaller balls, one each, at the places marked y (Figure 12–11c). Notice that the two x locations are criss-cross in relation to the two y locations. This is the fundamental structure of fluorite, with the large balls representing fluoride ions and the small balls representing calcium ions. To finish the model, put nine more balls on top: $27+4$ balls, total.

If we were to continue, piling up more layers, we would next add two smaller balls in the x positions, criss-cross from the y positions below; then another layer of larger balls, two smaller balls in the new y positions, and so on.

Of course, a ball model that more closely represents a real crystal of fluorite would not begin with a nine-ball layer. Instead, the first layer would contain millions upon millions of larger balls, with the smaller balls at the x positions, something like Figure 12–12. And the successive layers would be

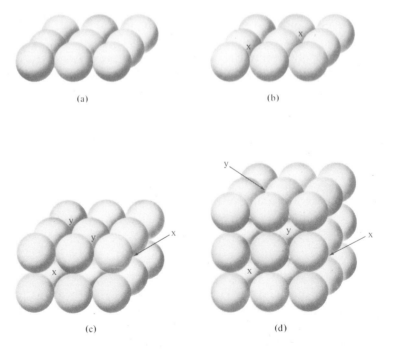

(a)   (b)

(c)   (d)

Figure 12–11.   Steps in the construction of a model of a fluorite structure.

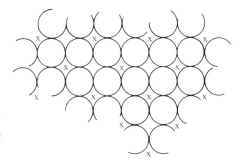

Figure 12–12.   Segment of first layer of fluoride ions; x's denote locations of calcium ions.

built up analogously to the way we built the smaller model for many, many layers. In that large structure, each fluoride ion would be near four calcium ions and would have a coordination number of 4 (counting calcium ions around a fluoride). Each calcium ion would have a coordination number of 8 (counting fluoride ions around a calcium).

You can see this in our $27+4$ ball model. The center fluoride ion ball in the middle layer has "below" it, at the x places, two calcium ion balls and two more "above" it, in the y locations, or four calciums around that central fluoride. In an extended structure, this effect repeats. On the other hand, each calcium ion at the x and at the y locations is near eight fluoride ions. This effect repeats in an extended, more realistic structure, also. (It is true, of course, that the fluoride ions and calcium ions at the edges, corners, and surfaces of a real crystal do not have their full compliment of nearest neighbors. But the number of these is so very small in comparison to the number of interior ions, which do have their full compliment, that the deficiency can be neglected.)

Molybdenite structure

In the modeled structure of molybdenite, molybdenum disulfide, MoS_2, we see an entirely different arrangement. In this case, the two species involved are each about the same size and are only partially ionic. We need 30 balls, all the same size, 10 to represent molybdenum and 20 to represent sulfur.

Begin with seven molybdenum balls, as in Figure 12–13a. Then add a layer of three sulfur balls, in the places as shown (Figure 12–13b). On top place a layer of seven more sulfur balls, each sulfur ball to be directly above the molybdenum balls in the first layer (Figure 12–13c). Follow with three molybdenum balls, each directly above the three sulfurs in the second layer (Figure 12–13d).

And so it goes, in an alternating ABABABABAB . . . structure. Each component of an A layer is directly above, or directly below, the component of each other A layer, whether it be molybdenum or sulfur in those layers. Similarly for the components in the B layers. In Figure 12–13d, if we call the bottom molybdenum layer an A layer, we see, going upward, a B sulfur layer, an A sulfur layer, a B molybdenum layer, an A sulfur layer, then at the top a B sulfur layer. If we continued, the structure would repeat over and over again.

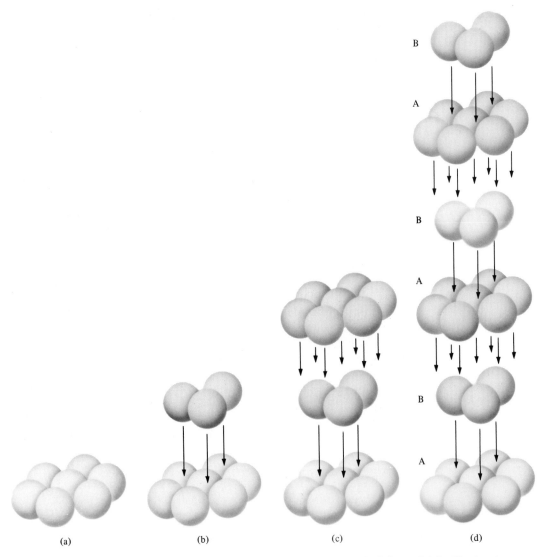

Figure 12–13.   Steps in the construction of a model of the molybdenite structure.

And, of course, instead of starting with only seven molybdenums, we would get a more realistic representation by starting with a first layer of millions upon millions of components, a few of which might look like Figure 12–14.

Other
crystalline
structures

Figure 12–15 shows other kinds of crystalline structures. In these, each illustrated sphere has been shown much smaller than it really is in relation to the other dimensions of the crystal so that you can see more detail about the

Figure 12–14. Segment of A layer showing more than seven molybdenums.

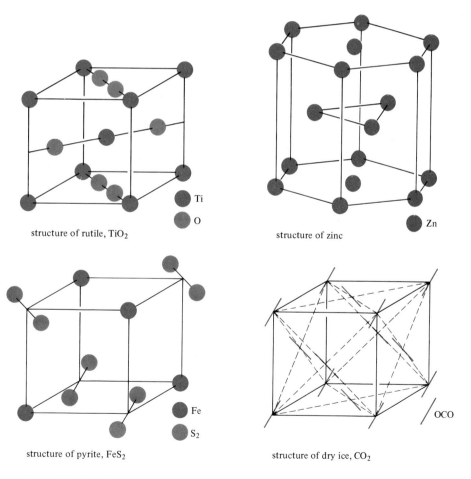

structure of rutile, TiO$_2$

Ti

O

structure of zinc

Zn

structure of pyrite, FeS$_2$

Fe

S$_2$

structure of dry ice, CO$_2$

OCO

Figure 12–15. Four crystal structures.

positions of the other components in the structure. The lines in each illustration are drawn by the artists to delineate the spatial relationships, and do not represent anything real. Some people have more difficulty than others in seeing these kinds of illustrations as three-dimensional pictures. If you have difficulty, usually a little extra attention will help to clarify the matter.

THE CLAYS

So far in this chapter our approach has been theoretical and esthetic. It is time to be practical. The clay minerals are essential to agriculture. They are a source of water in dry seasons and of nutrient ions, potassium, K^+, calcium, Ca^{2+}, iron, Fe^{2+}, magnesium, Mg^{2+}, and a host of others needed by plants as they grow. Without clay, there would be little plant life. It is interesting to note that, when this Earth's crust was formed, there was no clay. Except for Bentonite, which comes from volcanic ash, all the common clays are formed from feldspar. Plant root hairs or microorganisms growing near feldspar extract potassium ions by exchanging them for the much smaller hydrogen ion. The feldspar structure collapses, it dissolves slowly in water, and from those water solutions solid clays eventually precipitate. Of course, this fact alone is insufficient to demonstrate a Providential plan, as though the feldspars were designed to have the properties needed so they could ultimately form clays. But the converse argument requires an even weaker logic, which alludes to coincidence. Either way, clays have the properties necessary to support plant life.

Structure of clay

There are nine known groups of clays, and within each group there are several variations. Our description will be necessarily very much generalized. To begin, we note that the sizes of the magnesium ions, Mg^{2+}, and hydroxide ions, OH^-, are just right so that magnesium ions can be surrounded with six hydroxide ions, a coordination number of 6. Let's set this up using balls of two different sizes, larger ones for the hydroxide ions and smaller ones for the magnesium ions.

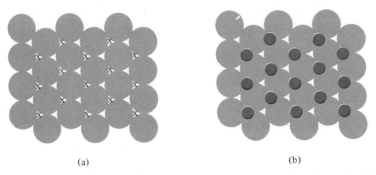

(a)   (b)

Figure 12–16. The first two steps in the construction of a model of a brucite structure.

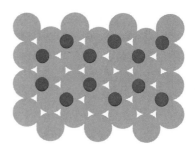

Figure 12–17. The second step in the construction of a model of a gibbsite structure. Compare with Figure 12–16b.

First, make a layer of hydroxide ions, as in Figure 12–16a. Then, at each x put a second layer, of smaller magnesium ions (Figure 12–16b). Third, another layer of hydroxide ions, placed so that these hydroxide ions are not exactly on top of the first layer hydroxide ions. That is, we have an ABC layering, with A and C hydroxide ion layers and the B layer for the magnesium ions. This triple layered structure is called **brucite**. Brucite layers are one of two kinds of layered structures that are found in all clays. Notice that, in the brucite structure, each magnesium ion has a coordination number of 6.

The other layered structure, **gibbsite** is almost the same. Begin with an A layer of hydroxide ions, as before. But for the B layer, we have aluminum ions, Al^{3+}; only two thirds as many aluminum ions as we had magnesium ions are necessary. So, fill up only two thirds of the available spaces in the B layer with the smaller balls, which represent aluminum ions in Figure 12–17. The third layer is a C layer, as before. The whole triple layered structure is gibbsite.

Now to make a clay, we need also a layer of silicate ions with the unit formula $Si_4O_{10}^{4-}$. These tetrahedra share three oxygens with adjoining tetrahedra, and look like Figure 12–18, but, as we know, form large sheets. One kind of clay is called **kaolinite**, it is excellent for making fine chinaware but not much good for agricultural purposes. It has an uncomplicated structure, however, so it is a good one to start with.

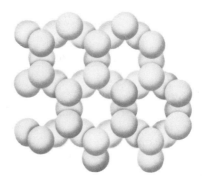

Figure 12–18. Segment of SiO_4^{4-} ion layer.

For kaolinite, take one gibbsite layered structure and set it on top of the silicate sheet. In various places on the bottom side of the gibbsite, the apical oxygens of the silicate sheet will bump into the hydroxide ions of the gibbsite. Wherever this happens, remove the hydroxide ion and let the oxygen of the silicate protrude into the hole you just made. The result is a combination silicate-gibbsite layered structure, called kaolinite. You can do the same thing with brucite and silicate. The resulting layered structure is called serpentine or **antigorite**.

Next, talcum powder, or **talc**. Put one brucite layered structure on top of a silicate layer, but this time only allow two out of every three hydroxide ions to be replaced with an oxygen from the silicate layer. Next, take another silicate layer, turn it upside down, and lay it on top of the brucite, again replacing only two out of every three possible hydroxides with oxygens from the top silicate layer. This is talc. Another clay is called **pyrophyllite**. It is exactly the same as talc except that the center of the sandwich is gibbsite.

Ionic exchanges in clays

Now, let us think about pyrophyllite a bit further. It is a sandwich, with silicate layers representing the bread and gibbsite the filling. The top and bottom of the sandwich is simply a layer of oxygens, at least somewhat negatively charged. If you had two pyrophyllite sandwiches, one on top of the other, between the top slice of one and the bottom slice of the other there would be a fairly high negative charge. Water molecules have two hydrogen atoms, each with a slight positive charge. If there is any water around, it will be drawn in and held fairly tightly by forming hydrogen bonds with the negatively charged oxygens on the bottom surface of the sandwich above it and with the negatively charged oxygens on the top surface of the sandwich below it. That is, water will form a bonding bridge between sandwiches, layer upon layer of them. Usually, this water will have ions dissolved in it, potassium ions, sodium ions, phosphate ions, nitrate ions, sulfate ions, calcium ions, and so on. As the water is drawn in, so also are the dissolved ions.

As the weather becomes dry in the countryside, the water will be drawn out of the clay little by little and be available to plant life along with the nutrient ions it holds in solution. When the climate is wet, the water that came out before is now replaced by other water from the falling rains. This clay structure of multiple sandwiched layers of pyrophyllite, is called **montmorillonite** and is important for plant life.

Other reactions can take place in the silicate layer. As it happens, an aluminum ion, Al^{3+}, can replace a silicon ion, Si^{4+}, in the silicate tetrahedra. Of course, every time this happens, we need an extra ionic charge of +1 to make up the difference. In some clays, we have a silicate layer in which some of the silicons have been replaced by aluminums. Those clays hold various positively charged ions outside the combined layers of talc structure, or pyrophyllite, or other structure. One K^+, one Na^+, or maybe one other $+1$ charged ion is held for each aluminum-silicon replacement in the silicate layer. Or, for each two aluminum-silicon replacements, those clays will hold one Ca^{2+}, or one Fe^{2+}, and so on.

Now, these clays are not going to be able to give up those ions just like that. The ions are held by a real negative charge in the silicate layer. They can trade, or exchange, say one H^+ ion, for a K^+ ion, or two H^+ ions for an Fe^{2+} ion. As you have guessed, usually many H^+ ions are around whenever there is some water around, in the form of H_3O^+ ions (one H^+ ion and one water molecule, H_2O). So, by exchange of ions, these clays will release the metal ion nutrients needed for plant life. Later on, perhaps years later, or next week if the farmer fertilizes his field with the proper ions, the clays will exchange their H^+ ions for metal ions, and hold them until they are again exchanged for H^+ ions. The cycle repeats, over and over again. Somehow, all this just seems to be a bit much to say that it is coincidental.

THE TECHNOLOGY OF METALS

Our last topic for this chapter is also on the practical side. How do we get metals from minerals for the use of mankind? We will begin with copper, the metal used by mankind longer than any other, and end with titanium, which well might be the metal of the future. In between we will look at a few other related materials.

Copper

Here and there copper occurs as the native metal, available simply by picking it up. Sometimes it is mechanically bound in crevices in surface rocks and can be obtained by heating the rocks and cooling them quickly with water. This cracks the rocks open, and the copper can be picked out. Most of these deposits have long since been exhausted because we have been using copper for at least 10,000 years.

Today we obtain copper from its ores, in which it exists combined with oxygen or sulfur (more often sulfur) in an oxidation state of $+2$. To obtain the metal, the copper must be reduced from the $+2$ oxidation state to zero. One reducing agent that could be used is carbon in the form of charcoal. Probably, this reaction first happened accidentally, when someone built a campfire on top of an outcropping of copper ore. The equation for the reaction, copper oxide was the ore, is

$$2\,CuO + C \rightarrow CO_2 + 2\,Cu$$

Notice that the oxidation state of the copper is reduced, from $+2$ to zero, and the oxidation state of the carbon, C, is oxidized or increased, from zero to $+4$; the oxidation state of the oxygen remains the same, at -2.

At present we have a little more trouble in obtaining copper. Although copper ore, largely copper sulfide, CuS, is plentiful and widely distributed geographically, it occurs mixed with a lot of rock. In principle, it would be possible to treat the whole mixture of rock and ore by some means and extract the copper; such a process would be very costly. The less expensive method is first to separate the copper ore from the rock and then to extract the copper from the ore. This is done in a device something like a giant milkshake mixer. The rock and ore from the mine is crushed into a fine powder of particles about the size of sand grains or a bit smaller. The giant mixer is partially filled with water and a little (not too much) oil, and a nozzle on the end of a pipe attached to a compressed air tank squirts air into the liquid at the bottom of the mixer. The mixer itself is operated at high speed so that an oily froth is produced on the top of the water. The powdered rock and ore is dumped into the mixing device. The ore particles collect in the froth and the unwanted rock particles, called gangue, collect on the bottom.

No one is yet able to state in complete detail exactly what happens, but we do know that the bubbles of air in the froth are coated with a film of oil and that the ore particles can be wet with oil better than they can be wet with water. The converse is true of the gangue particles; that is, they can be wet with water better than they can be wet with oil. The ore particles are fine enough so that, when they stick to the oil film around a bubble formed at the bottom of the mixer, they do not prevent that bubble from rising to the surface.

The process that occurs is not too much different from the action of soap in laundering. Oily or greasy dirt is preferntially wet by conglomerations of soap molecules (called **micelles**) in solution, and thus freed from the cloth. **Preferential wetting** is a concept that may be unfamiliar; some people think that water is water, for example, and wet is wet. However, you may have noticed drops of water on a very clean glass surface compared to water drops on a mildly dirty glass surface. If you looked very closely at those drops in the two circumstances, you could see a difference in the appearance of the drops. Depending upon the composition of the soil on the dirty glass surface, you would notice that the drops were more, or less, spread out on the soiled surface than on the very clean surface. When drops spread out, they are wetting that surface more. We can notice the same thing in comparing the behavior of a drop of water on a piece of newspaper to a drop of water on a piece of waxed paper. The water wets the newspaper preferentially compared to the waxed paper. That is, the drop spreads out, even disappears into the newspaper but stands up quite well rounded on the waxed paper. You could not use a piece of waxed paper to remove drops of water from a newspaper, but you could do the opposite.

Back to copper. The froth is scooped off the top from time to time; the gangue that collects on the bottom is removed similarly, and once in a while more oil, more water, and more ore-rock mixture is added; and the process continues. The gangue is left in a pile somewhere and saved. It still has some copper ore in it because the froth-flotation separation process is not perfect. Perhaps some time in the future we will have enough knowledge to be able to extract that residual copper at a price for which it can be sold. Unfortunately that pile of rock gets larger (or we get more smaller piles) as the years go by, and these are, at the least, not esthetically pleasing. That problem has not yet been solved, either.

The scooped off froth contains the CuS ore particles. These are collected and roasted. In extractive metallurgy terms, roasting means to heat in the presence of air. A chemical reaction takes place.

$$CuS + 1\tfrac{1}{2} O_2 \rightarrow CuO + SO_2$$

The CuO is a solid and the SO_2, sulfur dioxide, is a gas, a very unpleasant corrosive gas. In times past, it was allowed to disperse into the surrounding atmosphere; more recently this is being prohibited.

If the copper ore was roasted long enough, all of the CuS would be converted to CuO. However, the ore is only roasted long enough to convert about two thirds of it to CuO. Air is then excluded, and a different reaction occurs.

$$2\,CuO + CuS \rightarrow 3\,Cu + SO_2$$

The solid product obtained is called **blister copper** because the copper obtained has blisters of gas pockets and other debris all over it. The reaction between CuO and CuS is exothermic and takes place in an enclosed container, so most of the heat stays inside. As a result, the copper is molten as it forms; when the container is opened, the liquid copper runs out along with some of the not yet reacted materials, the bubbles of SO_2, and some other minor impurities from the ore. (These are mostly iron sulfide, FeS, and iron oxide, FeO, which were present in the ore to begin with but not mentioned in this story until now.) The molten mixture is cooled in shallow depressions or molds until it solidifies. The solidified blister copper is exceptionally beautiful. Here and there you can see the color of the reddish copper (like a new penny) decorated with irridescent blues, purples, greens, and even yellows, from the various impurities, all in a random fashion.

Before it can be used commercially, to make copper wire, pipe, pennies, nickels (a nickel coin has more copper in it than it has nickel) or used in other alloys, such as brass or bronze for marine hardware, machine bearings, or other uses, the blister copper must be purified. The final purification is done electrically, and involves an oxidation half-reaction,

$$Cu \rightarrow Cu^{2+} + 2\,e^-$$

That is, the impure blister copper is oxidized under conditions controlled so that only the copper, and nothing else present, oxidizes. The process is carried out in an electrical cell which is schematically shown in Figure 12–19a. The pile of debris is underneath the blister copper contains the impurities that are not oxidized, including a small amount of gold and other precious metals.

Now, only the copper is oxidized, and we have therefore purified the blister copper. The problem is that, in the act of purifying it, we have oxidized it again to a Cu^{2+} ion. It will be necessary to reduce it once more.

$$2\,e^- + Cu^{2+} \rightarrow Cu$$

The reduction is carried out on the other side of the cell (Figure 12–19b), where only copper atoms are formed, thus providing us with pure copper metal.

Today, copper is the third most widely used metal, outranked only by aluminum and iron. We see it mostly as pennies, but that is only a small amount of the total. Approximately half of the copper used today is in the form of copper wire because copper is a good conductor of electricity. (Aluminum is a pretty good electrical conductor too; silver is the best of all.) Other uses of copper involve alloys; copper alloyed with zinc, called brass, or with tin, called bronze, is a hard material, but easily worked, that is, easily formed into desired shapes, and these alloys resist corrosion. There are hundreds of varieties of brasses and bronzes, each with different amounts of zinc, tin, and other metals as necessary to impart the desired properties for particular uses.

In recent years, the annual world consumption of copper has been a little more than 6 million tons. In the United States, which currently produces about one fourth of this total, the mixture of copper ore and rock averages less

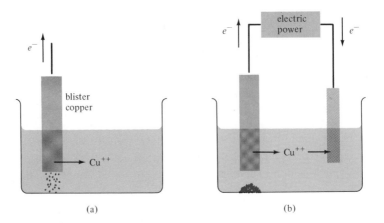

Figure 12–19.   Schematic diagrams, copper refining. (a) Formation of copper ions. (b) The complete cell. Note pile of impurities below the blister copper.

than 0.6% copper. For each ton of copper produced in the United States, it is necessary to remove 350 tons of "overburden," or stuff that lies on top of the copper-bearing rocks. After removing the overburden, approximately 150 tons of copper ore and rock mixture are processed by froth flotation, which yields about 3 tons of mostly copper ore. From that, we get 1 ton of copper metal.

Now, 6 million tons or more of copper each year is a lot of copper. How much longer can we go on using copper at this rate? The answer is complex, but we can look at the broad details. From careful and extensive analyses carried out by competent geologists and chemists, the total crust of the Earth is 0.0055% copper (in the +2 oxidation state). At ten times the present annual rate of consumption, this is enough copper to last us for 60,000,000 years, if we could extract all of it from the crust. This amount of copper is called the **resource base** for copper. It is the best estimate of the total amount present, including that which can be extracted by techniques we have today at reasonable cost, that which can be extracted by techniques we know will work but which are too costly at present or those which we think could be made to work but would be much too costly, and all the rest which probably could never (as far as we can hopefully guess now) be extracted.

On this basis, that 60,000,000 years at ten times our present rate of consumption, or 600,000,000 years at our present rate of consumption, is only a figure to start with. How much of that can we really every hope to get? That amount is called the **resources** for copper. It represents what we can extract with known and with probable future techniques, not considering the cost. It also includes copper deposits that have not yet been discovered but which we are all but certain are somewhere, deposited in eons past by the geological processes mentioned earlier in this chapter. Why not go out now and find these deposits? The answer again is economic. It costs money to carry out the exploration. If new deposits were found today, there would be no need to begin to extract the copper from them. At present, we have enough known copper deposits in the United States alone to last for 10 or perhaps 20 years at our present rates of world consumption. It would be economically foolish to invest large sums of money in finding new deposits because no possible return on that investment could be expected for at least 10, and probably many more, years. Saying it differently, the amount of copper that we know we can get from the Earth's crust is dependent upon how much has been invested in the search. The copper we know we can get at reasonable cost, because we know where it is and we know how much it will cost to extract it, is called the **reserves** for copper.

To summarize: For copper or any other metal, the total amount in the Earth's crust, which we believe to exist but some of which we can never extract, is the resource base. A lesser amount, representing some known deposits and some not yet located, from which copper can or could be extracted at present or higher cost, is the resources. A still lesser amount, known deposits sufficiently rich and available to be extracted at costs that permit sale today, is the reserves. The relation between the amount of reserves and of resources depends upon the rate of use and the willingness of investors to hire geologists to look for more. There is no forseeable shortage of copper, even though we know that there is a finite, limited amount available. Of course we cannot say what the situation will be like even a hundred years from now, not to mention a thousand years. On this basis, it is not reasonable to go to great lengths to conserve copper for future generations; it is equally unreasonable to use it profligately. As nearly as we can predict on the basis of what we know now, there will be sufficient copper for the future far, far beyond any forseeable time as long as we continue to use copper with prudence.

Iron and steel

Our next metal is iron, the most used metal ever known to man, accidentally discovered in Africa (and perhaps in China) by prehistoric man who built a campfire on an outcropping of iron oxide, Fe_2O_3.

$$2\,Fe_2O_3 + 3\,C \rightarrow 4\,Fe + 3\,CO_2$$

It takes a hotter fire to reduce the iron in iron oxide to the metallic state than it does for the corresponding process for copper. Perhaps this is why iron was discovered a long time after copper was made from its ore outcroppings. It is also interesting to note that we still prepare iron from its ores by using carbon mixed with the ore, whereas copper is no longer produced in any substantial amounts at all by the prehistoric method. Iron is useful to man in the form of alloys, not in the pure state. Iron that contains carbon is called steel. Usually other elements are also deliberately introduced, most often by mixing with molten iron and stirring vigorously. The mixture is then allowed to solidify. It can be further worked or shaped after solidification.

To generalize, pure iron is very soft, for a metal. With a little carbon, say about 0.1%, it becomes much harder and stronger. As we add more carbon, up to about 1% or perhaps 1.2% maximum, this effect increases. The steel in a "tin" can underneath the protective coating contains about 0.1% carbon. The steel in a watch or clock spring contains about 1% carbon. The steel in a knife blade (not a stainless steel blade) contains about 0.3 to 0.5% carbon.

Perhaps the major reason for our use of steel, though not the only one, is that its properties can be changed by heat treatment. You can see this for yourself by using an old clock spring or perhaps a bobby pin. If you attempt to bend the spring or bobby pin, you will notice that it breaks at the bend after only a couple of flexings back and forth. Now, take another spring or bobby pin and heat it in a flame until it is red hot; then allow it to cool, very, very slowly, by moving it upward and away from the flame as gradually as possible. After it is cool again, bend it back and forth at the place where it was heated and slowly cooled. You will notice that it is now much softer and can be bent back and forth many times before it breaks in two. Finally, if you have resisted the temptation to break the softened steel, put it back in a flame and heat it red hot again. This time, after it gets red hot, put it *immediately* under water. When you withdraw it from the water and try to bend it, you will notice that it is once again hard and brittle.

By varying the rate at which steel is cooled, we can control its hardness or brittleness, or stiffness, almost as we wish. Briefly, and in a general sense, iron (steel) can exist in different crystalline forms, depending upon the temperature. If red hot steel is cooled quickly, it does not have time to change from the form it had at the high temperature to the form that is stable at the low temperature. That is, when cooled quickly, the cool steel still has the same crystalline form it had at the high temperature. That particular form is hard. When cooled slowly, there is time to change to the crystalline form that is stable at cooler temperatures. That form is soft. The situation is complicated, however, because the temperatures at which each form is stable depend upon the per cent of carbon present and also upon the formation of an iron carbon compound, Fe_3C, which we haven't mentioned. The behavior of steel is, of course, further complicated by the presence of other alloying elements.

Nickel as an alloying element makes steel stronger; chromium imparts an increase in hardness when heat treated and, if present in amounts more than 5%, it imparts corrosion resistance. There are several different varieties of stainless steel. The kind used in cooking utensils, for example, is called "18-8' stainless;" it contains 18% chromium and 8% nickel. This alloy forms a tenacious coating, mostly chromium oxide, on its surface and thereby prevents further oxidation of the underlying metal. However, if heated too strongly, the protective oxide coating becomes porous; darker colored, thicker, nontenacious iron oxide forms. This can be scoured off, with effort, and a new protective coating will form almost immediately.

Manganese as an alloying element enhances the "hot-working" of steel by tending to prevent fracture when hot steel is forced by hammerlike blows to assume a desired shape. Vanadium is added to enhance the ability of steel to maintain a cutting edge. Tungsten is a particularly desirable alloying element when steel is to be used to make wood-working tools; it imparts toughness and the ability to keep a sharp edge. These and still other alloying elements make steel useful to us as structural members, in machinery and tools, and in electromagnets for the generation of electrical power and the conversion of that power by motors to kinetic energy. Approximately 90% of the total metal consumption in the world is in the form of steel.

Most of the iron ore we use today is iron oxide intimately mixed with silicon dioxide, SiO_2. The iron oxide can be thought of as either Fe_2O_3 or Fe_3O_4, depending upon the source. As we know, the essential idea is to reduce the iron to a zero oxidation state. The reducing agent is carbon monoxide, CO, instead of carbon as charcoal, as we shall see, but the main problem is to remove the SiO_2. Fortunately, there is a simple answer. The SiO_2 can be dissolved in some suitable solvent. Water will not do the trick fast enough, but molten, hot, liquid calcium oxide, CaO, works fine.

We need a source of carbon monoxide and a source of calcium oxide. We get our CO from coke. (Coke is made from coal by heating it and not letting it burn very much. This drives off the volatile constituents from the coal and what is left is coke, mostly carbon.) We get our CaO from limestone, $CaCO_3$, by heating it.

$$CaCO_3 \rightarrow CaO + CO_2$$

Now we are ready. Mix a quantity of iron ore, limestone, and coke together and put it in a big furnace, blow in a little air, and heat it a bit to get the whole thing started. The result is called a blast furnace operation. These reactions occur:

1. Limestone forms liquid calcium oxide and gaseous carbon dioxide

$$CaCO_3(s) \rightarrow CaO(l) + CO_2(g)$$

2. Calcium oxide reacts with the silicon dioxide to form a substance that is liquid at the high temperatures involved, calcium silicate.

$$CaO(l) + SiO_2(s) \rightarrow CaSiO_3(l)$$

3. Now the iron oxide is freed, ready to be reduced, and we need the carbon monoxide from the coke and oxygen.

$$2\ C(s) + O_2(g) \rightarrow 2\ CO(g)$$

$$3\ CO(g) + Fe_2O_3(s) \rightarrow 2\ Fe(l) + 3\ CO_2(g)$$

The last two reactions are exothermic, and provide enough energy to facilitate the other two reactions, especially the formation of calcium oxide from the limestone. All we need is to start the whole thing with an external source of heat.

There is seemingly another problem, but fortunately it solves itself. It appears as if the iron we have just made in the last reaction listed will react with the oxygen that we must have present to make the carbon monoxide. We need something to cover the iron, to protect it from that oxygen. The calcium silicate is just right for this trick. It is a liquid, and it will float on molten iron, as oil floats on water, and will keep the oxygen from reacting with the iron.

A blast furnace (Figure 12–20) is a large structure, generally circular, with a double tapered shape. It is smaller at the bottom, widens out to a

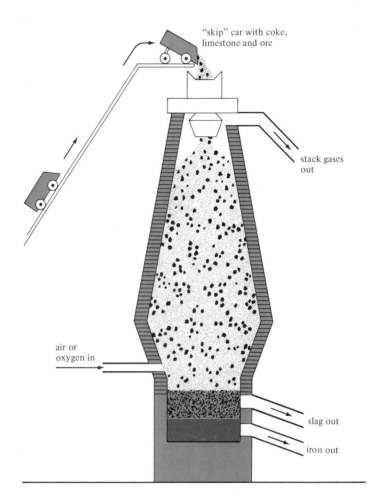

Figure 12–20.   Blast furnace, schematic.

diameter of 30 ft or more, and then narrows again; the whole thing may be 150 ft high or more. The mix of ore, coke, and limestone is put in at the top. The reactions listed (and some others) take place as the mix moves downward (about 15 ft/hr). Oxygen or air is forced in through holes in the bottom, and below that molten iron collects, protected by a layer of molten calcium silicate, called **slag**. In a typical modern blast furnace, the reactions, once started, continue for up to 5 years. Every so often during each day and night the molten iron and molten slag are drawn off through holes located for the purpose in the bottom part of the structure. Typically, from 3000 to more than 6000 tons of iron are produced daily by a single blast furnace.

The iron from a blast furnace is called **pig iron** because in times past it was poured into molds shaped like, and about the size of, the body of a pig and allowed to solidify before being processed further. Now, most of the molten iron is transported directly to a **basic oxygen converter** to be purified. Iron from the blast furnace contains much too much carbon and often some silicon and phosphorous that must be removed. A basic oxygen converter is similar to a flower vase in shape, sort of long and narrow, but much larger because it will hold tons of impure molten iron. A long pipe, water cooled (so it won't melt) is inserted into the mouth of the converter, down to just above the hot liquid metal surface. Then, oxygen is squirted from the pipe into the liquid metal as a supersonic jet for a few minutes. The oxygen reacts with the silicon, the phosphorus, and some of the iron, forming oxides, which float on top of the now purified iron (what is left of it). The reactions are

$$Si + O_2 \rightarrow SiO_2$$

$$4\,P + 5\,O_2 \rightarrow P_4O_{10}$$

$$4\,Fe + 3\,O_2 \rightarrow 2\,Fe_2O_3$$

Fortunately, the silicon and phosphorus react much more readily with oxygen than the iron (otherwise, this purification process would not work). Suitable alloying elements may be added to the purified, liquid iron in the amounts necessary, and then the alloy is poured into a mold to harden. After that it may be formed into useful shapes by passing between large double rollers or by forging (hammering) or other means.

At the present rate of annual consumption, world wide, the resources for iron will last for thousands of years.

Aluminum

Aluminum is the third most abundant element in the Earth's crust (after oxygen and silicon) existing there in a +3 oxidation state. Aluminum is the metal of modern times. Only 100 years ago, it was more precious than gold, and emperors had special tableware made of aluminum to impress their guests at formal banquets. (Both are gone now, emperors and aluminum plates at fancy dinners.) The reasons for the high cost were chemical, of course.

Aluminum is difficult to reduce from +3 to zero; it requires a good reducing agent. For example, metallic sodium, Na, will reduce aluminum from +3 to zero. Aluminum chloride reacts with sodium to form aluminum and sodium chloride.

$$AlCl_3 + Na \rightarrow Al + 3\,NaCl$$

But the problems of getting sodium in a zero oxidation state are formidable and costly. In effect, to get the aluminum metal, you had to also pay the cost of getting sodium metal. Cheaper reducing agents could not be found. Carbon as charcoal or coke is an inexpensive reducing agent but not strong enough to reduce aluminum. In fact, the reaction goes the other way around, aluminum (once you have it) will reduce carbon from +4 to zero.

$$4\ Al + 3\ CO_2 \rightarrow 2\ Al_2O_3 + 3\ C$$

The reaction goes rapidly in the presence of carbon dioxide at high pressures, although slowly or not at all (fortunately) if the pressure of CO_2 is as low as it is in ordinary air.

The problem could be solved if there were some way to dissolve the aluminum in the +3 state and then reduce it electrically, much as copper ions are reduced in the final purification we described previously. It would also be less expensive to use an aluminum compound found in nature rather than aluminum chloride, which must be made at some extra cost.

Aluminum occurs naturally in all clays and particularly in a special clay called bauxite, a combination of iron and aluminum oxides with silicon oxides. Bauxite is widely distributed over the surface of the Earth and mined by surface digging, once the overburden is removed. At the mine site, the bauxite is leached, that is, washed thoroughly with a strongly alkaline water solution, that is, a solution that contains hydroxide ions. We can summarize the reaction with this equation in which the aluminum oxide in bauxite is represented by Al_2O_3.

$$Al_2O_3 + 6\ OH^- + 3\ H_2O \rightarrow 2\ Al(OH)_6^{3-}$$

The aluminum hexahydroxy ion, $Al(OH)_6^{3-}$, formed as the product is soluble in the water used for leaching. Hence, in one major step, the aluminum is removed from the other unwanted species present in the original bauxite. This processing is performed at the site of the bauxite mine to save shipping costs on the unwanted components of the bauxite.

It would be desirable also to reduce the aluminum to the zero oxidation state at the mine site, and save the shipping costs for the oxygen too. This reduction requires large amounts of electrical power that are simply not available at the mine site. Instead, the water solution of aluminum hexahydroxy ions is evaporated, ultimately forming more or less dry, solid, pure aluminum oxide, Al_2O_3, which weighs less than half as much as the original bauxite. At a location, such as the Pacific Northwest in the United States or Canada, where hydroelectric power is inexpensive and abundant (more or less, at least) the process is continued.

Aluminum oxide will dissolve in molten cryolite, Na_3AlF_6, a naturally occurring mineral (which is now made synthetically also). The process is simply a reduction of the aluminum and a simultaneous oxidation of graphite that is forced to occur by the imposition of 5 or 6 volts at very high amperage (lots of electrons per second). To produce a ton of aluminum requires approximately the same electrical power as is needed for 14,000 1000-W light bulbs lit for 1 hr. The output of a typical aluminum refining plant is from 50 to 150 tons of aluminum per day, at full capacity.

Eventually, the bauxite mines will be exhausted although the reserves for aluminum will last for several decades even at ever increasing rates of consumption. (Current consumption in the United States, for example, is more than 50 lb of aluminum per person each year and is projected to exceed more than 150 lb per person per year by the end of this century.) It will be necessary to turn to less rich aluminum ores, but the technology for obtaining pure Al_2O_3 from such ores is already known, with relatively minor details yet to be worked out. There is no realistically forseeable time when we will ever run out of aluminum; the resources for aluminum are extraordinarily large.

The use of aluminum is growing. Aluminum itself is a light metal and, when alloyed, it becomes a strong, durable material that can be used to replace almost every other metal, depending upon the alloying element or elements. Aluminum is a good conductor of electricity; it forms a hard, nonporous, tenacious coating of aluminum oxide on its surface that resists further oxidation; it is an excellent reflector of light and heat; it is not magnetic in itself, which might limit its replacement for steel, but alloys that are magnetic can be prepared. It is safe for uses involving food storage and preparation. It can be formed and shaped by each of the metal forming processes already known and used by men.

Magnesium

The story for magnesium is parallel to that for aluminum in some ways. In the preceding paragraph, for example, we could substitute the word "magnesium" for the word "aluminum" and have an accurate description of its versatility. However, the use of magnesium is not growing nearly as rapidly as the use of aluminum, and magnesium is slightly toxic and therefore not suitable for use in food storage or preparation. On the other hand, magnesium is used in pyrotechnics; when it burns a very bright light is emitted, and that light can be colored by the inclusion of other metal salts. For example, in the presence of sodium chloride, the emitted light has an amber or yellowish orange color; strontium compounds, produce red; barium compounds, green; thallium compounds, blue.

Magnesium is very light; it has the lowest density (least weight for a given volume) of any usable metal. Therefore it is used in some furniture construction, as a casting in the casings of portable machines such as vacuum cleaners and sewing machines. Orbiting satellites contain more magnesium than any other metal. It is used in luggage frames for the same reason; it is both light and strong.

Since it is more easily oxidized than steel, magnesium can be used as a sacrificial anode in the cathodic protection of steel structures exposed to oxidation. Steel pipe is used in pipe lines buried in the ground, and, at intervals along the buried pipe, large pieces of magnesium connected to the pipe by a wire are also buried. Under these circumstances the buried magnesium is corroded in preference to the steel. It is much cheaper to replace a block of magnesium than a corroded steel pipe. In a similar manner, magnesium prevents the corrosion of steel bridges and ship hulls. For this latter use, the magnesium is fastened directly to the hull, of course.

Most of the magnesium produced in the United States comes from the sea, where magnesium occurs as the magnesium ion, Mg^{2+}, dissolved in the water. One cubic mile of sea water contains 10 billion lb of magnesium, enough for a 20 year world supply at the present rate of consumption. To obtain magnesium from the sea we need to separate it from the rest of the ocean content, and then reduce the ion to the metal. For the separation, we take advantage of the fact that magnesium ions will react with hydroxide ions, OH^-, to form an insoluble substance, magnesium hydroxide, $Mg(OH)_2$ (also called "milk of magnesia" when it is mixed with a little water).

The first step is to obtain the hydroxide ions from old oyster shells, which are handy from the sea also. Oyster shells are largely calcium carbonate, $CaCO_3$.

$$CaCO_3 \xrightarrow{\text{heat}} CaO + CO_2$$

(This is the same reaction as is used in the production of steel, but there limestone is the source of the calcium carbonate.) Now calcium oxide, CaO, will react with water to form calcium hydroxide, $Ca(OH)_2$, and this will dissolve to some extent in water, forming calcium ions, Ca^{2+}, which we don't care about, and hydroxide ions, which are what we want.

$$CaO + H_2O \rightarrow Ca^{2+} + 2\,OH^-$$

The hydroxide ions react with the magnesium ions to form insoluble magnesium hydroxide.

$$Mg^{2+} + 2\,OH^- \rightarrow Mg(OH)_2$$

At this point we have an awful lot of sea water with a little bit of fine, powdery, white magnesium hydroxide particles in it. The whole thing is filtered through a fine cloth, which retains the magnesium hydroxide while almost everything else passes through. We have separated the magnesium from the sea water.

Now, we have several choices for the reduction of the magnesium in the magnesium hydroxide to magnesium metal. We could, for example, reduce it with carbon monoxide (as we did reduce iron in the iron oxide ore). As it happens, the magnesium-oxygen bonds in magnesium hydroxide are quite strong, and reduction with carbon monoxide is difficult. Translated into practical terms, that reduction is expensive. We could melt the magnesium hydroxide and then reduce it electrolytically, much as was done in the refining of blister copper. But, magnesium hydroxide has a very high melting point, which, translated, means a high cost for melting. We could reduce the magnesium in the magnesium hydroxide with silicon; this is done industrially to some extent, although the process is somewhat costly.

First, magnesium hydroxide is heated, forming magnesium oxide and water.

$$Mg(OH)_2 \xrightarrow{\text{heat}} MgO + H_2O$$

Then, the magnesium, now in the form of its oxide (not hydroxide) is allowed to react with silicon at high temperatures.

$$2\,MgO + Si \rightarrow SiO_2 + 2\,Mg$$

At the high temperatures involved, the magnesium is produced as a gas, separated from all other solids present, the MgO, the Si, and the SiO_2. Magnesium produced in this way is very pure, and for some purposes it is worth the extra cost.

In the less expensive process, by which most magnesium is produced today, magnesium hydroxide is converted to magnesium chloride, $MgCl_2$, which melts at a much lower temperature. For this hydrochloric acid, HCl, is used.

$$2\,HCl + MgO \rightarrow MgCl_2 + H_2O$$

The magnesium chloride is carefully dried, heated until it melts, forming magnesium ions and chloride ions in the molten, hot liquid. Then, using an external source of electrical power, the magnesium ions are reduced and the chloride ions are oxidized.

$$Mg^{2+} + 2\,e^- \rightarrow Mg$$

$$2\,Cl^- \rightarrow Cl_2 + 2\,e^-$$

Although written separately, the two reactions occur simultaneously as electrons are forced into and out of the molten liquid.

$$Mg^{2+} + 2\,Cl^- \xrightarrow{\text{electrolysis}} Mg + Cl_2$$

The other product, chlorine, Cl_2, is collected as a gas and used to make more hydrochloric acid for the next cycle.

Magnesium is not used currently as much as aluminum for several reasons, although there is an ample supply of magnesium in the Earth's crust and in the seas for thousands of years. Both reasons are economically derived. Aluminum is produced from its ore by more than one metallurgical processor, whereas magnesium has been (except during World War II in the United States) produced by only one. Further, the process now used to make magnesium is more costly than that for aluminum.

Recently, a new and promising less expensive procedure for converting magnesium oxide to magnesium chloride, using salt instead of hydrochloric acid has been developed.

$$MgO + 2\,NaCl \rightarrow MgCl_2 + Na_2O$$

Salt is much less expensive than hydrochloric acid. This reaction has been known to metallurgical chemists for some time, but, although we can write the equation for the reaction easily, making the reaction practical is far more difficult. In fact, the problem may not yet be completely solved. When it is solved, the price of magnesium will be lowered, and we will see more of it in common use in the years to come.

Titanium

Titanium is an abundant element; there is more titanium, as rutile or titanium dioxide, TiO_2, and as ilmenite, $TiO_2 \cdot FeO$, in the Earth's crust than the combined total of copper, lead, tin, zinc, and nickel. The problem is in the reduction of the titanium from the +4 state to the metal. Titanium is extremely difficult to reduce; hence, the metal is costly. As a metal its strength is high, especially when alloyed with small amounts of aluminum, vanadium, molybdenum, manganese, iron, or chromium. Titanium has a low density, although not as low as aluminum. It is corrosion resistant for the same reason that aluminum is, an impervious coating of TiO_2 forms spontaneously on the surface of the metal when it is exposed to air.

To the small extent that titanium is used today, we find it in aerospace vehicles, in corrosive environments for some structures used in chemical processing, and in marine environment applications. By far the greatest use of titanium is as the dioxide. Titanium dioxide is a white, opaque solid that naturally forms as small, very thin, flat plates. Hence it is used as a white opacifying pigment in paint, paper manufacture, textiles, inks, and cosmetics.

To obtain the metal, the more abundant ilmenite ore is treated with sulfuric acid after which water is carefully added, precipitating the titanium as titanium tetrahydroxide, $Ti(OH)_4$. This separates it as an insoluble solid from other substances present in the original ore. Note that this step is not unlike the steps involved in the preparation of magnesium.

The titanium tetrahydroxide is heated, forming titanium dioxide and water.

$$Ti(OH)_4 \xrightarrow{\text{heat}} TiO_2 + 2\,H_2O$$

At high temperatures and in the presence of carbon, titanium dioxide will react with chlorine to form titanium tetrachloride and carbon dioxide.

$$TiO_2 + C + 2\,Cl_2 \rightarrow CO_2 + TiCl_4$$

The titanium tetrachloride obtained at this point is still not pure, but titanium tetrachloride can be purified by distillation. In fact, the whole point of the processing up to this stage is to obtain titanium in a form that can be purified. This is the least costly procedure now known, but it is very expensive.

The distilled, pure titanium tetrachloride will react with magnesium, reducing the titanium and oxidizing the magnesium, at high temperatures. Unfortunately, at these high temperatures, the titanium will react with oxygen and with nitrogen in the air. So, adding further to the expense, the reaction can only be allowed to take place when air is excluded.

$$TiCl_4 + 2\,Mg \rightarrow 2\,MgCl_2 + Ti$$

Notice that the use of magnesium, not the least expensive metal, is another necessary addition to the cost of preparation. Finally, titanium metal produced in this manner resembles a sponge, solid chunks with lots of holes. It has to be melted (in the absence of air, of course) and finally we get cooled, solid titanium metal for further fabrication. If a less expensive process can be devised, then indeed titanium may be the metal of the future.

SOME NONMETALS OF INDUSTRIAL IMPORTANCE
Sulfur

Everyone has seen, touched, and used iron (steel), copper, and aluminum. Few have seen sulfur, fewer yet have touched a piece of this yellow solid, but all have used it indirectly. Sulfur is used in the manufacture of sulfuric acid. The first step is the production of sulfur trioxide, SO_3, a gas under ordinary conditions. In the presence of a catalyst, sulfur and oxygen react to form sulfur trioxide.

$$2\,S + 3\,O_2 \rightarrow 2\,SO_3$$

Sulfur dioxide from ore refining can also form sulfur trioxide by reacting with oxygen, in the presence of a catalyst.

$$2\,SO_2 + O_2 \rightarrow 2\,SO_3$$

Or, hydrogen sulfide, H_2S, an odorous and very poisonous gas derived from petroleum refining, can react with oxygen under the proper conditions to form sulfur trioxide and water.

$$H_2S + 2\,O_2 \rightarrow SO_3 + H_2O$$

Sulfur trioxide from any of these sources or any other sources will react with water to form sulfuric acid, H_2SO_4, an oily appearing, high boiling acid which is, also, a good oxidizing agent.

$$SO_3 + H_2O \rightarrow H_2SO_4$$

No other substance possesses these three properties to as high a degree as sulfuric acid: acidic, high boiling point, oxidizing agent. Because of one or more of these three properties, sulfuric acid is used in the manufacture of fertilizer, pigments, detergents, explosives, storage batteries, and fibers; it is used in petroleum refining and in metallurgy, to name a few of thousands of different uses. Together, the use of sulfuric acid adds up to an annual consumption of 70,000,000 tons for the whole world. It is almost possible to state that either you are now touching an object or article of commerce that involved sulfuric acid in its manufacture or that you will do so within the next 30 sec, so pervasive is this single compound in our lives. In addition about 12,000,000 tons of sulfur is used annually in other ways, principally indirectly, in commercial processes involving paper, insecticides, dyes, bleaches, vulcanization of rubber, food preservatives, photography, textiles, and pharmaceuticals. It is probably correct to suggest that no other single element is used indirectly in as many varied ways in our daily lives as is sulfur, an element which many people have never even thought of.

Until near the end of the last century, 95% of the world's supply of sulfur came from Sicily as the only important exportable substance from that island. It was known that large deposits of elemental sulfur existed, not too deeply embedded, in the ground in Louisiana, Texas, Mexico, and offshore in the Gulf of Mexico. Of these, the deposits in Louisiana were the most accessible, but were considered unusable for two reasons. The deposits were overlaid by marshy ground, very expensive and difficult to dig through. The digging was made additionally hazardous because of the presence of a very poisonous gas down there with the sulfur, hydrogen sulfide, H_2S.

In 1891 Herman Frasch succeeded in persuading a few people to invest in what others thought was a foolish scheme: a way to mine sulfur from those deposits without digging a mineshaft at all. The Frasch process does seem so simple as to be unworkable at first sight. Simply drill a hole in the ground, a few inches in diameter, and keep pushing a pipe into the hole, immediately following the drill, to keep the marshy stuff from collapsing in on the drilled hole. Then remove the drill, put a little smaller pipe inside the first one, and inside that, a still smaller third pipe. Now, in the space between the big and the medium sized pipe, pump hot water (very, very hot superheated water) down to the bottom to melt the sulfur. Then, pump air under high pressure down the smallest pipe. Eureka! Up through the space between the medium sized and the smallest pipe will come melted sulfur, hot water, and air bubbles! It worked so well and so inexpensively that Sicilian sulfur could no longer be sold. Unfortunately our gain was the Sicilian's loss. A disruption of the economy of Sicily ensued from which those people have not yet recovered fully after more than 75 years.

Today, some of the originally exploited deposits in the Gulf of Mexico area have been exhausted and new drillings have been made. However it is doubtful that many more drillings will be made. Within a few years it is likely that we will have a glut of sulfur on the world market. For ecological reasons, it is desirable to remove sulfur from fossil fuels. If it is not removed, sulfur dioxide is formed during the burning of the fuel, is dispersed into the air where eventually it forms sulfuric acid or sulfurous acid, H_2SO_3 (not as unpleasant as sulfuric acid, maybe, but a close second or third). These substances have very deleterious effects. They burn your eyes, "dissolve" nylon stockings, and corrode buildings made of marble or limestone. Hence, the removal of sulfur from fossil fuels will yield more sulfur than we can foreseeably consume by using it in the manufacture of sulfuric acid or other substances. That is, the resources for sulfur are overwhelming.

Phosphorus

Phosphorus itself occurs in two forms, one as a slightly yellowish white, soft solid, which is quite poisonous and spontaneously flammable in air. The other form is a dark red solid, not as poisonous and not spontaneously flammable. The element phosphorus, P, has limited use, but in the form of phosphate ion, PO_4^{3-}, it is widely used in fertilizers, detergents, water softeners, baking powders, insecticides, beverages, photographic reagents, dental cements, ceramics, catalysts, and animal feed supplements. Phosphorus is absolutely necessary for life in plants and animals; as we know it is a constituent of adenosine triphosphate and adenosine diphosphate, the compounds most involved in energy transfer in living systems. It is also a component of bones and teeth.

Our teeth, for example, are composed in part of a network of proteinaceous structure, something like a sponge with microscopic pores. The pores are filled with hydroxyapatite, for the most part. This compound, $Ca_5(PO_4)_3OH$, forms the major phosphate constituent of our teeth. Although it is indeed a rock, found as such in nature, it is not as hard as a related compound, fluorapatite, $Ca_5(PO_4)_3F$. The topical application of fluoride compounds or the ingestion of fluoride containing food or water causes some of the hydroxyapatite to be converted to fluorapatite.

$$F^- + Ca_5(PO_4)_3OH \rightarrow OH^- + Ca_5(PO_4)_3F$$

Some of the controversial questions concerning the addition of fluoride ions to municipal water supplies were mentioned in Chapter 9.

Most phosphate containing rock is fluorapatite. It occurs in large and small deposits widely distributed in many geographical areas throughout the world. When this rock is treated with sulfuric acid (the equations are too complicated to bother with here) the product may be superphosphate, mostly a mixture of monocalcium phosphate, $Ca(H_2PO_4)_2$, and calcium sulfate, $CaSO_4$, or triple superphosphate, which is monocalcium phosphate, alone. The world reserves of fluorapatite and other apatites is sufficient to last for approximately 1000 years at the present rate of consumption. However, the world resources exist mostly as dissolved phosphate compounds in the sea and are sufficient to last into the unforseeable future, although at present we do not have the technology to extract phosphate from that source at low cost.

Calcium carbonate, calcium oxide, and cement

The most useful rock on Earth, and by far the most widely used, is limestone. We have noted its use in metallurgical processing earlier in this chapter. Here we can briefly consider the rock itself and its common derivative, cement.

Limestone deposits underlie almost one fifth of the United States and are widely distributed, appearing occasionally as outcropping rock as well. The limestone is used directly in road building, in soil conditioning, and in blast furnace operation. To a small extent it is used indirectly as lime, CaO (which we know as calcium oxide), and it is used in the manufacture of cement. The total current annual consumption of limestone in the United States is more than 500 million tons.

Cement, as we shall see, is a mixture of substances, produced as a darkish, off-white, or grayish finely divided powder. When cement is mixed with sand or fine aggregate and water, the mixture is called mortar. Concrete, technically speaking, is a mixture of cement, sand, coarse aggregate, and water. Mortar and concrete set to a hard, rocklike solid rather rapidly. However, while pliable, the mortar or concrete can be formed or molded into shapes and forms and reinforced by embedded steel wire or rod. It is used for a wide variety of structural purposes, as everyone knows. The current annual consumption of cement in the United States is more than 600 lb for each person in the nation, and even higher per capita consumption figures apply to Canada, Japan, and West Germany. Ninety percent of the cement produced in the United States is made from raw materials obtained within 160 miles of the location where the concrete and mortar are eventually used.

Portland cement, the most common kind made, is a mixture of calcium oxide, CaO, silicon dioxide, SiO_2, aluminum oxide, Al_2O_3, iron trioxide, Fe_2O_3, and calcium sulfate dihydrate, or gypsum, $CaSO_4 \cdot 2H_2O$. The calcium oxide is made from limestone by heating, and the other oxides are obtained from clay. More silicon dioxide is added as sand and more iron and aluminum oxide are added as bauxite or from other sources to produce a final mixture of the proper proportions. Depending upon the use to which the mortar or concrete will be put, the proportions vary.

The proper amounts of limestone and clay are mixed and crushed to a semicoarse particle size, mixed further, ground more finely, and roasted in a kiln, usually a long, almost horizontal, hollow cylindrical tube up to 600 ft long and 20 ft in diameter, which rotates slowly. At the lower end, intense heat is applied, while the ingredients are introduced at the upper end. Temperatures inside the kiln are as high as 1500° or 1600°C. After the roasting, gypsum is added and the entire mixture is very finely ground, so finely that 1 oz of the powder particles have a total surface area equal to that of 15 football fields.

To see this, imagine a piece of cheese shaped like a cube 1 in. on each side. Cut the cheese in half, right through the middle. We now have two smaller pieces. Before cutting, the surface area was 6 in.^{2}. By cutting it once, we made two new surfaces that were not there before, each with an area of 1 in.^{2}. Now the total surface area of the two (smaller) halves is 8 in.^{2}. If we now cut both halves into quarters, the surface area is increased 2 in.^{2} more to a total of 10 in.^{2}, and we now have four (still smaller) pieces of cheese. Each time we make the pieces smaller we increase the surface area. If the pieces could be cut very small, their surface area would be quite large. From the total surface area of the particles in 1 oz of cement, we can see that each particle must be very small indeed.

When portland cement is mixed with water a reaction begins within a few minutes, and the mortar or concrete starts to harden. The details are not well understood and are controversial, but one possible reaction is expressed by this equation:

$$3\ CaSO_4 + 3\ CaO + Al_2O_3 + 32\ H_2O \rightarrow (CaO)_6Al_2O_3(SO_3)_3(H_2O)_{32}$$

The complicated looking product is a crystalline substance that tends to form long, narrow, needlelike crystals. If we could see what happens submicroscopically, probably we would see the more or less roundish cement particles disappear and in their place interlocking, tangled, long, bent, and twisted needles would appear. Such a structure would be stiff, or hard.

As the concrete or mortar continues to harden, a second, slower reaction, is thought to occur:

$$10\ CaO + 4\ SiO_2 + 10\ H_2O \rightarrow 4\ Ca(OH)_2 + 2\ (CaO)_3(SiO_2)_2(H_2O)_3$$

The calcium hydroxide, $Ca(OH)_2$, also forms long, interlocking needlelike crystals, whereas the more complicated compound probably forms a gel, an amorphous material something like gelatin in its properties, which resides in the empty spaces among the two kinds of interlocking needles.

Because of the high temperatures required, the manufacture of cement is relatively expensive. Cement itself is dense, a small amount is heavy, so transportation costs for cement are high. Fortunately, the raw materials for the manufacture of cement can usually be found near the point of use so that the total transportation costs for such short hauls are not exorbitant. Unfortunately, therefore, the limestone quarries and mined clay banks are located near cities and towns, sometimes within the city limits, and are considered by some to be unsightly, dusty, undesirable blights on the urban environment.

Upside down mirror image helpful puzzle hint

about what you have just done.

for sixty-seven pennies in the square. To get the last penny in, think a little bit in any row touches six other pennies instead of four, and there is room the adjacent even unnumbered row. Now, except for pennies on the edges, each and then slide all seven into the "notches" between the two nearest pennies in move the remaining seven pennies a distance equal to the radius of a penny, starting at any edge. Take one penny, each, from the odd unnumbered rows? Answer to the 64/65 penny puzzle: Number the rows from 1 through 8,

FOR FURTHER INFORMATION

J. R. Heirtzler: Where the earth turns inside out. *Nat. Geogr.*, **147**:586 (May 1975). With pictures and diagrams of the mid-Atlantic ridge; awesome to contemplate.

P. J. Wyllie: The Earth's mantle. *Sci. Amer.*, **232**:50 (March 1975). The indirect evidence from which we think we know what is going on beneath us in regions that may or may not always remain inaccessible.

K. K. Turekian: *Geology of Oceans*. Prentice-Hall, Englewood Cliffs, N.J., 1968. Written for those who would like to know, authoratative, well-presented.

M. H. P. Bott: *The Interior of the Earth*. St. Martin's Press, New York, 1971. A textbook with lots of interesting information. Movement within the mantle, history of the Earth–Moon system, heat flow in the Earth, composition and properties of rocks, and many other topics.

R. B. Gordon: *Physics of the Earth*. Holt, Rinehart and Winston, New York, 1972. Something like the work by M. H. P. Bott, but with a physics emphasis. If both books are available, comparisons are instructive and informative.

P. J. Wyllie: *The Dynamic Earth*. Wiley, New York, 1971. A relatively advanced work. Do not read this one first. If you are first informed and then use this book, you will be glad you waited. Recommended, with this caveat.

R. S. Dietz: Geosynclines, mountains and continent-building. *Sci. Amer.*, **226**:30 (March 1972). If you did not know beforehand how to make mountains and continents, you would know after reading this handy set of "instructions." Marvelous!

J. A. Shimer: *Field Guide to Landforms in the United States*. Macmillan, New York, 1972. Just the thing to take with you on a short walk in the countryside, or on a long cross-continental trip to identify, from what you see, how this continent was built.

J. A. Shimer: *This Changing Earth*. Harper and Row, New York, 1968. Something like the same author's *Field Guide*, but without as many helpful illustrations.

C. S. Hurlbut: *Dana's Manual of Mineralogy*, 18th ed. Wiley, New York, 1971. Any book that has gone through seventeen previous editions has to be useful. One can hardly claim to be well educated unless this book at least resides on their library shelves, in a location where it can easily be retrieved, and read. This is perhaps the other book to take with you if you plan to be marooned on an uninhabited island, along with a book on "How to build a raft."

J. W. Feiss: Minerals. *Sci. Amer.*, **209**:128 (Sept. 1963). As the title states, this article is about minerals and written in the clear and interesting style for which this publication is well-known.

H. Boegel, edited and revised by J. Sinkankas: *A Collector's Guide to Minerals and Gemstones*. Thames and Hudson, London, 1971, translated from Germany by E. Fejer and P. Walker. This fine book is a sort of Dana's Manual, simplified, with superb illustrations. Recommended to all who would like to collect rocks.

D. K. Fritzen: *The Rock-hunter's Field Manual*, Harper and Row, New York, 1959. Take this book with you on a field trip, it is filled with detailed instructions about what to do when you wonder "What is this rock made of?"

M. H. Battey: *Mineralogy for Students*, Hafner Press, New York, 1973. Information useful to those who found out what the rock was made of, and now want to know still more.

A. Holden and P. Singer: *Crystals and Crystal Growing*, Doubleday, Garden City, N.Y., 1960. Another classic like Dana's Manual, but this one is for fun. Crystals are always beautiful, especially when you make your own. This book tells how. Enjoy!

E. A. Wood: *Crystals and Light*, Van Nostrand-Reinhold, New York, 1964. Like Holden and Singer, this author also was associated with Bell Laboratories, where scientists generally very much enjoy their work, and it shows in the writing of this gifted author.

A. Holden: *The Nature of Solids*, Columbia University Press, New York, 1968. Another book to complete the trilogy, this one more advanced than the other two.

P. Gay: *The Crystalline State*. Hafner Press, New York, 1972. An introduction to crystals, at about the same level of difficulty as Holden's work on solids.

K. K. Turekian: *Chemistry of the Earth*. Holt, Rinehart and Winston, New York, 1972. This is an elementary book if you already know some chemistry. Well presented, with many instructive illustrations (and some that are a bit obscure). But look at it anyway, you'll be glad you did.

H. J. Sanders: *Chemistry and the Solid Earth*. American Chemical Society (Special Sales Dept.), Washington, D.C., 1967. Something like Turekian's *Chemistry of the Earth* but much easier to read even though it has fewer illustrations. Originally, this work appeared as a Special Report in *Chemical and Engineering News*.

M. Hudson: *Crystals and Crystal Structure*. Longman Group, London, 1971. Professor Hudson has one of the finest personal collections of slides of minerals in Great Britain, and, though some parts of his book are to be read more than once for comprehension, his love of crystals shows through clearly and is infectious. Recommended.

P. E. Desautels: *The Mineral Kingdom*. Hamlyn House, New York, 1969. As Curator of the mineral collection of the Museum of Natural History, Smithsonian Institution, Mr. Desautels has under his cognizance the finest collection of minerals in the world. With photographs by an expert, Lee Boltin, this is an outstanding book. Absolutely great!

A. E. Schreck (ed.): *Minerals Yearbook*. U.S. Government Printing Office, Washington, D.C. (various years). Volume I, on metals, minerals, and fuels. At intervals, the staff of the U.S. Bureau of Mines prepares this statistical compilation of production, exportation, and importation of every metal, mineral, and fuel you can think of. This is *the* authoritative reference in the subject. Dry reading until you start to compare this datum with that one; and then it becomes like a detective story.

D. J. G. Ives: *Principles of the Extraction of Metals*. Royal Institute of Chemistry, London, 1969. Actually, this small book was first written in 1960. Since it treats the topic largely from the point of view of thermodynamic principles, it is both reasonably up to date and not quickly read. Try it anyhow, just to see what those principles are like.

O. A. Battista: Chromium, the metal that glitters. *Chemistry*, **42**:19 (Jan. 1969); Aluminum, featherweight champion of metals. *Chemistry*, **42**:14 (March 1969); Titanium, the Cinderella of metals. *Chemistry*, **42**:13 (May 1969). Clear, informative discussions about these three metals with some historical allusions. Recommended for easy and useful reading.

J. Skalny and K. E. Daugherty: Everything you always wanted to know about Portland cement, but did not ask. *Chemtech*, **2**:38 (1972). Exactly what the title says.

J. F. McDivitt and G. Manners: *Minerals and Men*. Johns Hopkins University Press, Baltimore, 1974. A very thorough, readable, important book, destined to be a classic (though some will disagree with that) on the exploration, development, production, processing stages and their technical, economic, political, and social considerations, of selected minerals. Absolutely must be read by any person who claims to be concerned with the future welfare of mankind. See also an article by D. B. Brooks and P. W. Andrews: *Science*, **185**:13 (1974) on the same topic.

H. Brown: Human materials production as a process in the biosphere. *Sci. Amer.*, **223**:194 (1970). In contrast with McDivitt and Manners, this author seems to take the same data (although not as extensively) and arrive at a very pessimistic conclusion about our future fate. Take your pick!

S. Brubaker: *In Command of Tomorrow*, Johns Hopkins University Press, Baltimore, 1975. To help you take your pick, try this one.

E. N. Cameron (ed.): *The Mineral Position of the United States, 1975–2000*, University of Wisconsin Press, Madison, 1973. The proceedings of a symposium, mostly attended by economic geologists, in Minneapolis in 1972. It is a review of the national mineral situation; it is timely and appropriate.

W. Alexander and A. Street: *Metals in the Service of Man*, 5th ed.

Pelican/Penguin, Baltimore, 1972. All about metals, where we get them from, how we get them, what we do to and with them. Very highly recommended!

Several years ago, Professor Erich Zimmermann wrote a book which turned out to be a classic reference on economic geography, resource economics and conservation. The first ten chapters of that work have been revised and edited by Professor Henry L. Hunker, published by Harper and Row, New York (1964); now even the revision is more than 10 years old. Look at it anyway, there are some stimulating ideas waiting to be examined. The title is *Eric W. Zimmermann's Introduction to World Resources*.

B. J. Skinner: A second iron age ahead? *Amer. Sci.*, **64**:258 (1976). Contrary to the position taken in this chapter, Dr. Skinner argues for a more severely limited availability of metallic ores, with interesting consequences.

J. Boyd, A. M. Weinberg, and D. L. Meadows: Resources and economic growth. *Wilson Quarterly*, **1**:35 (Autumn 1976). The proceedings of an "evening dialog" among a geologist, physicist, and systems analyst, accompanied by excerpts from essays by W. W. Rostow, H. C. Wallich, and E. B. Skolnikoff. No thoughtful person who is concerned for our future should miss this contribution from the first issue of what promises to be an influential quarterly.

C. Bunn: *Crystals: Their Role in Nature and Science*. Academic Press, New York, 1964. This book is addressed to natural philosophers; a description of natural phenomena, from the root of the word "crystal" to the use of crystals as gems, with a lot of excursions in between.

D. L. Anderson, G. Sammis, and T. Jordan: Composition and evolution of the mantle and core. *Science*, **171**:1103 (1971). A technical article on why we think we know a lot, but not all by any means, about the mantle and the core.

S. Moorbath: The oldest rocks and the growth of the continents. *Sci. Amer.*, **236**:92 (March 1977). If the Earth is more than 4 billion years old, why have no rocks ever been found that are as old as the Earth? A good question, with the answer in this article.

J. Sinkankas: *Gemstone and Mineral Data Book*, Macmillan, New York, 1974. Useful information for the mineralogist, with formulas and recipes for the hobbyist jeweler and collector. Sinkankas is a gifted writer and meticulous researcher; his book is invaluable. No library on mineralogy is complete if this book is missing.

J. E. Arem: *Man-Made Crystals*. Smithsonian Institution Press, Washington, D.C., 1974. Colorful illustrations, with commentary, of synthetic gems and other pretties. Informative and interesting.

C. Hall: On the history of Portland cement after 150 years. *J. Chem. Educ.*, **53**:222 (1976). A brief discourse with interesting anecdotes and 14 further references on the title topic.

chapter 13

The Chemical Industry

Almost all formal discussions of the chemical industry begin with a disclaimer to the effect that the "chemical" industry is not precisely definable. Like any other arbitrary industrial category, it is simple enough to permit clear identification of some members and complex enough to make other identifications nebulous. Thus, the manufacture of sulfuric acid is chemical manufacture, but what about soap, or paint, or drugs? These latter three are also usually classified as part of the chemical industry, but ceramics are not. Synthetic rubber is usually considered to be part of this industry, natural rubber is in a different category.

The United States Bureau of Census *Standard Industrial Classification* (S10) numbers 281 to 289, inclusive, is more widely used than any other and includes these as the major descriptive identifications:

SIC Number	Descriptor
281	Industrial inorganic and organic chemicals (sulfuric acid, ethyl alcohol)
282	Plastics materials; synthetic resins, rubber, fibers (polyethylene, nylon)
283	Drugs (aspirin, sulfathiazole)
284	Soap and detergents, perfumges, cosmetics
285	Paints, varnishes, lacquers, enamels, and allied products
286	Gum and wood chemicals (turpentine, "tall" oil, rosin)
287	Agricultural chemicals (fertilizers, insecticides)
288	Open for future developments
289	Miscellaneous chemical products (adhesives, explosives, inks, etc.)

However, whatever the chemical industry may comprise, it is not simply a list of descriptive phrases accompanied by a few examples. Other characteristics should be delineated. The chemical industry has several characteristics, some are unique to it, some are shared with a few other industries, and other characteristics are common to almost all industries. The treatment in this chapter is deliberately brief so that it will be quite clear that much is left unsaid. A thorough treatment is not possible; a more extensive discussion would necessarily omit some important details and include others that would be no more important, so an overview is appropriate here. For further, detailed information, the references cited at the end of this chapter are of course recommended.

PRODUCTS AND INDUSTRIES CONNECTED WITH THE CHEMICAL INDUSTRY

Probably the outstanding characteristic of the chemical industry is its pervasiveness. The following list of industries, businesses, services, and activities shows a fraction of those to which the chemical industry contributes goods and services, directly or indirectly: Agriculture, appliances, aviation, automobiles, cement, clothing, communication, construction, corrosion inhibitors, cosmetics, detergents, dyestuffs, elastomers, electrical, electronic, environmental control, explosives, fire prevention, fermentation, fertilizers, food preservation, food processing, footwear, fuels, furniture, leisure time, lubricants, medical products, nuclear energy, nutritional supplements, packaging, paints, paper, perfumes, personal care, pesticides, pharmaceutical materials, plastics, polymers, railroads, road building, rubber, sewage treatment, soaps, space, sports, steel, textiles, varnishes, vetinary products, water treatment, waxes, weed killers.

This pervasive characteristic is illustrated from a different point of view by a few numbers comparing the dollar sales volume of the chemical industry with chemical related industries and with all manufacturing, in Table 13–1.

TABLE 13–1. Sales by Groups of Industries, Compared to Total Assets* and Net Worth†, in the United States for 1976, in Billions of USA 1976 dollars

	Sales	Total Assets	Net Worth
The Chemical Industry:			
Drugs	14	16	10
Industrial Chemicals and Synthetics	50	46	24
All others	38	26	15
Totals	102	88	49
Chemical-related Industries:			
Petroleum and Coal Products	136	132	81
Rubber and Misc. Plastic Products	29	19	10
Primary Metal Industries	76	70	37
Stone, Clay, and Glass Products	29	22	12
Paper and Allied Products	39	31	16
Totals	309	274	156
All other manufacturing	792	491	257
All manufacturing	1203	853	462

*Sum of: current assets, investments, prepaid expenses, net plant equipment, and other tangible assets.
†Sum of: equity of all stock holders and other contingency and miscellaneous reserves not otherwise committed.

Multistep processing

The chemical industry is also characterized by the complexity of movement of tons of materials, within the industry. Stated differently, one industrial component manufactures an intermediate material from raw materials, ships it to another manufacturer who processes it further and ships it to still another, who processes it still more and so on. Finally, after a few or many such steps a finished product reaches the ultimate buyer.

Of all the items mentioned or implied previously, from agricultural products and appliances to waxes and weed killers, only a few per cent pass

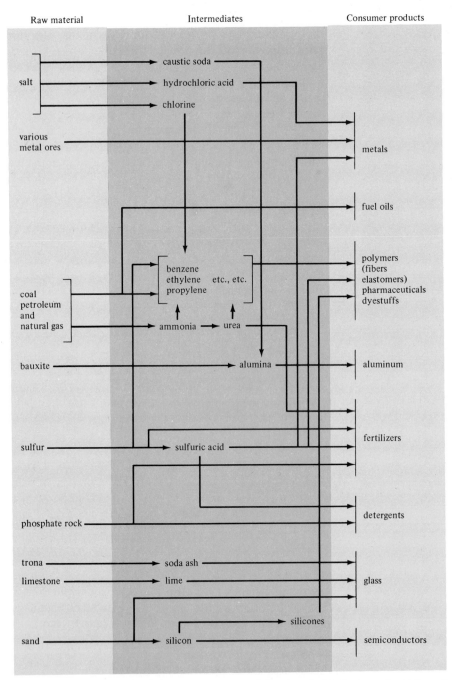

Figure 13–1. Flow of materials for a segment of the chemical industry, schematic and generalized.

through less than five or six such steps before they are finally displayed for consumer purchase and use, and many pass through dozens of intermediate steps. This complexity is indicated by Figure 13–1, which has been considerably simplified for the sake of clarity. If it were to show the complete details, even schematically, it would be illegible. A statistic demonstrates this same characteristic even more vividly. The sales volume of the chemical industry, measured in dollars, is approximately 8% of the total sales volume for all manufacturing, but the shipping costs within the chemical industry are 12% of the total for all manufacturing. That is, within the chemical industry, from one manufacturing plant to another, there is a great deal of internal shipping.

Two examples will illustrate this further. Ethylene production in the United States exceeds 25 billion lb/year. Two of the products made from ethylene are well known to everyone, polyethylene (discussed in an earlier chapter) and ethyl alcohol. There are also eight other major products made from ethylene: ethylene oxide, ethylene dichloride, ethyl benzene, acetaldehyde, ethyl chloride, ethyl dibromide, vinyl acetate, and chloroethylene. Few have heard of these eight, yet they comprise a substantial percentage of that several billions of pounds of ethylene used each year. These eight, and some others not mentioned, are intermediates; we see them only indirectly in other products. Thus, one industrial system (which may be one industrial plant or a complex) processes the ethylene oxide further; another plant or complex processes ethylene dichloride, and so on. Typically, these intermediates are further processed in a series of further steps, until the much transformed ethylene finally reaches the ultimate consumer in a finished product which might be antifreeze, a polymetric plastic, ethyl alcohol as "neutral spirits" in whiskey, or any one of dozens of other consumer products.

For the second example, most of us know that chlorine is used to make drinking water potable. This is a minor use for chlorine in terms of consumption; only 4% of the 11 million tons/year produced in the United States is used in this way. From the other 96% more than a thousand chemical intermediates are made. Their uses, which we see only indirectly, include chemicals for the manufacture of steel, aluminum, plastics, paper, synthetic fibers, food preservation and processing, dyestuffs, fabric processing, insecticides, detergents, explosives, and rocket fuels. It is evident that without these chlorine based intermediates, our lives would be lived differently; the same essential quality of industrial chlorine chemistry is illustrated by the fact that more than 35 different, independent companies, at more than 75 locations in the United States produce chlorine. It seems almost incidental to mention that chlorine is made, at those locations, from salt, for a total sales volume of $1.2 billion, annually.

Another view is presented by Table 13–2, which lists the annual production, in billions of pounds, of the top ten chemicals and of other selected chemicals. The point to notice is that most of these are intermediates; they are

TABLE 13–2.   1976 Production by the United States Chemical Industry of the Top Ten Chemicals and of Other Selected Chemicals, in Billions of Pounds

Sulfuric acid	68	Methyl alcohol	6.9
Lime	42	Ethylene oxide	5.0
Oxygen	35	Styrene-butadiene rubber	3.5
Ammonia	33	Carbon black	3.1
Ethylene	25	Ethyl alcohol	2.5
Sodium hydroxide	22	Phenol	2.3
Chlorine	22	Acetone	2.1
Nitrogen	21	Titanium dioxide	1.6
Nitric acid	16	Aniline	0.6
Sodium carbonate	16	Neoprene	0.3
Phosphoric acid	15	Butyl rubber	0.4
Benzene	12	Aspirin	0.03
Urea	8.9		

not purchased directly by the ultimate consumer, yet these are some of the major products of the chemical industry.

Although not directly evident from Table 13–2 the total (estimated) production of the chemical industry in the United States for 1976 was 480 billion lb, for a total sales volume of $100 billion. This averages at almost 21 ¢/lb. Since many chemicals cost much more than this (drugs serve as an outstanding and well known example), the selling price of many other chemicals is less than a few cents per pound. Chlorine, for example, costs not quite 6 ¢/lb.

Manufacturing complexity and variety

Another characteristic of the chemical industry is the apparent complexity of the processes that take place in any middle size to large manufacturing plant. This kind of complexity is illustrated by Figure 13–2, taken from photographs of chemical plants but with minor details slightly altered to show the complexity (large tank in the foreground deleted, an important tower structure moved out from an obscuring tangle of pipes, and so on).

Yet another characteristic of the chemical industry is its flexibility. Most of the products of this industry, whether intermediates or final products, can be made in more than one way, often by several possible routes. Each individual manufacturer chooses the route which is hoped to be the most economical for the particular circumstances. Most individual manufacturers support research and development laboratories; these are charged with the responsibility of finding further economies in the established processing steps for that particular manufacturer or of developing a new series of steps that will prove to be more economical.

The concluding characteristic of the chemical industry to be discussed is indicated by the word "innovation." Many of the products taken for granted

Figure 13–2. **Two views of a chemical manufacturing plant showing typical complexity of appearance.**

today were not in existence 10 years ago, many more were not even conceptually possible 75 years ago. Familiar examples of such products include ethyl gasoline, water based paints, cellophane, nylon, glass fiber reinforced plastics, and nonwoven fabrics. Not entirely to their credit, some members of the chemical industry have presented these developments as though they were altruistically conceived for the benefit of the consumer. Of course, as in any other industry, the return on investment under the pressure of competition was actually the primary motivating consideration. However, the consumer has benefited also, as is obvious from one glance at your surroundings compared to what it was like only a few decades ago.

THE GROWTH OF THE CHEMICAL INDUSTRY
History

A brief historical review will put this concluding characteristic of the chemical industry in perspective. Since prehistoric times man has relied on molecules with special properties to provide necessities and luxuries. Initially, all of these were natural products of plant and animal life obtained directly. Examples include cotton and wool fibers, wood, herbs and herb extracts. Other natural products from nonliving sources used in early times include clay, a few metals such as copper, zinc, and tin, and (a little later in time) iron from iron ore. By and large, most of the available natural products came from what we would today call agricultural sources.

Over the centuries, some of these natural products were reworked to produce molecules with other properties (although the processes used were not thought of in that context) that would serve useful needs. These also were largely agriculturally based and included many dyes, such as madder root and indigo, other substances such as shellac, rosin, turpentine, soap, and paper. Little by little sophisticated utilization of agricultural products increased, in degree and in kind. For example, although it does not sound sophisticated today, a very good way to bleach cotton (if you are in no hurry) is to soak the fabric in milk and let it stand, wet, in the sun, for a few days, repeating the procedure as often as necessary until the material is sufficiently white.

Within the past two centuries the importance of the agricultural base for useful molecules began to fade. Within the last century it has become rapidly less important. Instead of letting nature do it for us, man began to learn how to build or modify molecules for designed purposes, more or less independently of what happened to be available from plants and animals. Essentially, the variety of useful molecules are compounds of carbon, and it is only necessary (if you know the tricks and also know what you want to make) to find some source of carbon and construct your special molecules from that source adding other elements, such as chlorine, nitrogen, sulfur, and others, as necessary.

For the first third of this century, the source of carbon was coal. Coal is a complex substance, and different varieties of coal differ in many ways; all varieties, however, are sources of carbon for the purpose of constructing molecules with desired properties. Nylon, for example, was first made from coal, limestone, air, and water (through a series of intermediate steps, of course). More recently, petroleum has been used as the primary source of carbon for the construction of the desired molecules. Examples abound, and many have been cited in this book in earlier chapters; here, only polyethylene and synthetic drugs need be mentioned.

Current expansion

Now, how did all this happen? The brief answer alludes to the existence of research and development laboratories in the chemical industry, in the universities (neoprene was developed in the research laboratory of Rev. Julius A. Nieuwland, at Notre Dame), and in government establishments. Consider the story of neoprene as an example. We have already discussed the structure of rubber in Chapter 10; the essentials of this structure have been known for decades. Why not synthesize the same molecule, hopefully at less expense than is involved in harvesting natural rubber? Though many attempted this, none succeeded. It was conjectured that a related molecule, with one carbon atom replaced by a chlorine atom, might have properties similar enough to natural rubber to be useful. In 1929 this was finally accomplished, after more than 10 years of research work by Nieuwland. The product, then called chloroprene, was found to have many properties superior to those of natural rubber, and a few inferior (including cost) attributes. As it turned out, the higher cost was offset by the superior properties: this particular synthetic rubber was not as susceptible to ageing; it was resistant to gasoline and other solvents, and so on.

The story has been repeated many times, particularly within the chemical industry itself. With an average price of less than 21 ¢/lb, a fraction of a cent makes a difference. Any chemical offered for sale at a lower price by one manufacturer will bring that manufacturer a substantial increase in business. A new, heretofore nonexistent material, perhaps costing a little more but with improved properties, will have a similar effect upon sales volume. As sales of the new product increase, the manufacturer's capacity becomes insufficient to meet the demand. A new and larger plant is built once present sales and future expectations can justify the very high investment. Almost always, a plant with larger capacity can produce the product at lower unit cost. As the initial investment is paid off, the price falls, and sales increase still more. Thus the cycle continues.

Meanwhile, competitors within the chemical industry attempt to find ways to make the same product or another similar product at suitably low cost. This happened to neoprene. At one time it was the only synthetic elastomer; a glance at Table 13–2 shows that today, less than 50 years after neoprene was first synthesized, styrene-butadiene rubber is produced in ten times the quantity and butyl rubber in quantities equal to that of neoprene. In either event, a competitor with a lower price for the same product or with a reasonable price for a new product with similar or related properties often initiates another cyclic construction of larger manufacturing capacity, lower unit price, still larger plant capacity, and so on. Eventually, the product and its related competitors saturate the market and dramatic expansion no longer occurs. Most of the chemicals identified in Table 13–2 are in this category; ethylene is perhaps the only one in that list which is currently in a cyclic increasing stage, and it may well be nearing the end of that process.

There are other reasons for the growth of the chemical industry, of course. Nitrogen and nitric acid, sulfuric acid and phosphoric acid, and urea, all listed in Table 13–2, are directly or indirectly involved in fertilizer manufacture. In these times, fertilizer demand is up and steadily growing. So these chemicals will experience an increased production for reasons quite different from those that apply to ethylene. As with any article of commerce, once all the major uses have been identified and have become familiar to the consumer, growth occurs largely either by lowering the price, thus initiating uses that were formerly marginal or by increasing the number of customers.

There is, in the foreseeable future, a significant sign of decline or restructuring of the chemical industry that must be mentioned. We pointed out in chapter 8 that the end of the petroleum supply is foreseeable. Total depletion may be postponed if more petroleum is allocated for the exclusive use of the chemical industry as the source of the carbon for molecular design. But, eventually, the end will come. When it does, the chemical industry will be forced back, so to speak, to the utilization of either coal, which is also limited, or to agricultural sources, which is more or less where we came in.

FOR FURTHER INFORMATION

ACS Committee on Chemistry and Public Affairs, *Chemistry in the Economy*. American Chemical Society, Washington, D.C., 1973. A summary of the role of science, especially chemistry, in meeting human needs for clothing, shelter, food, health, energy transportation, and communication. Authorative and unbiased. Probably a classic in the published literature.

B. G. Reuben and M. L. Burstall: *The Chemical Economy*. Longman, New York, 1974. On the chemical industry, with a broad social and economic perspective to help nonspecialists see how the chemical industry works. Heavy going in some places; uses British examples for the most part, but applicable elsewhere. Also see F. R. Bradbury and B. G. Dutton: *Chemical Industry*, Butterworths, London, 1972.

For the chemical economy on a worldwide basis, see Anon.: The world chemical economy. *Chem. Eng. News*, April 16, 1973, pp. 19–64, *The Chemical Industry of Western Europe. ECN Chemical Data Services*, London, 1974; and N. Platzer: World Trade in chemical technology. *Chemtech*, **2**:402 (July 1972).

B. J. Luberoff: The business of chemistry. *Chemistry*, **40**: 8 (Sept. 1967). A brief introduction to the topic, illustrated with examples.

K. M. Reese: Economics in the chemical industry, parts I and II, *J. Chem. Educ.*, **46**:725 and 827 (1969). Written primarily for teachers, covering the title topic in more depth and sophistication and with equal clarity, compared to Luberoff.

A. J. Paradiso: The marketing of dyestuffs, present and future. *Chemtech*, **1**:292 (May 1971). A description of the chemical industry using only one small but important part of that industry as the example, with emphasis on marketing rather than on production.

J. Backman: *The economics of the chemical industry*. Manufacturing Chemists Assn., Washington, D.C., 1970. A compendium of facts, presented in graphs, tables, and clipped sentences. Complete, thorough, useful, and boring.

F. R. Bradbury and B. G. Dutton: *Chemical Industry, Social and Economic Aspects*. Butterworths, London, 1972. A lively picture of the social and economic significance of the chemical industry; the practice and problems met by the people who work in this industry. Recommended.

Anon: *Sources and Production Economics of Chemical Products*, Chemical Engineering McGraw-Hill, New York, 1974. This work is the first of a series. It now appears under the title *Financial Summary, Chemical Week 300,* and is published annually near the end of each year. It is written for the people in the chemical industry and is filled with facts related to the economics of production, with some data for some products covering the preceding ten years. An interesting publication, but it helps to know a bit about the subject before attempting to interpret the information that is provided.

For two insights into the operation of the chemical industry, see Kline, C.: Maximizing profits in chemicals. *Chemtech*, **6**:110 (Feb. 1976); and V. D.

Herbert, and A. Bisio: The risk and the benefit. *Chemtech*, **6**:174 (March 1976).

E. V. Anderson: Consumer chemical specialties. *Chem. Eng. News*, Dec. 1, 1969, pp. 37–64. A readable, well-written, authoratative, article on the chemical industry and freezer wrap, aluminum foil, antifreeze, waxes and polishes, you name it.

J. M. Gould and F. C. Katznelson: Input/output techniques in the CPI. *Chemtech*, **2**:405 (July 1972). "CPI" means "Chemical Process Industry"; "Input/output" means a grid consisting of horizontal rows, for sellers, and vertical rows, for buyers. Row intersections are filled with data signifying a supplier customer relationship; and so on. Technical, interesting to economists and managers; read it for yourself.

E. F. Schumacher: *Small is Beautiful.* Harper and Row, New York, 1973. The subtitle is "Economics as if people mattered." The content suggests that man has lost his sense of values in his involvement with technology and the time has arrived for us all to pay the price of this folly. Schumacher then suggests how this might be done at least cost and inconvience. Maybe so, maybe not; decide for yourself.

R. G. Ridker: To grow or not to grow: that's not the relevant question. *Science*, **182**:1315 (1973). In a way, an answer to Schumacher. We cannot afford to stop the ship, now surrounded by fog; the poorer majority of the world cannot wait, especially since it is not at all certain that what we might decide to do after waiting would be any wiser than what we might decide to do, now, instead of waiting.

Committee on Science and Public Policy, NAS, *Materials and Man's Needs*, National Academy of Sciences, Washington, D.C., 1974. A report outlining the place of materials, of science and of technology in a changing context, with 24 recommendations for action. Probably a milestone document.

V. K. Smith: *Technical Change, Relative Prices, and Environmental Resource Evaluation.* Johns Hopkins University Press, Baltimore, 1974. Presents an analytical-economic framework to interrelate the relative price behavior of amenity services of natural areas to measurable parameters, such as technical change rates, demands for goods, and substitutes for these.

These two articles form an interesting contrast. The first is a general look, written by H. H. Szmant: Science and the economy. *Chemtech*, **6**:152 (1976). The second is a specific look at a single class of product, by E. E. Magat and R. E. Morrison: The evolution of man-made fibers. *Chemtech*, **6**:702 (1976).

G. T. Seaborg: Chemistry—key to our progress. *Chem. Eng. News*, **31** (Sept. 13, 1976). A thorough, readable, and concise description of the progress of academic, research, and industrial chemistry during the first hundred years of the existence of the United States. The emphasis is on the organized application of chemistry to fill human needs.

J. Phelan and R. Pozen: *The Company State*, Grossman, New York, 1973. A Nader report on DuPont, in Delaware. Not totally objective, this book is a study of one well-known chemical company. However, all in all, a distinct attempt to tell it like it is fairly and forthrightly.

Anon.; *MCA*, 1872–1972; *A Centennial History*. Manufacturing Chemists Association, Washington, D.C., 1972. A celebrative document which describes the historical development, accurately but briefly, of the chemical industry in the United States over the past century.

J. H. Saunders: *Careers in Industrial Research and Development*. Marcel Dekker, New York, 1974. This work was written for beginning chemists to inform them of what life is like after they graduate; it also discloses to others what it is like to be a chemist in industry.

There are several references to consult that should be mentioned: "Facts and Figures for the Chemical Industry" and annual feature in *Chemical and Engineering News*, usually appearing in one issue of this weekly publication in May or June. Many of the data cited in this chapter were taken from or derived from this source. "Chemical Week 300" appears annually in the weekly publication, *Chemical Week*. The data in *Chemical Week* are the same as in *Chemical and Engineering News* (since both come from the same primary sources) but *Chemical Week*'s treatment is more diversified in some respects. For a well-rounded annual picture, both sources are helpful. Two compendia must be noted: the *Encyclopedia of Chemical Technology*, Wiley, New York, and the *Kirk–Othmer Encyclopedia of Chemical Technology*, McGraw-Hill, New York. Both of these works are authoratative, useful, and informative; part V of *Basic Organic Chemistry* by J. M. Tedder, Wiley, New York, 1975 is a minature version of these compendia.

In addition, several other sources of information should be at least mentioned in this general context. These are price/cost indexes, such as *Chemical Marketing Reporter*, a weekly publication; statistical compilations, such as the *United Nations Statistical Yearbook*, packed with data; buyer's guides, such as *Chemical Week, Buyer's Guide Issue*, every October, McGraw-Hill, New York, with interesting advertisements that complicate the extraction of desired information (unless that information is in an ad); and trade publications, for which see *Scientific, Technical and Engineering Societies Publications in Print 1974–1975* by J. M. Kyed and J. M. Matarazzo (R. R. Bowker, New York, 1974) and *Encyclopedia of Associations*, 9th ed., by Margaret Fisk (Gale Research, Detroit, 1975). These references plus several more are described and some of their uses outlined by Barbara Lawrence: *Chemtech*, **5**:678 (1975).

It is usually interesting to get behind the screens, so to speak, and learn what is really happening. For this, you will find the *Chemical Safety Data Sheets* helpful. There are more than 100 of these 15 to 40 page books, each one dealing with the procedures used throughout the industry for handling a particular chemical, all the way from acrolein to vinyl acetate with hydrogen sulfide, sodium cyanide, and several others alphabetically in between. They are published at intervals by the Manufacturing Chemists Association, Washington, D.C.

W. J. Reader: *Imperial Chemical Industries. A History*. Oxford University Press, Vol. 1, New York, 1970; Vol. 2, New York, 1975. Multinational corporations will have an increasing influence on economic developments throughout the world. This history is an accurate and complete account of one of perhaps the dozen largest non-American corporations in the world. It is also comprehensible and interesting.

For a quick but thorough introductory treatment of industrial chemistry, see *Industrial Horizons* by Margaret Walker, edited by D. J. Daniels (Bath University Press, Bath, 1974). This interesting work consists of five small booklets plus three maps. It begins with the chemistry of oil refining and ends with the social aspects of the siting of a refinery, treating history, economics, and geography in between. The collection makes it quite clear that for any component of the chemical industry a multitude of factors are involved and a few of these are chemical. Highly recommended.

All industry deals with materials, one way or another. A comprehensive look at materials, from policy considerations, energy and the environment, to resources now and in the future, as well as other topics, is presented in the 20 February, 1976, issue of *Science*, vol. 191. Highly recommended.

chapter 14

Chemistry, Pollution, and Environment

In general for each of the other chapters in this book it has been possible to focus on a single theme, addressing some of the implications and applications showing their chemical relationship to matters that interest those who are not primarily interested in chemistry, as such. However, the topic of this chapter is so pervasive and so varied in its details as to preclude even an introductory treatment of these details. Instead, your attention is invited to several themes and to the annotated bibliography at the end of this chapter. As you examine one or more of the cited references or other references on the same topic, look for explicit and implicit presentations of one of the themes identified in the following paragraphs. Additionally, since there are several other applicable points of view, try to identify other thematic foci in the references you choose to read.

In a very real sense, this chapter is the most important chapter in this book. Paradoxically, because of this it is not possible for an author to lead the interested reader through the content to be encompassed. Instead, each must find his/her own path, guided by his/her own comprehension; an author can only say, "There it is; this is a crude and not guaranteed accurate map; it is your task during the years to come to construct a more useful diagram of the roads and detours and swamps and difficult terrain." To emphasize the point that now you are indeed on your own, I have deliberately turned to other chemistry colleagues for the preparation of the rough map. As you use what you choose of their annotated bibliography, keep the following themes in mind:

THEMES IN ECOLOGY

1. The only physical resources we have come from the crust of the Earth, the water in rain and snow, lakes, rivers, oceans, and glaciers, the air in our atmosphere, and radiant energy from the Sun. Some of these resources are finite and limited, such as crude oil and coal; some are finite but not limited in the same sense, such as silicon dioxide and solar energy; some are renewable, involving the use of energy to renew them, such as fresh water and cellulose.

2. A relatively small, undramatic, event can produce awesome consequences. Two examples: A slight change in the temperature of a lake or shallow ocean waters can result in significant increases or decreases (depending upon the needs of various aquatic species) in the harvesting of fish and other hard and soft shell organisms. The relatively small amount of DDT which by now is widely distributed over the surface of the Earth has markedly altered the balance of life in many areas.

3. Different systems are not independent. Catalytic converters for automobile exhaust systems have increased the amount of sulfur dioxide in the air. This sulfur dioxide, along with sulfur dioxide from other sources, is conveyed by wind currents from North America to Scandanavia where it adversely affects the production of timber, grown there as an important national resource. As the amount of carbon dioxide increases in the atmosphere, from various sources such as living animals and the burning of fuel, some is taken up by the waters of the oceans, ultimately precipitating largely as calcium carbonate; later, should the amount of carbon dioxide in the atmosphere decrease, some is released (slowly) from the calcium carbonate deposits.

4. The Earth, air, and water constitute a "sink" of limited capacity. Continued discharge of unusual contaminants, planned or accidental or incidental to some other necessary activities, can potentially destroy or seriously

alter a system that is necessary or at least beneficial to life on this Earth. The case of DDT has been mentioned. Discharges of phosphate and other substances into lakes and rivers are yet another.

There is no need to view these themes pessimistically, although their portent, should we ignore them, is indeed pessimistic. The fact that we have become more aware of these concepts during the past few decades is in itself an optimistic sign. Further, the four remaining themes are in themselves hopeful in their scope.

5. In principle, almost anything can be recycled and reused, over and over again. Except for a bit of hydrogen from the dissociation of water vapor by the Sun, high in the atmosphere, no significant amount of substance is removed from the whole Earth system. The atoms of carbon and hydrogen originally in the form of available crude oil are still largely present somewhere within reach. To return them to their original condition as useful hydrocarbon molecules, should it be found desirable to do so, requires only the expenditure of energy and lots of imagination and ingenuity.

The restrictions on recycling everything are economic. The energy we would need is plentifully available in solar energy. Its inaccessibility resides in the present high cost of making it accessible. Ultimately, doubtless, the cost will become reasonable by the application of knowledge that we now possess, but cannot recognize as applicable, or that will be acquired. Currently, aluminum is being successfully recycled in very large amounts. Scrap iron and steel, on the other hand, is only partially recycled. The reason for the difference is cost. It is cheaper to recycle aluminum than it is to obtain it from its ores; this is not the case for some instances in the steel industry. Some would suggest, in consequence, that our economic priorities are wrong in the case of iron and steel, that we are not now paying the real cost for iron ore and its processing into steel.

Continuing with this fifth theme, there are other economic (and sociological) consequences when one insists on strict application of ecological principles. In a sense, this involves the third theme, on dependent interacting systems. Thus, vinyl chloride is a known carcinogen; workers in the manufacture of polyvinyl chloride have been exposed to this substance, and the end results have been fatal. Yet, if the manufacture of this polymer were stopped by legislative or other action, thousands of people would be immediately unemployed. The economy of the market place would be seriously disrupted. For these kinds of reasons, it is not desirable to stop making polyvinyl chloride. Fortunately, faced with this dilemma, a solution has been found: Alter the manufacturing process so as to all but completely prevent the exposure of the workers to vinyl chloride in the manufacture of the polymer. The alteration of the manufacturing process is costly; the price you and I pay for garden hose, raincoats, phonograph records, floor tile, and many other products have increased accordingly.

6. Specific treatments are possible is our next theme. Symptoms of pica serve as an example. A person afflicted with pica eats unusual substances such as dirt or paint flakes. Some flaking paint contains lead, a severe and cumulative poison. The solution is to repaint flaking surfaces with paint that does not contain lead and to prohibit by legislation the further manufacture and sale of paint that does contain lead. Again, there is a cost to the consumer. Lead containing paints are much more durable, when properly applied, than other paints. Now we must repaint more often. The example cited for minimizing workers' exposure to vinyl chloride is another example of applying a specific treatment.

7. The final theme is perhaps the most hopeful of all. Substitutions are possible. Sometimes at lower, sometimes at higher costs. Plastics can in many instances be substituted for scarce metals; many small parts in a modern automobile, for example, that were formerly made of metal are now made of plastics. Of course, since plastics or polymers are made from crude oil as the primary raw material, this may not seem like much of an improvement. There are two factors to keep in mind: First, if all of the crude oil now left were reserved exclusively for the manufacture of polymers, there would be plenty of crude oil until the unforeseeable future. Second, since this exclusive utilization of crude oil is most unlikely, it is reassuring to note that we have substitutes for crude oil as the raw material: coal, wood, corn cobs, almost any plant growth (in principle), can be used instead.

INTRODUCTION TO THE BIBLIOGRAPHY The bibliography that follows has been compiled by Professor John W. Moore and Elizabeth A. Moore.* It is divided into eight sections, the first on General References, followed by these major topic headings:

> Energy-related pollution problems
> Air pollution
> Resources and recycling
> Water pollution
> Pollution and health
> Population
> Science, philosophy, and environment

References are numbered sequentially; in some cases several closely related references are cited under the same numerical entry. Entries are differentiated

---

* Some of the entries in the bibliography are taken from J. W. Moore and E. A. Moore: *J. Chem. Educ.*, **52**:288 (1975); **53**:167 and 240 (1976); and from the text by the same authors, *Environmental Chemistry*, Academic Press, New York, 1976. The kind permission of the copyright owners to reproduce that material here is gratefully acknowledged.

by small capital letters, A, B, and C shown in square brackets according to the following scheme:

> A—of general interest and relatively easy reading. Should probably be consulted first before going on to other references under the same heading.
>
> B—more specialized but still readily readable. Gives specific information on a particular problem or problems. Should probably be consulted second.
>
> C—more difficult. Usually quite worthwhile but may require some effort to understand completely. Best consulted after other references in a category have been studied.

In other words, if the references are read in order A, B, C one ought to be able to develop a quite detailed knowledge of any of the topic in the bibliography.

GENERAL REFERENCES

Except for the first (a relatively short article), the next to last (a comprehensive survey) and the last (not yet available), the references cited in this section are all either textbooks or collections of articles that give a broad general coverage of the entire field of environmental chemistry.

1. D. R. Brower: The third planet: operating instructions. *New York Times Magazine*, March 18, 1975. A concise description of pollution and energy problems, in a clever scenario [A].

2. R. E. Lapp: *The Logarithmic Century*. Prentice-Hall, Englewood Cliffs, N.J., 1973. The dramatic use of graphs and a readable text give an excellent overall view of current problems [A].

3. J. C. Giddings and M. B. Monroe (eds.); *One Chemical Environment*. Canfield Press, Harper and Row, New York, 1972. A selection of previously published articles encompassing all of the subjects of environmental pollution, including that portion which is energy related [B].

4. H. S. Stoker and S. L. Seager: *Environmental Chemistry: Air and Water Pollution*. Scott-Foresman, Glenview, Ill., 1972. Provides good treatment of the two areas covered [B].

5. L. Hodges: *Environmental Pollution. A Survey Emphasizing Physical and Chemical Principles*. Holt, Rinehart and Winston, New York, 1973. A survey covering most of the areas of this bibliography, with the exception of energy [C].

6. S. A. Manahan, *Environmental Chemistry*. Willard Grant Press, Boston, 1972. Provides a good treatment of air and expecially water pollution; also one chapter devoted to soil chemistry [C].

7. J. W. Moore and E. A. Moore: *Environmental Chemistry*. Academic Press, New York, 1976. Provides thorough coverage of all the areas in this bibliography [C].

8. W. H. Matthews, F. E. Smith, and E. D. Goldberg (eds.): *Man's Impact on Terrestrial and Oceanic Ecosystems*. The MIT Press, Cambridge, Mass., 1971. A comprehensive collection of well-written papers [B].

9. N. I. Sax (ed.): *Industrial Pollution*. Van Nostrand Reinhold Co., New York, 1974. This is a collection of articles by experts on a broad range of environmental topics. It contains many facts, presents problems clearly, and suggests solutions in a very readable way [C].

10. W. W. Lowrance: *Of Acceptable Risk*: *Science and the Determination of Safety*. Wm. Kaufmann, Los Altos, Calif., 1976. On the issue of risks that are necessary and the roles of scientists and others in evaluating the factors that determine, politically and scientifically, which risks and how much to accept and which to reject in favor of alternative actions [C].

In the reorganization of the U.S. Senate in 1977 several new committees were formed. Among these are three with these new names: Committee on Commerce, Science, and Transportation, Committee on Energy and Natural Resources, and the Committee on Environment and Public Works. To be well informed on many of the topics identified in this chapter, it will be useful to refer to records of the proceedings of these three Committees as they become available, Probably A and B.

ENERGY-RELATED POLLUTION PROBLEMS

Most of the energy resources currently being consumed in the United States and throughout the world are in the form of fossil fuels—coal, petroleum, and natural gas. A great many environmental problems are the direct result of burning of these fuels. On a worldwide basis, combustion of wood still outpaced consumption of nuclear fuel in 1974, but the special, long-term hazards of nuclear power have drawn much scientific and public attention. Should the world eventually come to depend on nuclear fission or fusion power, energy storage and conversion to other forms than electrical will become even more important than they are now. We have included several references to systems based on synthetic fuels (H_2 and CH_3OH) and electrochemical cells. Each of the references in the general category at the beginning of this section gives an excellent overview of energy-environment problems.

General

11. For information about the environmental effects of energy usage over a broad range, the following may be useful: (a) G. T. Miller: *Energy and the Environment*: *Four Energy Crises*. Wadsworth Publishing Co., Belmont, Calif. 1975 [A]; (b) H. S. Stoker, S. L. Seager, and R. L. Capener: *Energy*: *From Source to Use*. Scott, Foresman and Co., Glenview, Ill., 1975 [B]; (c) C. E. Steinhart and J. S. Steinhart: *Energy. Sources, Use and Role in Human Affairs*. Duxbury Press, Belmont, Calif., 1974 [B].

The effects of fossil fuels— coal

12. For information about the possible environmental effects of coal mining, see: (a) Shall we strip mine Iowa and Illinois to air condition New York? *Atlantic*, **232**:84 (Sept. 1973) [A]; (b) two books by H. Caudill tell in a moving way of the effects of coal mining on Appalachia: *Night Comes to*

the Cumberlands, Little Brown, and Co., Boston, 1963 [A]; and *My Land is Dying*, E. P. Dutton, New York, 1971 [A]; (c) L. A. Sagan: Health costs associated with the mining, transport, and combustion of coal in the steam-electric industry. *Nature*, **250**:107 (1974) [C].

13. Acid-mine drainage resulting from coal mining: (a) J. E. Biesecker and J. R. George: Stream quality in Appalachia as related to coal-mine drainage, 1965. In W. A. Pettyjohn (ed.): *Water Quality in a Stressed Environment*. Burgess Publishing Co., Minneapolis, 1972, and the references contained therein [B]; (b) Ion exchangers sweeten acid water. *Environ. Sci. Technol.*, **5**:25 (1971) [B]; (c) T. Aaronson: Problems underfoot. *Environment*, **12**(9):17 (1970) [A]; see also chapter 8 of reference 11(b).

14. Gasification and other developments in coal research: (a) Harry Perry: The gasification of coal. *Sci. Amer.* **230**:19 (March 1974) [B]; (b) J. P. Henry, Jr. and B. M. Louks: An economic study of pipeline gas production from coal. *Chemtech*, **1**:238 (1971) [C]; (c) G. R. Hill: Some aspects of coal research. *Chemtech*, **2**:292 (1972), both liquefaction and gasification of coal are discussed in this article [C]; (d) G. A. Mills: Gas from coal—fuel of the future. *Environ. Sci. Technol.*, **5**:1178 (1971) [B]; (e) E. F. Osborn: Coal and the present energy situation. *Science*, **183**:477 (1974), a good summary of the ways in which coal can help alleviate the current energy crisis [B]; (f) A. M. Squires: Clean power from dirty fuels. *Sci. Amer.*, **227**:26 (Oct. 1972) [B] and clean power from coal. *Science*, **169**:821 (1970) [C].

15. About power plants: (a) D. E. Abrahamson: *The Environmental Cost of Electric Power*, Scientists' Institute for Public Information, Washington, D.C. 1970 [A]; (b) Energy and the environment—electric power. Council on Environmental Quality, Washington, D.C. 1973 [B]; (c) D. L. Scott: *Pollution in the Electric Power Industry, Its Control and Costs*. D. C. Heath, Lexington, Mass., 1973); written by an economist, this small volume gives a good discussion of balancing costs and benefits [B]; (d) L. B. Young and H. P. Young: Pollution by electrical transmission. *Bull. Atom. Sci.*, **30**(10)35 (1974) [B]; (e) Magnetohydrodynamics. *Chemistry*, **46**:21 (June 1973), generation of electrical current by passage of partially ionized combustion gases through a magnetic field may permit power plant efficiencies as high as 50% [A].

16. Air pollution from power plants: (a) *Man's Impact on the Global Environment. Assessment and Recommendations for Action*, Report of the Study of Critical Environmental Problems, sponsored by MIT, MIT Press, Cambridge, Mass., 1970 [B]; (b) W. W. Kellogg, et al.: The sulfur cycle. *Science*, **175**:587 (1972) [B]; (c) for information on acid rain see refs. 32 and 33 in the section on air pollution; (d) D. H. Klein and P. Russell: Heavy metals: fallout around a power plant. *Environ. sci. technol.*, **7**:357 (1973); the metals discussed are: Ag, Ca, Co, Cr, Cu, Fe, Hg, Ni, Ti, and Zn [B]; (e) K. K. Bertine and E. D. Goldberg: Fossil fuel combustion and the major sedimentary cycle. *Science*, **173**:233 (1971); combustion of

fossil fuels can mobilize many elements into the atmosphere and waters potentially at rates comparable to natural processes [C]; (f) O. I. Joensuu: Fossil fuels as a source of mercury pollution. *Science*, **172**:1027 (1971) [B]; (g) C. E. Billings and W. R. Matson: Mercury emissions from coal combustion. *Science*, **176**:1232 (1972) [B].

17. Waste heat from power plants: (a) Thermal discharges: ecological effects. *Environ. Sci. Technol.*, **6**:224 (1972) [B]; (b) R. D. Woodson: Cooling towers. *Sci. Amer.*, **224**:70 (May 1971) [B]; (c) D. Furlong: The cooling tower business today. *Environ. Sci. Technol.*, **8**:712 (1974) [B]; (d) The following are similar articles in that both deal with waste heat released into the Connecticut River by a (nuclear) power plant, but they come to opposite conclusions. J. R. Clark: Thermal pollution and aquatic life. *Sci. Amer.*, **220**:19 (March 1969) [A] and D. Merriman: The calefaction of a river. *Sci. Amer.*, **222**:42 (May 1970) [B]; (e) Temperature and aquatic life. *Laboratory Investigations Series # 6*, Federal Water Pollution Control Admin., U.S. Department of the Interior, Cincinnati, Ohio, 1967. The succinct chapter on chemical reactions is especially useful; also contains references and an extensive selected bibliography [B].

The effects of fossil fuels— petroleum and natural gas

18. The use of petroleum for transportation: (a) Eric Hirst: Transportation energy use and conservation potential. *Bull. Atom. Sci.*, **29**(9):36 (1973) [B]; (b) U.S. Environmental Protection Agency, *A Report on Automotive Fuel Economy*. February, 1974. Excellent, well-written, concise, full of useful data, free [A].

19. Alternative sources of power for automobiles: (a) R. U. Ayres and R. P. McKenna: *Alternatives to the Internal Combustion Engine. Impacts on Environmental Quality*, Resources for the Future, Johns Hopkins University Press, Baltimore, 1972 [C]; (b) A series of articles in *Road and Track*, a publication aimed at automobile enthusiasts, reports on the alternatives to the internal combustion engine. They are very well written and technically correct. [A]. (1) Principles and promises of the Wankel (February 1971); (2) The gas turbine (April 1971); (3) Honda's new CVCC engine (February 1973); (4) Stirling engine (March 1973); (5) Emission standards finalized (July 1973); and (6) The diesel engine (September, 1973); (c) G. Walker: *Stirling-Cycle Machines*. Clarendon Press, Oxford, 1973 [C]; also a briefer treatment by the same author, The Stirling engine. *Sci. Amer.* **229**: 80 (Aug. 1973) [B]; (d) D. E. Cole: The Wankel engine. *Sci. Amer.* **227**:14 (Aug. 1972) [B].

20. About oil and/or natural gas (LNG) spills: (a) C. E. Steinhart and J. S. Steinhart: *Blowout, A Case Study of the Santa Barbara Oil Spill*. Duxbury Press, Belmont, Calif., 1972 [A]; (b) What happens when LNG spills? *Chemtech* **2**:210 (1972) [C]; (c) J. McCaull: The black tide. *Environment*, **11**(9):2 (1969) [A]; (d) Oil spills: An environmental threat. *Environ. Sci. Technol.*, **4**:97 (1970) [B]; (e) N. Mostert: *Supership*. Alfred A. Knopf, New York, 1974 [A].

*Possible
environmental
problems from
the use of
nuclear energy*

21. Information about nuclear safety, ethics, and nuclear wastes: (a) A. M. Weinberg: Social institutions and nuclear energy. *Science*, **177**:27 (1972) [B]; (b) W. H. Jordan: Nuclear energy: benefits versus risks, *Phys. Today*, May 1970, p. 32 [B]; (c) E. C. Tsivoglou: Nuclear power: the social conflict, *Environ. Sci. Technol.*, **5**:404 (1971) [B]; (d) L. A. Sagan: Human cost of nuclear power. *Science*, **177**:487 (1972) [B]; (e) A series of articles about the safety of nuclear reactors appeared in *Science* between 1971 and 1973 [B]; *Science*, **172**:918; **176**: 492; **177**:771, 867, 970, 1080; **178**:482; **179**:360; (f) S. Novick: *The Careless Atom*, Houghton Mifflin, Boston, 1969 [A]; (g) M. Eisenbud: *Environmental Radioactivity*. 2nd ed. Academic Press, New York, 1973. A classic covering all aspects of the topic [C]; (h) G. Bethlendy: Environmental aspects of nuclear power stations. *Environ. Lett.* **4**:151 (1973). A readable and thorough account [B].

*Energy storage
and conversion*

22. The hydrogen economy: (a) D. P. Gregory: The hydrogen economy. *Sci. Amer.*, **228**:13 (Jan. 1973) [B]; (b) W. E. Winsche et al.: Hydrogen: its future role in the nation's economy. *Science*, **180**:1325 (1973) [C]; (c) L. W. Jones: Liquid hydrogen as a fuel for the future. *Science*, **174**:367 (1971) [C].

23. Methanol as a fuel: (a) T. B. Reed and R. M. Lerner: Methanol: a versatile fuel for immediate use. *Science*, **182**:1299 (1973) [B]; (b) E. E. Wigg: Methanol as a gasoline extender: a critique. *Science*, **186**:785 (1974) [B].

24. Electrochemical cells: (a) New batteries. *Chemtech*, **1**, 487 (1971) [B]; (b) A. B. Hart and G. J. Womack: *Fuel Cells*, Chapman and Hall, Ltd., London, 1967 [B]; (c) J. O'M. Bockris and Z. Nagy, *Electrochemistry for Ecologists*. Plenum Press, New York, 1974 [B].

25. Other means of energy transmission and storage: (a) Mechanochemical energy conversion. *J. Chem. Educ.*, **50**:753 (1973) [C]; (b) R. F. Post and S. F. Post: Flywheels. *Sci. Amer.*, **229**:17 (Dec. 1973) [B]; (c) G. Y. Eastman, The heat pipe. *Sci. Amer.*, **218**:38 (May 1968) [B].

**AIR
POLLUTION**

That air pollution has been a problem for a lot longer than one might expect· can be seen by reading reference 26. The U.S. Environmental Protection Agency (EPA) has divided the problem into five major categories: sulfur oxides, particulates, nitrogen oxide, photochemical smog and oxidants, and carbon monoxide. In addition the EPA is concerned with pollution of the stratosphere (a region that does not readily cleanse itself) and the emission of trace substances, especially heavy metals, into the atmosphere. Information on these problems will be found in the references that follow.

*Air pollution
in general*

26. An early history and discussion of air pollution problems can be found in J. P. Lodge, Jr. (ed.): *The Smoake of London*, Maxwell Reprint Co., Elmsford, N.Y., 1969, [A]. It contains a reprint of John Evelyn's *Fumifugium*, (written in 1661) and of Robert Barr's *The Doom of London*.

27. A general, overall treatment of all aspects of air pollution, at the introductory level: (a) E. Inglauer: The Ambient air, *New Yorker*, April 13, 1968, [A]; (b) V. Brodine: *Air Pollution*. Harcourt Brace Jovanovich, New York, 1973, [A]; (c) D. A. O'Sullivan: Air pollution, a *C & E News* special report, *Chem. Eng. News*, June 8, 1970, [B]; (d) N. Hinch: Air pollution, *J. Chem. Educ.*, **46**:93 (1969) [B].

28. Information about the economic aspects of air pollution can be found in U.S. Environmental Protection Agency, *The Economic Damages of Air Pollution*, *EPA*-600/5-74-012, U.S. Government Printing Office, Washington, D.C., 1974, [B].

29. A general treatment of air pollution, but with some technical information in addition to descriptive material: (a) R. E. Newell: The global circulation of atmospheric pollutants. *Sci. Amer.*, **224**:32 (Jan. 1971) [B]; (b) W. Bach: *Atmospheric Pollution*, McGraw-Hill, New York, 1972. This book, written by a geographer, contains a good discussion of meteorology, health effects, economic aspects, technology and control, and public action relating to air pollution [B]; (c) *Air Conservation*. American Association for the Advancement of Science, Washington, D.C., 1965. A classic, basic reference despite its age [B]; (d) A. T. Rossano, Jr. (ed.): *Air Pollution Control. Guidebook for Management*. Environmental Science Service Div., ERA, Inc., Wilton, Conn. This covers the chemical, biologic, engineering, meteorologic, law and administrative aspects of air pollution. It is written in a practical, straightforward manner [B]; (e) *Cleaning Our Environment*. American Chemical Society, Washington, D.C. 1969, contains much useful information [B]; (f) W. H. Matthews, W. W. Kellogg and G. D. Robinson (eds.), *Man's Impact on the Climate*. MIT Press, Cambridge, Mass., 1971. Prepared from background materials for MIT's Study of Critical Environmental Problems, 1970 [B].

The atmosphere and the stratosphere's importance

30. About the atmosphere: (a) H. J. Sanders: Chemistry and the atmosphere. *Chem. Eng. News*, Special Report, 1967 [B]; (b) H. E. Landsberg: Man-made climatic changes. *Science*, **170**:1265 (1970) [B]; P. V. Hobbs, H. Harrison and E. Robinson: Atmospheric effects of pollutants. *Science*, **183**:909 (1974) [B]; (c) R. M. Garrels, A. Lerman, and F. T. Mackenzie: Controls of atmospheric O_2 and CO_2: past, present, and future. *Amer. Sci.*, **64**:306 (1976) [B].

31. The importance of the stratosphere in protecting life on earth: (a) A. K. Ahmed, Unshielding the sun by diminishing ozone . . . Human effects. *Environment*, **17**(3):6 (1975) [B]; and J. Eigner; Unshielding the sun . . . Environmental effects. *Environment*, **17**(3):15 (1975) [B]. The first of these articles is a thorough and readable treatment of causes of ozone depletion and the possible resulting effects, especially on humans; the author is a staff scientist with the Natural Resources Defense Council. The second article describes possible effects on plants and animals; (b) Climatic Impact Committee, National Research Council/National Academy of Sciences: *Environmental Impact of Stratospheric Flight*. National Academy of Sciences, Washington, D.C., 1975. This report is

up-to-date, thorough, readable and has an excellent summary of findings [C]. A more concise summary also appears as, Jetliners in the stratosphere: concerns for consequences. *News Report* (of the National Academy of Sciences) XXV, no. 3 (March 1975) [B]; (c) A. L. Hammond and T. H. Maugh, II, Stratospheric pollution: Multiple threat to Earth's ozone. Research News, *Science*, **186**:335 (1974) [B].

Sulfur oxides

32. About sulfur compounds in the atmosphere: (a) U.S. Environmental Protection Agency, *Air Quality Criteria for Sulfur Oxides*, AP-50, U.S. Government Printing Office, Washington, D.C., 1970. A thorough, well-written and understandable description of the properties, sources and toxicology of sulfur oxide air pollutants [B]; (b) S. K. Hall, Sulfur compounds in the atmosphere. *Chemistry*, **45**:16 (March 1972) [A]; (c) F. Leh and K. M. Chan: Sulfur compounds: pollution, health effects, and biological function. *J. Chem. Educ.*, **50**:246 (1973) [B].

33. Acid rain: (a) G. E. Likens, F. H. Bormann, and N. M. Johnson: Acid rain. *Environment*, **14**(2):33 (1972). The authors present a thorough, well-documented account [A]; (b) G. E. Likens and F. H. Bormann: Acid rain: a serious environment problem, *Science*, **184**:1179 (1974) [B]. See also *Science*, **188**:957 (1975) for a further discussion of points raised in this article; (c) W. W. Kellogg, et al.: Acid rain: a serious regional environmental problem. *Science*, **175**:587 (1972) [B]; (d) E. M. Winkler, *Stone: Properties, Durability in Man's Environment*, Springer-Verlag, New York, 1973. This book is primarily about stone, but contains some excellent photographs of damage caused by acid rain in Germany [A].

Particulate air pollutants

34. For information about particulates in air: (a) U.S. Environmental Protection Agency, *Air Quality Criteria for Particulate Matter*, AP-49, U.S. Government Printing Office, Washington, D.C., 1970. Provides good overall coverage of particulate air pollution; well written and understandable [B]; (b) V. J. Schaefer: Auto exhaust, pollution and weather patterns. *Bull. Atom. Sci.*, October 1970, p. 31. Considers the danger of smaller, less visible particulate pollutants; easily understandable [A]; (c) P. W. West and S. L. Sachdev: Air pollution studies. The ring oven technique. *J. Chem. Educ.*, **46**:96 (1969). An interesting experiment [B].

Nitrogen oxides

35. About nitrogen oxides as air pollutants: (a) U.S. Environmental Protection Agency, *Air Quality Criteria for Nitrogen Oxides*, AP-84, U.S. Government Printing Office, Washington, D.C., 1971. Provides the usual complete and understandable summary that characterizes this series [B]; (b) H. J. Hall and W. Bartok: NO_x control from stationary sources. *Environ. Sci. Technol.*, **5**:320 (1971). Various techniques are described for control of NO_x at power plants and other stationary sources (which produce 50% more total nitrogen oxides than automobiles) [B]; (c) T. R. Hauser and C. M. Shy: Position paper: NO_x measurement. *Environ. Sci. Technol.*, **6**:890 (1972). Describes the difficulties encountered by EPA

when it was discovered that the standard analytical method for NO_2 contained large determinate errors [B].

36. Experiments: (a) J. H. McFarland and C. S. Benton: The oxides of nitrogen and their detection in automobile exhaust. *J. Chem. Educ.*, **49**:21 (1972) [B] and E. R. Stephens and M. Price: Analysis of an important air pollutant: peroxyacetyl nitrate. *J. Chem. Educ.*, **50**:351 (1973) [B].

Photochemical smog

37. For information about photochemical smog: (a) A. J. Haagen-Smit, Theory and practice of air conservation. In W. Beranek, Jr. (ed.): *Science, Scientists and Society*, Bogden and Quigley, Belmont, Calif., 1972 [B]; this chapter appears in a somewhat abbreviated version as, The light side of smog. *Chemtech*, **2**:330 (1972) [B]. Well written articles by a pioneer in photochemical smog chemistry; (b) U.S. Environmental Protection Agency, *Air Quality Criteria for Photochemical Oxidants*, AP-63, U.S. Government Printing Office, Washington, D.C., 1970. A complete and well-written summary of sources, characteristics and effects of these pollutants [B]; (c) R. D. Cadle and E. R. Allen: Atmospheric photochemistry. *Science*, **167**:243 (1970). A good general treatment of chemistry in the atmosphere, with emphasis on photochemical smog reactions [C].

38. A computer model of smog formation: B. J. Huebert: Computer modeling of photochemical smog formation. *J. Chem. Educ.*, **51**:644 (1974) [C].

Carbon monoxide

39. For information about the introduction of carbon monoxide into the atmosphere: (a) U.S. Environmental Protection Agency, *Air Quality Criteria for Carbon Monoxide*, AP-62, U.S. Government Printing Office, Washington, D.C., 1970. The usual thorough and well-written treatment of this series [B]; (b) Committee on Effects of Atmospheric Contaminants, *Effects of Chronic Exposure to Low Levels of Carbon Monoxide on Human Health, Behavior and Performance*, National Academy of Sciences/National Academy of Engineering, Washington, D.C., 1969. Well written and concise [B]; (c) L. D. Bodkin: Carbon monoxide and smog. *Environment*, **16**(4):34 (1974) [B]; (d) T. H. Maugh, II, Carbon monoxide: natural sources dwarf man's output. *Science*, **177**:338 (1972). A good overall view [B].

The pollution of the air by lead, mercury, and other heavy metals

40. Lead in the atmosphere: (a) T. J. Chow: Our daily lead. *Chem. Brit.*, **9**:258 (1973). Facts and figures on environmental distribution of lead [B]; (b) J. J. Chisholm, Jr.: Lead poisoning. *Sci. Amer.*, **224**:15 (Feb. 1971) [B]; (c) Committee on Biologic Effects of Atmospheric Pollutants, *Lead: Airborne Lead in Perspective*, National Academy of Sciences, Washington, D.C., 1972 [B]. A thorough report by a panel which some have accused of being biased toward industry; see also *Science*, **174**:800 (1971); (d) U.S. Environmental Protection Agency, *EPA's Position on the Health Effects of Airborne Lead*, Washington, D.C., 1972 [B]; (e) J. P. Day, M. Hart, and M. S. Robinson: Lead in urban street dust. *Nature*, **253**:343 (1975) [B]

and T. M. Roberts et al.: Lead contamination around secondary smelters: estimation of dispersal and accumulation by humans. *Science*, **186**:1120 (1974) [B]. Both of these articles deal with the subject of dustfall; some researchers have contended that this dustfall can be ingested by young children and is of major importance in lead poisoning.

41. For imformation on air pollution by mercury and other heavy metals, see references 16(d), 16(f), and 16(g) of this bibliography.

Air pollution by trace substances; monitoring/ analysis

42. Trace substances: (a) R. A. Duce, G. L. Hoffman, and W. H. Zoller: *Science*, **187**:59 (1975) [B]; R. A. Carr and P. E. Wilkniss: *Science*, **181**:843 (1973) [B]; and H. V. Weiss, M. Koide, and E. D. Goldberg: *Science*, **172**:261 (1971) [B]. These papers and other references within them are part of a series of analyses of trace substances in remote areas aimed at determining the role of man in contributing to global biogeochemical cycles; (b) M. J. Prival and F. Fisher: Fluorides in the air. *Environment*, **15**(3):25 (1973) [A]. A lengthy article with numerous references; easily readable; (c) Committee on Biological Effects of Atmospheric Pollutants, *Asbestos*, National Academy of Sciences, Washington, D.C., 1971 [B]. Concludes that "it would be imprudent to permit additional contamination of the public environment with asbestos."

43. Monitoring: R. L. Chapman: Continuous stack monitoring, *Environ. Sci. and Technol.*, **8**:520 (1974). This is a feature article that describes the latest industrial monitoring techniques; useful schematic diagrams [B].

RESOURCES AND RECYCLING

World and United States supplies of a number of nonrenewable mineral resources are small relative to current and projected demand. Fertile soils also constitute a very slowly renewable resource, as well as serving to detoxify numerous air and water pollutants. Conservation of such resources is a major plank in most environmentalists' platform for an ecologically sound society. Even with conservative practices, however, large amounts of solid waste will be produced in any industrialized country. Disposal of or preferably recovery of the resources from such wastes will therefore be a major problem facing chemical and other technologists in coming years. Of particular importance in all plans for recycling is the consumption of energy resources, since these are the only truly nonrenewable or nonrecyclable ones in a scientific sense. The quantity of energy resources consumed by various recycling methods is discussed in the last group of references in this section.

Supplies of resources

44. A general treatment of the supply of all types of resources: (a) H. Brown, J. Bonner, and J. Weir: *The Next Hundred Years*, Viking Press, New York, 1957 [A]; (b) Committee on Geological Sciences, *The Earth and Human Affairs*, National Academy of Sciences, Canfield Press, Harper and Row, New York, (1972) [B].

Supplies of metal/mineral resources and their distribution on Earth

45. C. F. Park, Jr.: *Affluence in Jeopardy: Minerals and the Political Economy*, Freeman, Cooper & Co., San Francisco, 1968. Presents material over a broad range in a manner which points out interconnections; readable [B]; (b) L. H. Ahrens, *Distribution of the Elements in Our Planet*. McGraw-Hill, New York, 1965 [C]; (c) B. J. Skinner: *Earth Resources*, Prentice-Hall, Englewood Cliffs, N.J., 1969. In 150 pages the author presents a clear and concise picture of the abundance and distribution of the Earth's resources [B]; (d) H. Brown, Human materials production as a process in the biosphere. *Sci. Amer.*, **223**:195 (Sept. 1970). The author discusses the consumption trends of metals and minerals, tracing their use through history up to the present time and the obvious need for recycling [B]; (e) A. L. Hammond, Manganese nodules (I): mineral resources on the deep seabed. *Science*, **183**:502 (1974) and Manganese nodules (II): prospects for deep sea mining. *Science*, **183**:644 (1974). A good, concise coverage of a new source of minerals, which is not without problems, both in recovery and in international competition [B]; (f) Committee on Resources and Man, National Academy of Sciences, *Resources and Man. A Study and Recommendations*, W. H. Freeman and Co., San Francisco, 1969, a classic [B]; (g) Committee on Mineral Resources and the Environment, *Mineral Resources and the Environment*, National Academy of Sciences, Washington, D.C., 1975. An excellent report that covers materials conservation through technology, estimation of supplies of fossil fuels and copper, the environmental implications of the extraction and use of coal, and the demand for fuel and mineral resources. Data are easy to find and understand in both the body of the report and in the summaries [B]; (h) V. E. McKelvey, Mineral resource estimates and public policy. *Amer. Sci.*, **60**:32 (1972). A comprehensive article by the Director of the U.S. Geological Survey [B]; (i) T. W. F. Russell and M. W. Swartzlander: The recycling index. *Chemtech*, **6**:32 (1976). A numerical evaluation of the potential for recycling several resources, from sulfuric acid to dodecyl benzene sulfonic acid [B].

46. Reference books with information about mineral supplies: (a) D. A. Brobst and W. P. Pratt (eds.): *United States Mineral Resources*, U.S. Geological Survey Professional Paper 820, U.S. Government Printing Office, Washington, D.C., 1973. The first comprehensive assessment by the Geological Survey of the nation's mineral resources in more than 20 years. Concludes that known deposits of raw materials are seriously depleted [C]; (b) U.S. Bureau of Mines, *Mineral Facts and Problems*, U.S. Government Printing Office, Washington, D.C., 1970. Estimates of reserves, explanations of technologies used, discussions of environmental effects, and projections of demand to the year 2000 are included for almost every element and many compounds of commercial importance [C].

Soil resources

47. The nature of soil resources: (a) R. G. Gymer, *Chemistry: An Ecological Approach*, Harper and Row, New York, 1973. The chapter on soil is useful [B]; (b) P. Sears; An empire of dust. In J. Harte and R. H. Socolow:

Patient Earth, Holt, Rinehart and Winston, New York, 1971, the authors write of the American Dust Bowl of the 1930's and the lessons to be learned from it from a personal perspective [A]; (c) H. D. Foth and L. M. Turk, *Fundamentals of Soil Science*, 5th ed. Wiley, New York, 1972. An excellent reference on all aspects of soil chemistry [C]; (d) R. C. Plumb: Trace elements make Australia fertile. *J. Chem. Educ.*, **51**:675 (1974) [C].

48. The ability of soils to absorb "wastes": (a) J. Allen: Sewage farming. *Environment*, **15**(3):36 (1973) [A]; (b) L. T. Kardos: A new prospect. *Environment*, **12**(2):10 (1970) [A]; (c) J. O. Evans: The soil as a resource renovator. *Environ. Sci. Technol.*, **4**:733 (1970) and Recycling sludge and sewage effluent by land disposal, *Environ. Sci. Technol.*, **6**:871 (1972) [B]. These articles deal with the ability of the forest and range land, as well as cultivated soils, to be disposal sites for sewage sludge and effluent with good results; (d) F. B. Abeles, L. E. Craker, L. E. Forrence, and G. R. Leather: Fate of air pollutants: removal of ethylene, sulfur dioxide, and nitrogen dioxide by soil. *Science*, **173**:916 (1971) [C].

The production of solid wastes

49. Information about solid wastes and their production: (a) D. H. M. Bowen (ed.): *Solid Wastes*, American Chemical Society, Washington, D.C., 1967–71. This collection of 25 articles previously published in *Environmental Science and Technology* provides a useful survey of the field [B]; (b) National Center for Resource Recovery, Inc. has three recent books which are up-to-date, readable, state-of-the-art studies. They are well written and abound with figures, tables, schematic drawings, etc. *Incineration*; *Sanitary Landfill*; and *Resource Recovery from Municipal Solid Wastes*, Lexington Books, D. C. Heath & Co., Lexington, Mass. 1974 [B]; (c) Chapter 13, Solid wastes, of reference 5 is a useful source of information; (d) The section on solid wastes of reference 29 (e) is quite useful, although some aspects of the problem have been omitted; (e) W. E. Small, *Third Pollution—The National Problem of Solid Waste Disposal*, Praeger Publishers, New York, 1971. A small, readable volume [A].

Plastics waste and biodegradability

50. (a) F. Rodriguez, The prospects for biodegradable plastics. *Chemtech*, **1**:400 (1971). A good exposition of developments in a field of chemical research that seems unlikely to be of much use in solving solid waste problems [B]; (b) G. Scott: Improving the environment: chemistry and plastics waste. *Chem. Brit.*, **9**:267 (1973). A details excursion into the role of plasticizers, antioxidants and activators in biodegradability of plastics [C].

Recovery of used resources; recycling; energy from solid wastes

51. (a) W. C. Kasper: Power from trash. *Environment*, **16**(2):34 (1974) [A]; (b) C. B. Kenahan: Solid waste. *Environ. Sci. Technol.*, **5**:594 (1971). A general description of a number of techniques developed for solid waste recovery by the U.S. Bureau of Mines [B]; (c) H. Ness: Recycling as an industry. *Environ. Sci. Technol.*, **6**:700 (1972). A description of the operation of the recycling industry with special emphasis on recovery of

copper and paper [B]; (d) R. Grinstead: *Environment*, **12**(10):2 (1970); *Environment*, **14**(3):2 (1972); and *Environment*, **14**(4):34 (1972). In this series of articles a research chemist describes current practices and available technology for resource recovery, giving greatest attention to paper recycling [A]; (e) Council on Environmental Quality, *Resource Recovery*, U.S. Government Printing Office, Washington, D.C., 1973. An assessment of technologies for resource recovery from municipal solid wastes [B].

Recycling, reclamation, and reuse from an energy standpoint

52. (a) D. J. Rose, J. H. Gibbons, and W. Fulkerson: *Phys. Today*, Feb. 1972, p. 32. A general approach to the problem of waste management with emphasis on physical processes, this paper gives some criteria by which the best solutions to solid waste problems may be identified [B]; (b) B. M. Hannon: Bottles, cans, energy. *Environment*, **14**(2):11 (1972). This paper discusses the energy costs of various beverage container systems [A]; (c) R. S. Berry: Recycling, thermodynamics and environmental thirft. *Bull. Atom. Sci.*, **27**(5):22 (1971); also appears in *The Energy Crisis*, R. S. Lewis and B. I. Spinrad (eds.), Education Foundation for Nuclear Science, New Haven, Conn., 1972. The production and consumption of automobiles could be made more efficient by increased recycling, extended lifetime of the product, and especially by new, more efficient technology of manufacture [B].

WATER POLLUTION

Water has unusual properties, many of which have had considerable influence on the evolution of life on earth. Water is also the chief agent by which modern cities, industries, and individual dwellings rid themselves of unwanted wastes. The large number and variety of substances found suspended or dissolved in rivers, lakes and oceans makes water pollution an extremely difficult problem to solve. In this area not even a bibliography of limited length can be exhaustive in its coverage. Insofar as possible we have attempted to choose introductory articles and books that contain further references to more advanced materials. Those who wish to obtain highly detailed information may therefore have to do additional library research on their own.

The nature of water

53. Water in general: (a) H. L. Penman: The water cycle. *Sci. Amer.*, **223**:99 (Sept. 1970). The unusal and unique properties of water are described and provide a background for understanding its importance in the biosphere [A]; (b) The wonder of water. *Chemistry*, **47**:6 (June 1974) [A]; (c) U.S. Environmental Protection Agency, *National Water Quality Inventory*, 2 vols., U.S. Government Printing Office, Washington, D.C., 1974. The first systematic survey of U.S. water quality. Also sets goals for 1977–1983 [C]; (d) *Oceanus* is a periodical that is issued four times per year by the Woods Hole Oceanographic Institute. It regularly contains timely and readable articles about all aspects of water, and is highly recommended [A].

54. Drinking water and drinking water standards: (a) J. Crossland and V. Brodine: Drinking water. *Environment*, **15**(3):11 (1973). A survey of problems relating to contamination of drinking water supplies [B]; (b) J. L. Marx: Drinking water: another source of carcinogens? *Science*, **186**:809 (1974). A report of studies done on the lower Mississippi River which detected the presence of carcinogens [B]; (c) U.S. Environmental Protection Agency: Interim primary drinking water standards. *Federal Register*, **40**:11990 (March 15, 1975). These are the EPA's proposed interim standards under the Safe Drinking Water Act; the enactment of standards is especially important in view of the previous report on carcinogens in drinking water [C].

Bodies of water

55. Rivers, bays, oceans: (a) H. Van Der Schalie: Aswan dam revisited. *Environment*, **16**(9):18 (1974) [A]; (b) W. Bascom: The disposal of waste in the ocean. *Sci. Amer.*, **231**:16 (Aug. 1974). After describing the nature of the ocean and the wastes that are being discharged into it, the author discusses the ocean's ability to accept some wastes [B]; (c) L. J. Carter: Galveston Bay: test case of an estuary in crisis. *Science*, **167**:1102 (1970). A good example of problems which can occur in the biologically essential area where fresh and salt water meet [B]; (d) G. P. Howells, T. J. Kneipe, and M. Eisenbud: Water quality in industrial areas: profile of a river. *Environ. Sci. Technol.*, **4**:26 (1970). A thorough report on the biological status of a major industrial river—the Hudson—which has been carefully studied [B].

*Water polution
from various
sources*

56. General treatment of subject: (a) T. L. Willrich and N. W. Hines (eds.), *Water Pollution Control and Abatement*, Iowa State University Press, 1967. Good coverage of all aspects of water pollution in a largely description treatment [B]; (b) G. C. Berg: *Water Pollution*. Scientists' Institute for Public Information, 1970 [A]; (c) J. McCaull and J. Crossland: *Water Pollution*. Harcourt Brace Jovanovich, 1974. Written at a popular level by two members of the staff of *Environment* [A]; (d) W. A. Pettyjohn, *Water Quality in a Stressed Environment*, Burgess Publishing Co., Minneapolis, 1972. A collection of readings dealing with ground water and surface water contamination [B]; (e) *Environ. Sci. Technol.*, Vol. 8, 1974. This special issue devoted to water pollution contains much useful information [B].

57. The economic aspects of water pollution: (a) *Chem. Eng. News*, Water pollution controls to cost a bundle. October 15, 1973, p. 13 [A]; (b) EPA sees no economic blocks to clean water. *Chem. Eng. News*, February 4, 1974, p. 16 [A]. Both of these articles contain reports on the economic impacts of water pollution control.

Eutrophication

58. (a) J. Crossland and J. McCaull: Overfed. *Environment*, **14**(9):30 (1972). Good, general coverage of the problem of eutrophication in a mainly descriptive manner, good references [A]; (b) G. E. Hutchinson: Eutrophication. *Amer. Sci.*, **61**:269 (1973). A thorough treatment of the scientific

background of the problem [C]; (c) R. D. Grundy, Strategies for control of man-made eutrophication. *Environ. Sci. Technol.*, **5**:1184 (1971). Scientific and legislative aspects of control are treated in this report [B]; (d) National Academy of Sciences, *Eutrophication: Causes, Consequences, Correctives*, Washington, D.C., 1969. The introduction, summary and recommendations are readable and concise sources of information [B].

Industrial water pollution

59. General treatment of industrial sources: (a) N. L. Nemerow, *Liquid Waste of Industry*, Addison-Wesley Publishing Co., Reading, Mass., 1971. Contains a useful catalog of emissions of various industries and abatement techniques for them [B]; (b) G. V. Cox: Industrial waste effluent monitoring. *Amer. Lab.*, July 1974, p. 36. A summary, in table form, of the methods of wastewater analysis which meet Federal standards and references for information about each method [C]; (c) J. B. Cox: Re: water. Where are we? *Chemtech*, **6**:566 (1976). A history of EPA efforts to enforce the Federal Water Pollution Control Act [A].

60. The periodical *Environmental Science and Technology* regularly carries articles about industrial and municipal water pollution and treatment [B]. Some representative topics covered recently include: metal finishing industries, **7**:209 (1973) and **4**:381 (1970); plastics manufacturing, **4**:637 (1970); farms and animal feedlots, **7**:797 (1973) and **8**:985 (1974); textiles, **6**:37 (1972); organic chemical industry, **8**:621 (1974); oil refinery, **5**:1099 (1971); photographic processing, **5**:1085 (1971); and fertilizer manufacturing, **6**:693 (1972).

61. Mercury from industrial sources: (a) N. Grant: Mercury in man. *Environment*, **13**(4):2 (1971) [A]; (b) T. Aaronson: Mercury in the environment. *Environment*, **13**:(4):16 (1971) [A]; (c) J. M. Wood: A progress report on mercury. *Environment*, **14**(1):33 (1972) [A]. These three articles should provide a good introduction to the subject; they are readable, contain a great deal of information, and have good references if more information is needed. (d) L. J. Goldwater, Mercury in the environment. *Sci. Amer.*, **224**(5):15 (1971). Describes the problem of mercury pollution from a conservative view [B]; (e). J. Gavis and J. F. Ferguson: The cycling of mercury through the environment. *Water Res.* **6**:989 (1972). A thorough review [C].

62. Information about the pulp and paper industry: (a) H. Gehm: *State-of-the-Art Review of Pulp and Paper Waste Treatment*, U.S. Environmental Protection Agency, U.S. Government Printing Office, Washington, D.C., 1973. A useful report of paper-mill waste treatment, from a somewhat industry-oriented view [C]; (b) Changes are in store for pulping technology. *Environ. Sci. Technol.*, **9**:20 (1975). A descriptive treatment of paper pulping techniques without the use of sulfur [B].

Water treatment

63. Standard methods for treatment of municipal wastewater are described in Chapter 14 of reference 4, Chapter 9 of reference 5, and Chapter 15 of reference 7. More advanced techniques, currently being tested for possible adoption in the future, are described in: (a) Nutrient removal in

wastewater treatment. *J. Coll. Sci. Teaching*, **3**:36 (1973) [B]; (b) L. B. Luttinger and G. Hoche: Reverse osmosis treatment with predictable water quality. *Environment. Sci. Technol.*, **8**:614 (1974) [B]; (c) I. R. Higgins: Ion exchange: its present and future use. *Environ. Sci. Technol.*, **7**:1110 (1973) [B]; (d) Ozonation seen coming of age. *Environ. Sci. Technol.*, **8**:108 (1974) [B].

Deep well disposal of wastes

64. Information about the technique of deep well disposal and the possible problems arising from it can be found in: (a) D. M. Evans and A. Bradford, *Environment*, **11**(8):3 (1969) [A]; (b) Deep-well disposal continued. *Environ. Sci. Technol.*, **7**:1106 (1973) [B].

POLLUTION AND HEALTH

Many of the forms of pollution described in previous references have harmful effects on human health. In this section we concentrate first on food supply and the effects of modern agricultural techniques on the environment. One of the consequences of such techniques is release to the environment of large quantities of toxic, synthetic pesticides. These have achieved sufficient notoriety that a separate section is devoted to them, but they are not the only toxic substances of concern to the general public. We have therefore included a separate section on toxic substances in general.

Food

65. The supply of food, in light of increasing population: (a) G. Borgstrom: *The Hungry Planet*, Macmillan, New York, 1965 [A]; (b) L. R. Brown and G. W. Finsterbusch: *Man and His Environment: Food*, Harper and Row, New York, 1972, considers how to produce enough food and the environmental consequences of trying to do so [A]; (c) L. R. Brown: Human food production as a process in the biosphere. *Sci. Amer.*, **223**:160 (Sept 1970) [B]. Similar to Brown's book in coverage; (d) N. Wade: Green revolution (I) and (II). *Science*, **186**:1093 and 1186 (1974). These two news articles delineate both social and ecological problems associated with increased food production [A]; (e) President's Science Advisory Committee, *The World Food Problem*, Vols. *I, II, III*, The White House, Washington, D.C., 1967. Presents a tremendous amount of useful information over a broad range; contains a myraid of figures and tables [C]; (f) *Science*, **188**, no. 4188 (9 May, 1975) has devoted the entire issue to food and provides excellent coverage over a broad area of topics [B].

66. Articles that deal with protein and nutrition include: (a) F. M. Lappé, Protein from plants. *Chemistry*, **46**:10 (Nov. 1973) [A]; (b) A. M. Altschul: Food: protein for humans. *Chem. Eng. News*, November 24, 1969, p. 68 [B]; (c) D. Spurgeon: The nutrition crunch: a world view. *Bull. Atom. Sci.*, **29**(8):50 (1974) [B]; (d) M. Hamdy: The nutritional value of vegetable protein. *Chemtech.*, **4**:616 (1974) [B].

67. The amount of energy necessary to produce food is discussed in the following references: (a) J. S. Steinhart and C. E. Steinhart: Energy use in

the U.S. food system. *Science*, **184**:307 (1974) [B]; and Chapter Six of, *Energy: Sources, Use, and Role in Human Affairs*, Duxbury Press, Belmont, Calif., 1974 [B]; (b) E. Hirst: Food-related energy requirements. *Science*, **184**:134 (1974) [B]; (c) D. Pimental et al.: Food production and the energy crisis. *Science*, **182**:443 (1973) [B] and correspondence relating to this in *Science*, **187**:560 (1975).

Environmentally-related illnesses/toxic substances

68. A general treatment of environmental illnesses and toxic substances causing these illnesses may be found in: (a) L. J. Carter and R. Gillette: Cancer and the environment (I) and (II). *Science*, **186**:239 (1975) [A]. Aldrin, dieldrin, benzidine and vinyl chloride are included in this report of chemical carcinogens; (b) G. L. Waldbott: *Health Effects of Environmental Pollutants*, C. V. Mosby, St. Louis, Mo., 1973. A readable and very useful book which has emphasis on airborne pollutants; good references [B]; (c) Council on Environmental Quality, *Toxic Substances*, Superintendent of Documents, U.S. Government Printing Office, Washington, D.C., 1971. An excellent, well-written summary of the problems of toxic substances in the environment [B]; (d) D. H. K. Lee, Specific approaches to health effects of pollutants. *Bull. Atom. Sci.*, **29**(8):45 (1973). A brief account of what is currently known about the effects of environmental agents on man which is presented in a clear manner [B]; (e) *The Toxic Substances List*, U.S. Department of Health, Education and Welfare, Annual Editions, U.S. Government Printing Office, Washington, D.C. This annual reference book by the National Institute for Occupational Safety and Health of HEW presents a wealth of data on all substances known to be toxic [C].

69. Information about asbestos as a toxic substance may be found in: (a) A. K. Ahmed, D. F. MacLeod, and J. Carmody: Control of asbestos: *Environment*, **14**(10):16 (1972). This is a very thorough yet readable treatment of the subject, which spans the period of its early history as a suspected cause of occupational disease to the possible effects on the community. Contains many notes and references [A]; (b) See also reference 42(c) of this bibliography (air pollution section) for more information about asbestos.

70. Vinyl chloride and its toxic properties are covered in: (a) J. W. Moore: The vinyl chloride story. *Chemistry*, **48**:12 (June 1975) [A]; (b) I. R. Tabershaw and W. R. Gaffney: Mortality study of workers in the manufacture of vinyl chloride and its polymers. *J. Occupational Med.*, **16**:509 (1974) [B].

71. The study of environmentally-related illnesses and their connection to the work-place environment is the subject of several recent books and articles, including: (a) A. Anderson, Jr.: The hidden plague. *New York Times Magazine*, October 27, 1974, p. 20 [A]; (b) W. Greene, Life vs. livelihood. *New York Times Magazine*, November 24, 1974, p. 17 [A]; (c) P. Brodeur, *Expendable Americans*, Viking, New York, 1973 [A]; (d) R. Scott, *Muscle and Blood*, E. P. Dutton and Co., New York, 1974 [A]; (e) J. A. Page and M. O'Brien, *Bitter Wages: Ralph Nader's Study Group Report on Disease and Injury on the Job*, Grossman Publishers, New York,

1973 [B]; and (f) J. M. Stellman and S. M. Daum, *Work is Dangerous to Your Health. A Handbook of Health Hazards in the Workplace and What You Can Do About Them*, Pantheon, New York, 1973 [B].

Pesticides

72. A good introductory treatment of pesticides and their effect on human and animal life can be found in: (a) D. L. Dahlsten et al.: *Pesticides*, Scientists' Institute for Public Information, 1970 [A]; (b) O. L. Loucks, The trial of DDT in Wisconsin. In J. Harte and R. H. Socolow: *Patient Earth*, Holt, Rinehart and Winston, New York, 1971 [A]; (c) C. F. Wurster: Aldrin and dieldrin. *Environment*, **13**(8):33 (1971) [A].

73. For information about why pesiticides are needed in agriculture and some alternatives to their use, consult: (a) L. Chiarappa, H. C. Chiang, and R. F. Smith: Plant pests and diseases: assessment of crop losses. *Science*, **176**:769 (1972) [B]; (b) G. W. Irving, Jr.: Agricultural pest control and the environment. *Science*, **168**:1419 (1970) [B]; (c) Council on Environmental Quality, *Integrated Pest Management*, U.S. Government Printing Office, Washington, D.C., 1972. Describes methods other than broad-scale application of chemical pesticides for control of agricultural pests [B]; (d) R. L. Metcalf: Pests and pollution. Challenge of modern insect control. In K. L. Rinehart, Jr., W. O. McClure, and T. L. Brown (eds.): *Wednesday Night at the Lab.*, Harper and Row, New York, 1973; surveys the uses and properties of DDT and substitutes for it as an insecticide [B]; (e) W. Worthy: Integrated insect control may alter pesticide use pattern. *Chem. Eng. News*, April 23, 1973, pp. 13–19. A good summary [B]; (f) C. A. Edwards: *Persistent Pesticides in the Environment*, 2nd ed., CRC Press, Cleveland, Ohio, 1973. This is a well-written and comprehensive treatment and is highly recommended. The section comparing various models of DDT movement in the environment is especially recommended [B].

74. An alphabetical listing of all pesticides, their common names, formula, use, and manufacturer is given in: S. S. Epstein and M. S. Legator: *The Mutagenicity of Pesticides*, MIT Press, Cambridge, Mass., 1971 [B].

75. PCB's are related to DDT and other chlorinated hydrocarbon insecticides and present some dangers to man and animals through their use as plasticizing agents. Basic information about these compounds may be found in: (a) K. P. Shea: PCB. *Environment*, **15**(9):25 (1973) [A]; (b) K. P. Shea: The new-car smell. *Environment*, **13**(8):2 (1971) [A]; (c) C. G. Gustafson: PCB's—prevalant and persistent. *Environ. Sci. Technol.*, **4**:814 (1970) [B].

POPULATION

There are some who would argue that the recent, rapid increase in human population is directly or indirectly responsible for nearly all environmental problems. Others disagree. It is certainly true, however, that supporting a large population requires greater consumption of energy resources than supporting a smaller one at the same standard of living. The concept of an optimum level of population is an appealing one, but arguments over its actual size are sure to continue for some time. The references that follow attempt to present many facets of this problem. Those who are mathematically inclined may enjoy reference 79, which illustrates the dire consequences (as well as the impossibility) of continuous exponential population growth.

Population statistics

76. Current information about population levels and growth rates for the United States and the world can be found, respectively, in: (a) U.S. Department of Commerce, Bureau of the Census, *Estimates of the Population of the United States and Components of Change: 1973*, U.S. Government Printing Office, Washington, D.C., 1974 [B]; (b) United Nations Secretariat, *World and Regional Population Prospects*, The United Nations, New York, 1974 [B].

The population problem

77. Introductory information about increases in population and its effect on the environment and social problems may be found in: (a) H. H. Hart, *Population Control: For and Against*, Hart Publishing Co., New York, 1973. Contains a selection of essays covering various points of view [A]; (b) S. F. Singer (ed.), *Is There An Optimum Level of Population?* McGraw-Hill Book Co., New York, 1971. Another series of essays, with emphasis on the relation of population levels to supplies of resources; education, health and welfare services; and life styles and human values [B]; (c) G. Hardin (assembler), *Population, Evolution, and Birth Control. A Collage of Controversial Ideas*, 2nd ed. W. H. Freeman, San Francisco 1969. The subtitle is an apt description of these writings selected from a wide range of literature and science [A]; (d) L. R. Brown, *In the Human Interest*, W. W. Norton and Co., (1974). Brown presents overpopulation as a problem and some ideas about stabilizing the world level [B].

78. More specialized information about population, with emphasis on the growth of population as it relates to environmental and world problems, may be found in: (a) The human population. *Sci. Amer.* (Special Issue), **231** (Sept. 1974) [B]; (b) G. Hardin: The tragedy of the commons. *Science*, **162**: 1243 (1968). A classic treatment of the population-resources-competition problem [B]; (c) J. P. Holdren and P. R. Ehrlich: Human population and the global environment. *Amer. Sci.*, **62**:282 (1974). The authors present their thesis that population is the sole and central issue in both the cause and solution of environmental problems [B]; (d) B. Commoner: The environmental cost of economic growth. *Chem. Brit.*, **8**:52 (1972) [B], and *The Closing Circle*, Alfred A. Knopf, New York, 1971 [A]. Commoner is in sharp disagreement with Holdren and Ehrlich and sees population as only one factor in environmental problems; he cites the wrong kind of technology as being at least as important as population.

79. A model to simulate global population change is present in: E. G. Rochow, G. Fleck, and T. R. Blackburn: Mechanisms of reactions. Chapter 11. In *Chemistry. Molecules That Matter*, Holt, Rinehart and Winston, New York, 1974 [B].

80. Birth control methods are discussed in: J. L. Marx: Birth control: current technology, future prospects. *Science*, **179**:1222 (1973) [B].

**SCIENCE,
PHILOSOPHY,
AND
ENVIRONMENT**

81. Listed here are samples of some provocative opinions on the causes and cures of the environmental problems presented in this chapter: (a) R. L. Heilbroner: What has posterity ever done for me? *New York Times Magazine*, January 19, 1975, p. 14. Questions why we should protect the environment for future generations [A]; (b) E. F. Schumacher: *Small is Beautiful*, Harper Colophon Books, New York, 1973. The subtitle, "Economics as if People Mattered," is apt [B]; (c) J. Platt: What we must do. *Science*, **166**:1115 (1969). Platt proposes that there should be a large-scale reassignment of scientists to work on problems such as ecology and pollution [B]; (d) K. E. Boulding, *Beyond Economics*; *Essays on Society, Religion, and Ethics*, University of Michigan Press, Ann Arbor, Mich. 1968. Boulding is an excellent economist who transcends his discipline; an example of this is in his placing an economic value on the environment [B]; (e) R. J. Dubos: Humanizing the Earth. *Science*, **179**:769 (1973). Dubos believes that with ecological wisdom and scientific knowledge the Earth will continue to be a place of continued growth of civilization [B]; (f) R. E. Train: The quality of growth. *Science*, **184**:1050 (1974). Train cites the need for political leaders who can take the long-term views needed for solving environmental problems [B]; (g) A. M. Weinberg: Science and trans-science. *Minerva* **10**:209 (1972) [B] and *Science*, **177**:211 (1972) [B]; **174**:546 (1970) [B]; **180**:1123 (1973) [B]. Weinberg's argument that certain questions which appear scientific have no scientific answers is presented and commented upon; (h) C. P. Snow, *Public Affairs*, Schribner's New York, 1971. As a former scientist who now participates in public affairs, Snow makes a strong case for the necessity of scientists to be involved in politics, especially since they are uniquely able to understand future-oriented affairs [B]; (i) A. W. Eipper: Pollution problems, resource policy, and the scientist. *Science*, **169**:11 (1970). The author describes some methods for dealing with environmental problems, from a first-hand point of view [B]; (j) K. Lorenz, *Civilized Man's Eight Deadly Sins*, Harcourt Brace Jovanovich, New York, 1974. Lorenz lists and describes eight "pathological disorders of mankind," some of which are related to environmental decay [C]; (k) A. Wolman: Ecologic dilemmas. *Science*, **193**:740 (1976). The challenge is to manage the environment, and to do this means first to set priorities, and then to implement the solutions.

A word in conclusion: Note that many of the references cited here as pertinent to the topic of this chapter are also cited for other chapters as pertinent to those topics. Indeed, this supports the thematic notions expressed in our opening paragraph.

chapter 15

Science and . . .

The final chapter in a book that has attempted to show the delights and the utilities of one branch of science, chemistry, probably should be addressed to the whole field of science. Further, the final chapter in a book addressed to people who have primary interests other than science ought to show that science is closely related to each and all of those other fields. The problem is that there are a very large number of other fields. A further problem is that each person must, in one way or another, individually determine that science, or history, or medicine, or cabinet-making (you name what is not really interesting to you) is instead distinctly pertinent and closely related to poetry, or glass-making, or paper clip chaining, or attorneying, or mountain climbing (or whatever *is* very interesting to you). In a few words, all and each are intimately related and directly relevant, one way or another.

Now, that may not seem so to you presently; it is not always possible to find obvious useful relationships between and among all conceivable fields of interest in the space of one lifetime particularly while you are busy doing things. However, the idea is worth some attention nevertheless. Part of the difficulty in relating one field with another is, and obviously, that they are not related. If they were related, then they would be practically similar in many ways; there would be no problem. Yet, and this is the paradox, any two conceptually separate fields *are* related.

How so? The best way is to demonstrate. Consider the fine arts, religion, and economics. Each of these can be juxtaposed with science by identifying commonalities and contrasts. The purposes of science and any one of the three can be stated conjointly to show their relatedness. And for the other half of the paradox, their mutual differences can be presented.

RELATIONS AND DIFFERENCES IN HUMAN ENDEAVOR

To begin, it will be helpful to state the purpose of each of the four in a way that demonstrates the unique character of each:

> The purpose of science is to satisfy our curiosity about natural phenomena.
>
> The purpose of the fine arts is to communicate to others the joy (or other emotion) the artist feels.
>
> The purpose of religion is to enable each to love their Creator. (This encroaches upon a delicate matter; some feel that religion is inane, opining that there is no Creator. This point will be touched upon later.)
>
> The purpose of economics is to put into order the production, distribution, and consumption of goods and services.

Science and the fine arts

Based on these definitions, the relatedness between science and any of the other three can be epitomized by a few key words for each pair. First, science and the fine arts: these words, imagination, critical judgment, and esthetics, form a bridge joining science and the fine arts. That is, both science and the fine arts considered together can be said to share this purpose:

To provide a means for contemplating the physical world with an ever new, ever deeper, admiration and awe as a result of the display of the power of the imagination of the scientist or artist.

Both the scientist and the artist begin by observing their environment, carefully distinguishing between fact and illusion in what they observe. The observed facts are then arranged in a kind of order and certain facts are selected from this arrangement. The selection is based upon a discernible connection between or among the selected facts, or upon an intuitively felt connection that is perhaps not clearly understood at the moment. The other, unselected, facts are relegated to a secondary position. Note that in these actions the critical judgment of the scientist or artist is used. Further, except for the initial observations of the facts and illusions, imagination is used as the tool to order, to select, and to discern obvious and intuitive relations between and among the facts.

In a second step, the scientist or artist uses both the faculty of critical judgment and the faculty of imagination to bring forth some kind of unifying statement (in words, in an equation, in a poem, a painting, a sculpture) that serves to show how all of the selected facts form an integrated, noncontradictory, whole. This second step is probably more critical than the first. Those who are neither scientist nor artist can observe, eliminate most illusory information, and put the result into a selective order. If all of the facts selected can readily be identified as consistent with each other, this is the end of action. If some facts are apparently inconsistent, an artistic or scientific creative contribution is necessary to show that the inconsistency is only apparent. The creative act, which is common to both art and science, consists in recognizing that there is a contradiction and in the further discovery of a unifying idea that demonstrates the contradiction is only apparent.

Some examples will help here. Consider first a masterpiece from the fine arts, Rembrandt's "Aristotle Contemplating the Bust of Homer." The contradictory facts, in this case, can be said to consist of the problem of handling light in the work. It is a fact that the pigments in oil paint used by this artist only reflect light; they do not emit light of their own accord. Yet Rembrandt used those pigments in such a way as to "force" those pigments to glow with their own light. This is clearly impossible, yet when you look at the work itself it has been done.

Or, from the history of chemistry, consider the problem presented by chlorine. Many years ago it was "known" that all acids contain oxygen (the word, oxygen, is derived from the French for "essential" and "acid"). This is true, of course, for such acids as nitric, acetic, sulfuric, and so on; they do contain oxygen. It is not true for hydrochloric acid, we now know, but it was not known in 1800; at that time it was thought that hydrochloric acid was composed of hydrogen, oxygen, and some other element, X. It was possible, then, to separate hydrochloric acid into two components, hydrogen and chlorine. Obviously, in 1800, chlorine was composed of oxygen and X. It was a "fact" then that chlorine was not a simple substance. The ultimate answer, first asserted about 1810, that all acids do not contain oxygen, was made in the face of an apparent contradiction that was quite real to thoughtful chemists of those days.

Or consider a more recent contradiction. At low temperatures, 2 K, or lower (about −271°C or −465°F) liquid helium behaves in an unusual manner. For example, it flows up the sides of its container, over the edge, and down the outside, dripping off the bottom. How is one to account for this behavior? Today, one explanation involves the emphatic statement that, at temperatures of about 2 K or lower, liquid helium is not composed of atoms; instead it is a continum, *one* "thing," at those low temperatures. If this theoretical explanation is correct, this perhaps apparent contradiction awaits a creative act that will show that there is no contradiction after all in saying that liquid helium is both atomic and continuous simultaneously. (At present, there are some likely resolutions of this which involve mathematical and abstract statements about the meaning of a quantum state. However, a full resolution has not yet been achieved.)

To find a unifying explanation that demonstrates an apparent contradiction is not contradictory is one thing. This alone is useless unless the finding can be communicated to others. So, in the third step, still involving the imagination but in a different way, scientists or artists must find a means of relating their unifying ideas to some other ideas that are already well known. In science, this is often accomplished by using a mathematical equation. In the fine arts this is usually accomplished by the disposition of pigment on canvas, or of sounds and pitches in music, or by choice of words in poetry.

Finally, in the fourth step, what has been communicated to others is seen to be esthetically pleasing. Unless it is delightful in some way or other, it is clearly not a work of fine art. Even in science where the esthetic beauty is sometimes discernible only to other scientists, there is an esthetic aspect which, if not present, renders the accomplishment suspect.

So far, the similarity between science and the fine arts has been presented. The other side of the paradox can be indicated in parallel contrasts:

Science	The Fine Arts
Discoveries are often made by intuition and then later found to be logical.	Masterpieces are often conceived intuitively and then materialized by expert technical skill, logically executed.
Mathematics is used as a tool.	The corresponding tools are rhythm, proportion, harmony.
Work of average quality is often useful.	Work of average quality is relatively sterile.
The esthetic aspects are usually abstract, the beauty is expressed indirectly to the senses.	The esthetic aspects are often abstract, but they are based upon a physical object (except in most poetry); the beauty is seen directly by the senses.

Science	The Fine Arts
An important scientific contribution can be paraphrased later by others with no semantic loss; only initially is it uniquely associated with the style of the initiating scientist.	A great work of art cannot be paraphrased later and still present the same semantic content; it is forever uniquely associated with the style of that artist.
Seeks truth about the material universe, knowing that such truth cannot be identified as such with certainty.	Makes suggestive statements about truth, hoping to stimulate others toward an internal certitude.
Uses symbols to represent reality, manipulating these in precise ways in order to provide new insights.	Uses symbols to represent reality, permitting their manipulation in a variety of ways in order to provide new insights.
A work of science is first understood and then experienced by one other than the scientist.	A work of art is first experienced and then understood by one other than the artist.
An important idea usually arises during the period of active concentration that follows the period of intensive data gathering.	An important idea usually arises during the period of relative inactivity that follows active saturation with the problem. (In this respect, mathematics is like the fine arts.)
The value of a scientific statement depends upon its precision of expression.	The meaning of a statement in the fine arts (that is, the work itself) depends upon its richness of expression.
The symbols used must mean the same to all. Subjective feelings are eliminated, at least in the formal description of the finished work.	The symbols can mean different things to different people. Emotions, passions, faith, hope, all play a part in the completed statement.
Is concerned with fewer, less intricate relationships between and among facts, often knowingly oversimplified, so that they can be handled.	Is primarily concerned with complex interrelationships, though often knowingly simplified, are only rarely made extremely simple.
Insists on a definite result.	Requires a satisfactory solution, a pursuit of the spirit.
Posits that, in the ultimate, there is only one harmonious way to explain the physical world.	Demonstrates that there are an endless number of different ways to see the physical world.

Science and religion

Except perhaps in books and articles that attend directly to the subject, in one way or another, published discussions of science and religion are usually restrained in tenor. In the first place, the substance of religion is in many respects a private matter, and should be discussed publicly with respect for other views. Secondly, and applicable here, the interrelationships between science and religion are more subtle than those concerning science and the fine arts. Thus, scientific statements have no direct bearing upon moral considerations or ethical conduct or subservience to any higher being (although scientific statements are at least indirectly related to esthetics, as we have seen).

However, the two can be bridged. For example, science and religion can be said to have this common purpose:

To provide a basis on which unifying concepts can be achieved. Or, less precisely, to provide a means of ascertaining what might be true in order that this knowledge can be used.

The word, truth, implies something about faith. Oversimplified, faith can be defined as a personal, interiorly recognized, certitude. Thus, a scientist has faith that all of the material universe is ordered in some way, that it contains no essential contradictions. There are few scientists who doubt this, though there are some who indeed question it; we might call them agnostic scientists. In religion, as is well known, many have a faith that there exists a living, infinite being on whom we are utterly dependent. Others have a faith in a kind of superior being who has (or beings who have) other attributes. A second group, as is also well known, have a faith that there exists no such being or beings. And a third, agnostic, group questions the faith of both other groups. The point here is to note that, in a discussion on science and religion, the notion of faith plays an important role, either positively or agnostically or negatively (the atheists).

In addition to faith, another key word that can serve as a bridge between science and religion is "logic." That is, it ought to be possible to consider what is held by faith from a logical, reasonable point of view. There are consequences that follow upon a statement of belief, and those consequences should follow logically from the statement. Thus, in a religious sense, if I hold on faith that mankind is utterly dependent upon his Creator and that the Creator is triunally infinite, this demands that I behave as though I were dependent and that I make some kind of intellectual attempt to understand how there can be three divine Persons and yet one God. If, speaking scientifically, I have certitude that the physical world is internally consistent, then I must as a logical consequence of this faith in principle seek to satisfy myself that helium can be nonatomic under certain conditions and atomic under other conditions. In this case, incidentally, the understanding is surely going to be difficult to attain, and I must be patient while those who are qualified to grapple with this problem find a solution. Of course, the same requirement for patience applies to my conception of my Creator and the mysterious aspects of my dependence on Him, for example.

To summarize, the key words that bridge science and religion are "faith" and "logic" (and application of the logic). Since the contrasts between the two are well known, only a few need to be listed here.

Science	Religion
Activity in science arises from a challenge posed by externally derived sources.	Activity in religion arises from a striving within.
Controllable experiments can be conducted. The principles, the theories, can be tested.	No experiments are possible, or, if possible, not controllable. The tenets of religion cannot be tested by experiments.
Scientific faith is based upon observable facts.	Our reliance on faith has an emotional as well as an intellectual basis.
Claims to deal with immediate realities.	Claims to deal with ultimate realities.
Many theories in science are thoroughly tested, and have a factlike status, the atomic theory, for example. These theories and their "certitude" yield a certain kind of pleasure, or satisfaction.	The certitude in religion is achieved in a different way. And, instead of satisfaction, or pleasure, these yield a kind of inner peace.

Science and economics

On the one hand it should be a simple matter to relate science and economics because economics is, at least potentially, a science. On the other hand, since economics is a young and developing science it is not too easy to discern exactly what economics is and, therefore, it is not clear how to relate what seem to be its essential characteristics to the corresponding and well-established characteristics of science as a whole. There are presently many different, and differing, schools of economic thought; for the purposes of this discussion a kind of melding of a few roughly similar schools is presented as though it were the orthodox position in economic thought. This may well not be the case, at least as viewed by those who favor the positions taken by responsible economists who represent points of view not apparent here.

With this disclaimer, this purpose is common to both science and economics:

To develop an ordered set of known, or knowable, interactions between mankind and the environment.

The similarities between science and economics are closely related to the similarities identified for science and the fine arts. In this case also two of the key ideas are "critical judgment" and "imagination." The third key word is "practical." Thus, both the scientist and the economist begin with observations, separating fact from illusion, and selecting facts that are obviously or intuitively related from which to pose a question involving apparent contradictions between or among some of the selected facts. The resolution of the apparent contradictions, as before, involves a kind of creativity, and in all of this critical judgment and imagination are intimately involved.

Then, in comparing science and economics, the concluding actions involve practical rather than esthetic considerations. The economist hopes to be able to determine that the recommended economic action will have practical results in the market place, bringing more order to the production, distribution, and consumption of goods and services. The scientist aims to achieve practical results from theory, in order to enable others to develop and test them further. (If engineering or technology is included as a part of science, another practical result could be mentioned: the efforts of the engineer to apply the theories from science in a practical way, to produce or distribute, or both, goods and services in a more efficient manner. This activity is clearly closely related to economics.)

As in the other pairs, the contrasts between science and economics are listed in parallel columns.

Science	Economics
The principles or theories of science are changed by scientists when newly discovered facts infer such changes. As an example, consider the phlogiston theory (discussed at some length in Chapter 3).	The principles or theories of economics are changed by economists when changing patterns of national or international culture alter the "ground rules." For example the currently burgeoning emphasis upon consumerism, compared to the older cultural acceptance of producerism, has already affected current economic theory and practice. (For one thing, it may soon be incorrect to relate economics only to goods and services; the interests of the general public even now cannot be ignored in rendering goods and services to a smaller segment.) Earlier changes in cultural patterns, such as the development of large corporations and the rise of trade unions demonstrated that the economic principle of lassez-faire had become erroneous.
Ultimately, science relies on rigorous mathematics.	Economics tends to rely on mathematics in the same way, but currently the principles of economics are not well enough established to permit as thorough a reliance on abstract mathematics.
Scientists tend to be most active in their work on problems that are interesting to them.	Economists tend to work most actively on problems that are of greater importance to society.
Science is indirectly, though intimately, related to most other areas of human endeavor.	Economics is directly related to almost all areas of human endeavor, sociology, political science, medicine, to name a few (perhaps excepting the fine arts).

Science	Economics
Scientists are interested in what can be known, in what might be called a natural harmony.	Economists are interested in what can be done, in a social harmony.
The use of experiment is rarely dangerous; even when it is, the danger is usually known in advance and usually forseeably detrimental to only a few individuals.	The use of experiment is often risky and almost always involves unknowable and uncontrollable effects, with potentially forseeable harm to a large number of individuals. Hence, a real experimental investigation is often precluded. As an alternative the economist sets up a simulated mathematical model (with known but unavoidable defects compared to reality) and performs a hypothetical experiment under innocuous conditions. (Unfortunately, due to the defects in the simulation, the results are not always of immediate practical value.)

FOR FURTHER INFORMATION

If you would like to try to develop your own bridging key words for any two areas along with a description of that commonality and a list of parallel contrasts, one place to start is with the essays in the Propaedia Volume of the *Encyclopaedia Britannica*, 15th ed., Encyclopaedia Britannica, Inc., Chicago, 1974.

R. M. Pirsig: *Zen and the Art of Motorcycle Maintenance*. Morrow, New York, 1974. On the relation between science and art using quality as the bridging word. Buddistic in its emphasis on good as superior to true. Pirsig errs in stating that the purpose of science is to achieve truth and that this is achievable (the purpose is to explain predictively, with theories that stand, fall, or are modified depending upon the accuracy of their prediction). Otherwise, a fine, enjoyable fantasy with an interesting moral worth pondering, with lengthy reflections.

M. Kranzberg and W. H. Davenport (eds.): *Technology and Culture*. Schocken Books, New York, 1972. A collection of previously published papers on sociotechnology that provide a varied perspective on this topic.

L. L. White: Letter from a scientist. *Leonardo*, **6**:351 (1973). One of the last essays by a philosopher, scientist, historian, and investment broker, suggesting that the perspectives of the artist are nearer to the real world than those of the scientist.

A. Eddington: *The Nature of the Physical World*. Michigan University Press, Ann Arbor, 1958 (paperback ed.). This is a classic, highly recommended.

H. Margenau: *The Nature of Physical Reality*. McGraw-Hill, New York, 1950. Another classic, similarly highly recommended.

J. Bronowski: *Insight: Ideas of Modern Science*, Harper, New York (1965); and *The Ascent of Man*, Little, Brown, Boston (1973).

Eddington and Margenau are scientists and this shows in their works; Bronowski is a humanist well acquainted with science, and this shows in his classic works. Highly recommended.

J. B. Conant: *Science and Common Sense*. Yale University Press, New Haven, 1951 and *On Understanding Science*, Yale University Press, New Haven, 1947.

Some would say that unless you have read, and enjoyed, at least one work by Eddington, Margenau, Bronowski, or Conant, you cannot validly claim to be fully educated. Anyone who has done so would be strongly inclined to agree.

Many people are properly concerned about the effects of scientific and technological developments in our culture; some have adopted a pessimistic view by the simple expedient of rejecting opposing facts. A balanced presentation is available from several sources. To get started, see T. R. LaPorte and D. Metlay: *Science*, **188**:121 (1975); D. Bell: *The Coming of the Post-Industrial Society*. Basic Books, New York, 1973; and V. F. Weisskopf: The significance of science. *Science*, **176**:138 (1972) and *Bull. Amer. Acad. Arts Sci.*, **27**:15 (March 1975).

E. W. Hobson: *The Domain of Natural Science*. Dover, New York, reprint 1968. First published in 1923, on the position which the scientific view of the world should occupy in relation to other factors of human experience, particularly religion and philosophy. Some of the scientific details may be out of date, but the thrust is as valid now, more than 50 years later, as it was then.

J. Neyman (ed.): *The Heritage of Copernicus-Theories "Pleasing to the Mind."* MIT Press, Cambridge, Mass., 1974. Copernicus was born five centuries ago. After the first chapter, which is an account of Copernicus' life and his desire to understand the motions of the planets in terms of an intellectually simple and clear theory, the remainder of this book is a series of essays that explore whether modern science has or has not understood contemporary complexities with the same clarity and simplicity. This query is not really answered, nor perhaps could it be; but this book is a treasury.

R. J. Seeger and R. S. Cohen (eds.): *Philosophical Foundations of Science*, Reidel, Boston, 1970. A collection of essays on the topic of the title; volume six of the series *Boston Colloquim for the Philosophy of Science*.

G. S. Stent: Prematurity and uniqueness in scientific discovery. *Sci. Amer.*, **227**:84 (Dec. 1972). A persuasive exposition of the assertion that creative works in art and science are unique to the particular artist or scientist; without Picasso, no one else would ever have made that contribution, without Copernicus we might only now be recently rid of geocentrism. Maybe so, maybe not; decide for yourself.

W. H. Davenport: Resource letter TLA-1 on technology, literature, and art since World War II. *Am. J. Phys.*, **38**:407 (1970). An annotated bibliography, not unlike this one, but somewhat preoccupied with "the Bomb." However, this defect does not obscure the value of the listings to a significant degree.

J. Benthall: *Science and Technology in Art Today*. Praeger, New York, 1972. An almost absolutely magnificent book, loaded with provocative delights, including for example a computer controlled gerbil environment.

Three on poetry and science: Wadsworth's preface to his second edition of *Lyrical Ballads*, W. H. Auden's *Secondary Worlds*, Random House, and Morris Bishop, *A Bowl of Bishop* with one poem, $E = mc^2$, Dial Press, New York, 1954.

R. G. Powers: The cold war in the Rockies, etc., *Art J.*, **33**:304 (1974). The title refers to, and the article discusses, interactions between architecture and technology as exemplified (or worse perhaps, in Powers' opinion) by the Air Force Academy. Not an objective discussion but definitely worth notice.

Bull. Atomic Sci., **15**:50–93 (1959). Eleven articles by artists and scientists on, of course, science and art. Stimulating.

A. V. Shubnikov and V. A. Koptsik (G. D. Archard and D. Harker, trans.): *Symmetry in Science and Art*. Plenum, New York, 1974. An extraordinary book, treating symmetry in its multifaceted aspects, written for the non-specialist and for the specialist by two Russian authors who know what they are talking about. The illustrations enhance the work considerably. This book belongs in every library and not on the back shelf.

W. G. Pollard: *Physicist and Christian*. Seabury, New York, 1970; *Science and Faith: Twin Mysteries*. Nelson, New York, 1970; *Chance and Providence*. Chas. Scribner's Sons, New York, 1958. Dr. Pollard is a well known physicist turned cleric but remaining a physicist. If it is written by Pollard, it is worth reading.

Several scientists have felt impelled to affirm their religious faith, usually in an autobiographical or anecdotal style. Of the several that are available, and written by individual scientists, four, each by an outstanding scientist, are listed here: L. du Nuoy, *Between Knowing and Understanding*. D. Mackay, New York, 1966. Written twenty years after du Nuoy's better known *Human Destiny*, on how he came to write that classic.

M. Polanyi: *Science, Faith, and Society*. University of Chicago Press, Chicago, 1964. Many who ought to know consider Michael Polanyi one of the half-dozen most distinguished scientists who ever lived. For various interesting reasons, he has not received the formal recognition he deserves.

E. Schrodinger: *Science and Humanism*. Cambridge University Press, Cambridge, 1951. An affirmation of faith by a distinguished scientist.

C. A. Coulson: *Science and Christian Belief*. University of North Carolina Press, Chapel Hill, 1955. On the unity of science and faith by a distinguished mathematician.

You may wish to browse through recent issues of *ISIS*, the journal of the History of Science Society, an international review published by the Museum of History and Technology, Smithsonian Institution, Washington, D.C. Many articles in this journal are related to the content of this chapter.

There are several collections of essays by various scientists and theologians, some with autobiographical or anecdotal slant, some with philosophical overtones. Six of the several follow:

C. P. Haskins (ed.): *The Search for Understanding*. Carnegie Institution, Washington, D.C., 1967.

F. J. Crosson (ed.): *Science and Contemporary Society*. University of Notre Dame Press, Notre Dame, Ind. 1967.

Anon. (ed.): *Science and Society, a Symposium*. Benjamin, Menlo Park, Calif., 1965.

R. W. Burhoe (ed.): *Science and Human Values in the 21st Century*. Westminster Press, Philadelphia, 1971.

I. G. Barbour: *Science and Religion*. Harper and Row, New York, 1968.

E. P. Booth (ed.): *Religion Ponders Science*. Appleton Century, New York, 1964.

W. Weaver: *Science and Imagination*. Basic Books, New York, 1967. A collection of Weaver's essays originally published elsewhere, here collected conveniently.

I. G. Barbour: *Issues in Science and Religion*. Prentice-Hall, Englewood Cliffs, N.J., 1966. A scholarly look by a historian who knows a great deal about all three: history, science, and religion.

J. M. Aubert: *A God for Science*. Newman, Westminster, Md., 1964. A bit smug, perhaps, but it has some nice spots.

L. Eiseley: The cosmic orphan. *Saturday Review/World*, 2 February, 1974, p. 18. This one is on the theme that we are not alone, but is cited here for two reasons: First, this article is highly recommended, and second, so is anything else Eiseley has written.

F. Dostoyevsky: an essay on the "chemistry" of life in *Letters from the Underground*. Everyman's Library, Dutton, New York, 1971.

T. Dobzhansky: *The Biology of Ultimate Concern*. New American Library, New York, 1967. One volume in a series, *Perspectives in Humanism* planned and edited by Ruth N. Anshen, designed "to affirm that the world, the universe, and man are remarkably stable . . . . Man is an organism, a whole . . . . The lawfulness of nature, including man's nature, is a miracle defying understanding." Amen! (and Hurrah!)

Two on economics for the uninitiated: A series of articles in the *Saturday Review*, January 22, 1972, pp. 33–57 (with advertisements intervening); and D. B. Brooks (ed.): *Resource Economics*, Johns Hopkins University Press, Baltimore, 1974 comprising selected works by O. C. Herfindahl on the theme that mineral resources are economic goods. A presage of future developments in economic theory demanded by the cultural change we are now somewhat unwillingly, but necessarily and gloriously, undergoing. Without doubt, this is what it will be (must be) like in 2084.

D. J. Rapport and J. E. Turner: Economic models in ecology. *Science*, **195**:367 (1977). On the relevance of economic principles in the study of ecology, another example of a relationship between two fields that may be thought by some to be unrelated.

S. L. Jaki: *Science and Creation*; *from Eternal Cycles to an Oscillating*

Universe. Neale Watson, New York, 1974. On the effect of religious belief on scientific progress, with the theme that the notion of rational physical law requires, first, the notion of a rational, transcendent Lawgiver.

C. N. Hinshelwood: *Fifteenth Eddington Memorial Lecture*. Cambridge University Press, Cambridge, 1961. A chemist's view: our understanding of nature is enriched by both the sciences and the arts.

J. Maddox: *The Doomsday Syndrome*. McGraw-Hill, New York, 1972. The history of the human race shows clearly that survival is dependent upon the spirit, not upon the availability nor upon the worship of material goods.

Anon: Turning points for America. *Wilson Quarterly*, **1**:10 (Autumn 1976). A book review in which the adverse effects of economic decisions suggest an eventual economic solution of either oligopolistic control of our life-styles or an equivalent control by labor unions. This contribution is cited here as another example of interactions between two apparently disparate concepts, economics and constitutional freedoms, in this case.

G. M. Lyons: United States of America. *International Social Science Journal*, **1** No. 1, 1976, UNESCO, Paris. Science influences political decisions, as evidenced by scientist-advisors to Presidents and Prime Ministers. But politics also influences science; to learn why and how, read this article.

G. Kepes (ed.): *Structure in Art and in Science*. George Braziller, New York, 1965. Fourteen contributed chapters, by as many well known scientists, philosophers, and artists, plus the editor's own contributions have produced a remarkably well illustrated book, intended to help the reader discern harmony in science through art, and to discern harmony in art through science. The emphasis is upon vision, structure, and whole systems.

Another kind of compilation can be found in *Stacks*, a periodical publication of the Libraries of the Polytechnic Institute of Brooklyn in the three issues: The great conversation. No. 12, May 1968; The sciences and the other humanities. No. 7, November 1967; and The continuing dialog. No. 28, May 1970. Each of these is an annotated bibliography, extraordinarily well selected and executed, on the topics of this chapter, plus selected detailed abstracts of other publications.

The bibliography for this chapter could not be complete if the works by René Dubos were not mentioned; their importance is emphasized by the placement of these comments here, in last place where they are more likely to be noted. To get started try *The Torch of Life*, Simon and Schuster, New York, 1962 (also available as a paperback printed in 1970) and *A God Within*, Chas. Scribner's Sons, New York, 1972. Dubos is a theistic scientist, as these cited works will show. Barbara Ward is a political economist and a theist. You will enjoy and profit from their collaboration in *Only One Earth* on the "care and maintenance" of this small planet, W. W. Norton, New York, 1972. More recently, Dubos has written an optimistic epilogue to the earlier works, in *Science*, **193**:459 (1976). Creative stewardship of this Earth is possible and humankind will strive to achieve this end.

S. Bellow: Literature in the age of technology. *Chemtech*, **7**:16 (1977). That a Nobel Laureate in Literature has written an article published in a journal edited by a chemist almost in itself demonstrates a relationship between literature and science, and science and literature. For the details, read the article.

O. T. Benfey: The limits of knowledge. *Chemistry*, **50**:2 (March 1977). An editorial to show that science has no certitude, nor did it ever have, nor can it ever. A relatively simple molecule such as benzene, with only 12 atoms, can never be fully comprehended even if the entire universe were one single, complex, intelligence which was devoted exclusively to this task. For the truth, apply to the poets, not to science.

Index